Kotlin 语言实例精解

[美] 艾亚努·阿德勒肯 著

王福会 译

清华大学出版社

北京

内 容 简 介

本书详细阐述了与 Kotlin 开发相关的基本解决方案，主要包括俄罗斯方块游戏、设计并实现 Messenger 后端应用程序、在数据库中存储信息、Android App 的安全和部署、Place Reviewer 后台应用程序、Place Reviewer 前端设计等内容。此外，本书还提供了相应的示例、代码，以帮助读者进一步理解相关方案的实现过程。

本书适合作为高等院校计算机及相关专业的教材和教学参考书，也可作为相关开发人员的自学教材和参考手册。

Copyright © Packt Publishing 2018.First published in the English language under the title
Kotlin Programming By Example.

Simplified Chinese-language edition © 2018 by Tsinghua University Press.All rights reserved.

本书中文简体字版由 Packt Publishing 授权清华大学出版社独家出版。未经出版者书面许可，不得以任何方式复制或抄袭本书内容。

北京市版权局著作权合同登记号 图字：01-2018-4394

本书封面贴有清华大学出版社防伪标签，无标签者不得销售。
版权所有，侵权必究。侵权举报电话：010-62782989　13701121933

图书在版编目（CIP）数据

Kotlin 语言实例精解/（美）艾亚努·阿德勒肯（Iyanu Adelekan）著；王福会译. —北京：清华大学出版社，2019
书名原文：Kotlin Programming By Example
ISBN 978-7-302-51655-2

Ⅰ. ①K… Ⅱ. ①艾… ②王… Ⅲ. ①JAVA 语言-程序设计 Ⅳ. ①TP312.8

中国版本图书馆 CIP 数据核字（2019）第 257374 号

责任编辑：贾小红
封面设计：刘　超
版式设计：魏　远
责任校对：马子杰
责任印制：李红英

出版发行：清华大学出版社
网　　址：http://www.tup.com.cn, http://www.wqbook.com
地　　址：北京清华大学学研大厦 A 座　　　邮　编：100084
社 总 机：010-62770175　　　邮　购：010-62786544
投稿与读者服务：010-62776969, c-service@tup.tsinghua.edu.cn
质 量 反 馈：010-62772015, zhiliang@tup.tsinghua.edu.cn

印 装 者：清华大学印刷厂
经　　销：全国新华书店
开　　本：185mm×230mm　　印　张：26.25　　字　数：523 千字
版　　次：2019 年 1 月第 1 版　　　　　　　印　次：2019 年 1 月第 1 次印刷
定　　价：129.00 元

产品编号：078569-01

感谢父母对我的信任和挚爱。此刻,语言无法描述我对你们的感激之情。

译 者 序

Kotlin 是一种新型语言且具有较好的稳定性，并可在所有 Android 设备上运行，同时还解决了 Java 无法处理的许多问题。Kotlin 为 Android 开发平台引入了许多已被证实的编程概念，使得开发过程变得更加轻松，并可生成更具安全性、表现力和简洁的代码。

同时，也希望读者具备开阔的头脑，以及对新技术的渴望之心，这对于程序设计学习来说十分有益。针对于此，本书精心挑选了与 Kotlin 语言相关的开发实例，涉及俄罗斯方块游戏、设计并实现 Messenger 后端应用程序、数据库中的信息存储、Android App 的安全和部署、Place Reviewer 后台应用程序、Place Reviewer 前端设计等内容。这里，我们也建议读者重点考查相关代码，并理解其所执行的任务。除此之外，还需要亲自实现、运行书中的每一个程序。

在本书的翻译过程中，除王福会之外，周建娟、李秋霞、程晓磊、黄丽臣、于鑫睿、刘祎、张骞、李垚、张颖、张弢、刘君、李强、李伟、李姣姣、沈旻、翟露洋、刘洋、蔡辉、张博、杨崇珉、刘璋、刘晓雪、张华臻、刘颙、张满婷等人也参与了本书的翻译工作，在此一并表示感谢。

译　者

前　　言

自Google宣布Kotlin为官方支持的Android语言以来，该语言的受欢迎程度大幅上升，这也反映了Kotlin是一种设计良好的现代编程语言，并适用于多个开发领域，包括Web、移动开发以及原生开发。由于受欢迎程度的不断提高，多年以来，Kotlin用户一直保持着稳定的增长。

适用读者

本书适用于各种年龄层以及不同水平的读者。也就是说，本书面向初学者以及具有一定开发经验的程序员，他们想要学习Kotlin语言方面的知识。

在本书的编写过程中，我特别注意到了以下一个事实：初学者需要轻松地理解相关主题和概念。为此，本书各章是按照难度递增的书顺序编写的。如果读者恰好是一名初学者，本书可使您快速融入学习过程中，同时保持学习的连贯性。

相比较而言，具有一定开发经验的读者则会更加流畅地阅读本书——一切都是平等的。如果读者具有应用程序开发的相关经验，那么，可以选择先浏览本书的示例代码，以了解所涵盖的主题和所期望的内容。特别是Java开发人员，他们可以直接阅读书中更高级的内容。

无论属于哪种类型的读者，请放心，我们依然为您撰写了相关的主题。

本书内容

第1章讨论了如何利用Kotlin语言编写简单的应用程序，包括构建Android项目、学习开发Android应用程序所需的基础知识，并以此与Web服务器进行通信。

第2章介绍一款相对简单的游戏作品，即俄罗斯方块，以使读者能够快速进行Android项目开发。

第3章介绍了如何生成视图、利用模型实现应用程序逻辑，并实现数据的视图化操作。

除此之外，本章还将学习 UI 事件处理方面的内容。

第 4 章将探讨如何设计和实现后台程序，进而向客户端应用程序提供 Web 资源。

第 5 章涉及模型-视图-表示模式的应用，从而编写一个可与 Messenger 后端程序通信的 Messenger 应用程序。

第 6 章则在第 5 章的基础上，进一步完善 Messenger 应用程序的开发。

第 7 章解释了 Android 框架所支持的各种数据存储方法。除此之外，本章还将学习如何使用这些方法存储/获取有效的应用程序信息。

第 8 章逐步分析了 Android 应用程序的部署问题；此外，本章还涵盖了较为重要的 Android 应用程序安全方面的问题。

第 9 章利用 Spring MVC 详细讨论了后台程序的设计和实现过程，即 Place Reviewer Web 应用程序。

第 10 章分析了如何创建一个 Web 定位程序，并学习使用强大的 Google Places API。另外，本章还将学习如何针对 Web 应用程序编写测试程序。

对于初学者来说，希望读者秉承一种开放、主动的学习态度。在学习一门新语言时，开始阶段可能会遇到种种问题，但只要坚持不懈，终将会获得成功。这里也建议读者逐章阅读本书，确保掌握书中的全部内容。特别需要指出的是，应重点考查相关代码，并理解其所执行的任务。同时，还需要亲自实现、运行书中的每一个程序。

资源下载

读者可访问 http://www.packtpub.com 并通过个人账户下载示例代码文件。另外，读者在购买本书后，可访问 http://www.packtpub.com/support，注册成功后，我们将以电子邮件的方式将相关文件发与读者。

读者可根据下列步骤下载代码文件：

❑ 访问并注册我们的网站（对应网址为 http://www.packtpub.com）。
❑ 选择 SUPPORT 选项卡。
❑ 单击 Code Downloads & Errata。
❑ 在 Search 文本框中输入书名。

当文件下载完毕后，确保使用下列最新版本软件解压文件夹：

❑ Windows 系统下的 WinRAR/7-Zip。

- ❑ Mac 系统下的 Zipeg/iZip/UnRarX。
- ❑ Linux 系统下的 7-Zip/PeaZip。

同时，读者还可访问 GitHub 获取本书的代码包，对应网址为 https://github.com/PacktPublishing/Kotlin-Programming-By-Example。

此外，读者还可访问 https://github.com/PacktPublishing/以了解丰富的代码和视频资源。

下载书中的彩色图像

我们还提供了相关 PDF 文件，其中包含了本书中与屏幕截图、示意图相关的彩色图像，读者可访问 https://www.packtpub.com/sites/default/files/downloads/KotlinProgrammingByExample_ColorImages.pdf 下载。

本书约定

代码块则通过下列方式设置：

```
release {
 storeFile file("../my-release-key.jks")
 storePassword "password"
 keyAlias "my-alias"
 keyPassword "password"
}
```

代码中的重点内容则采用黑体表示：

```
release {
 storeFile file("../my-release-key.jks")
 storePassword "password"
 keyAlias "my-alias"
 keyPassword "password"
}
```

命令行输入或输出如下所示：

```
./gradlew assembleRelease
```

🛈 图标表示较为重要的说明事项。

💡 图标则表示提示信息和操作技巧。

读者反馈和客户支持

欢迎读者对本书的建议或意见予以反馈，以使我们进一步了解读者的阅读喜好。

反馈意见对于我们来说十分重要，以便改进我们日后的工作。对此，读者可向 feedback@packtpub.com 发送邮件，并以书名作为邮件标题。

尽管我们在最大程度上做到尽善尽美，但错误依然在所难免。如果读者发现谬误之处，无论是文字错误抑或是代码错误，还望不吝赐教。对于其他读者以及本书的再版工作，这将具有十分重要的意义。对此，读者可访问 http://www.packtpub.com/submit-errata，选取对应书籍，单击 ErrataSubmissionForm 超链接，并输入相关问题的详细内容。

若读者在互联网上发现本书任意形式的副本，请告知网络地址或网站名称，我们将对此予以处理。关于盗版问题，读者可发送邮件至 copyright@packtpub.com。

若读者针对某项技术具有专家级的见解，抑或计划撰写书籍或完善某部著作的出版工作，则可访问 authors.packtpub.com。

评论本书

欢迎读者对本书的建议或意见予以反馈，以进一步了解读者的阅读喜好。

读者可访问 packtpub.com 并获取与 Packt 相关的更多信息。

目 录

第1章 基础知识 1
1.1 开始 Kotlin 之旅 1
1.1.1 安装 JDK 3
1.1.2 编译 Kotlin 程序 4
1.1.3 运行第一个 Kotlin 程序 6
1.1.4 在 IDE 中工作 8
1.2 Kotlin 编程语言基础知识 10
1.2.1 Kotlin 知识 10
1.2.2 面向对象程序设计 28
1.3 Kotlin 的优点 30
1.4 利用 Kotlin 开发 Android 应用程序 31
1.4.1 设置 Android Studio 31
1.4.2 构建第一个 Android 应用程序 34
1.5 Web 基础知识 42
1.5.1 Web 的含义 42
1.5.2 超文本传输协议 42
1.5.3 客户端和服务器 43
1.5.4 HTTP 请求和响应 43
1.5.5 HTTP 方法 43
1.6 本章小结 44

第2章 构建 Android 应用程序——俄罗斯方块游戏 45
2.1 Android 概述 45
2.1.1 活动 46
2.1.2 意图 46
2.1.3 意图过滤器 47
2.1.4 片段 47
2.1.5 服务 47

- 2.1.6 加载器 47
- 2.1.7 内容提供商 47
- 2.2 理解俄罗斯方块游戏 48
- 2.3 创建用户界面 49
    - 2.3.1 ConstraintLayout 51
    - 2.3.2 定义尺寸资源 54
    - 2.3.3 视图 56
    - 2.3.4 视图组 57
    - 2.3.5 定义字符串资源 62
    - 2.3.6 处理输入事件 65
    - 2.3.7 与 SharedPreferences 协同工作 70
    - 2.3.8 实现游戏活动布局 75
- 2.4 App 清单文件 78
    - 2.4.1 <action> 81
    - 2.4.2 <activity> 81
    - 2.4.3 <application> 81
    - 2.4.4 <category> 82
    - 2.4.5 <intent-filter> 83
    - 2.4.6 <manifest> 83
- 2.5 本章小结 83

第 3 章 俄罗斯方块游戏的逻辑和功能 84
- 3.1 实现游戏体验过程 84
    - 3.1.1 图块建模 85
    - 3.1.2 构建应用程序模型 100
    - 3.1.3 创建 TetrisView 111
- 3.2 MVP 模式简介 122
    - 3.2.1 MVP 的含义 122
    - 3.2.2 MVP 实现 123
- 3.3 本章小结 123

第 4 章 设计并实现 Messenger 后端应用程序 ... 124
- 4.1 设计 Messenger API 124

目录

- 4.1.1 应用程序编程接口 ... 124
- 4.1.2 REST ... 125
- 4.1.3 设计 Messenger API 系统 ... 125
- 4.2 实现 Messenger 后端 ... 128
  - 4.2.1 PostgreSQL ... 128
  - 4.2.2 创建新的 Spring Boot 应用程序 ... 129
  - 4.2.3 Spring Boot 概述 ... 132
  - 4.2.4 限制 API 访问 ... 154
- 4.3 将 Messenger API 部署至 AWS 上 ... 173
  - 4.3.1 配置 AWS 上的 PostgreSQL ... 173
  - 4.3.2 向 Amazon Elastic Beanstalk 部署 Messenger API ... 176
- 4.4 本章小结 ... 178

第 5 章 构建 Messenger Android App（第 1 部分）... 180
- 5.1 开发 Messenger App ... 180
  - 5.1.1 纳入项目依赖关系 ... 180
  - 5.1.2 开发登录 UI ... 182
  - 5.1.3 设计注册 UI ... 209
- 5.2 本章小结 ... 219

第 6 章 构建 Messenger Android App（第 2 部分）... 220
- 6.1 创建主 UI ... 220
  - 6.1.1 创建 MainView ... 220
  - 6.1.2 创建 MainInteractor ... 222
  - 6.1.3 创建 MainPresenter ... 225
  - 6.1.4 封装 MainView ... 227
  - 6.1.5 创建 MainActivity 菜单 ... 238
- 6.2 创建聊天 UI ... 239
  - 6.2.1 创建聊天布局 ... 239
  - 6.2.2 准备聊天 UI 模型 ... 241
  - 6.2.3 创建 ChatInteractor 和 ChatPresenter ... 242
- 6.3 应用程序设置 ... 249
- 6.4 Android 应用程序测试 ... 259

6.5	执行后台操作	260
	6.5.1　AsyncTask	260
	6.5.2　IntentService	260
6.6	本章小结	261

第 7 章　在数据库中存储信息 ... 262

7.1	与内部存储协同工作	262
	7.1.1　向内部存储中写入文件	262
	7.1.2　从内部存储中读取私有文件	263
	7.1.3　基于内部存储的示例程序	263
	7.1.4　保存缓存文件	277
7.2	与外部存储协同工作	277
	7.2.1　获得外部存储许可	277
	7.2.2　媒介的有效性	278
	7.2.3　存储共享文件	279
	7.2.4　利用外部存储缓存文件	279
7.3	网络存储	279
7.4	与内容提供商协同工作	295
7.5	本章小结	306

第 8 章　Android App 的安全和部署 ... 307

8.1	Android 应用程序安全	307
	8.1.1　内部存储	307
	8.1.2　网络安全	308
	8.1.3　输入验证	309
	8.1.4　与用户凭证协同工作	309
	8.1.5　代码混淆技术	309
	8.1.6　广播接收器的安全性	309
	8.1.7　动态加载代码	309
	8.1.8　服务的安全性	310
8.2	启用和发布 Android 应用程序	310
	8.2.1　理解 Android 开发者程序策略	311
	8.2.2　设置 Android 开发者账号	311

- 8.2.3 本地化规划 311
- 8.2.4 规划同步版本 311
- 8.2.5 根据质量标准进行测试 311
- 8.2.6 构建可发布的 APK 312
- 8.2.7 规划应用程序的 Play Store 列表 312
- 8.2.8 将应用程序包上传至 alpha 或 beta 测试 312
- 8.2.9 设备兼容性定义 312
- 8.2.10 启用前报告评估 312
- 8.2.11 定价和应用程序分发配置 312
- 8.2.12 分发选项的选取 313
- 8.2.13 应用程序内产品和订阅设置 313
- 8.2.14 制定应用程序内容评级 313
- 8.2.15 发布应用程序 313
- 8.2.16 发布 Android 应用程序 320
- 8.3 本章小结 324

第 9 章 创建 Place Reviewer 后台应用程序 325
- 9.1 MVC 设计模式 325
  - 9.1.1 模型 325
  - 9.1.2 视图 325
  - 9.1.3 控制器 326
- 9.2 设计并实现 Place Reviewer 后台程序 326
  - 9.2.1 用例标识 326
  - 9.2.2 标识数据 327
  - 9.2.3 设置数据库 327
  - 9.2.4 实现后台应用程序 328
  - 9.2.5 将后台程序连接至 Postgres 330
  - 9.2.6 创建模型 330
  - 9.2.7 创建数据存储库 333
  - 9.2.8 Place Reviewer 业务逻辑实现 334
  - 9.2.9 Place Reviewer 后台应用程序的安全问题 336
  - 9.2.10 基于 Spring MVC 的 Web 内容服务 340
- 9.3 利用 ELK 管理 Spring 应用程序日志 343

9.3.1 利用 Spring 生成日志 343
9.3.2 安装 Elasticsearch 343
9.3.3 安装 Kibana 345
9.3.4 Logstash 346
9.3.5 配置 Kibana 347
9.4 本章小结 349

第 10 章 实现 Place Reviewer 前端 350
10.1 利用 Thymeleaf 生成视图 350
 10.1.1 实现用户注册视图 351
 10.1.2 实现登录视图 365
 10.1.3 Google Places API Web 服务 369
 10.1.4 实现主视图 371
 10.1.5 生成评论 382
10.2 Spring 应用程序测试 395
 10.2.1 添加测试依赖关系 395
 10.2.2 定义配置类 396
 10.2.3 利用自定义配置设置配置类 396
 10.2.4 编写第一个测试程序 397
10.3 本章小结 400

后记 401

第 1 章 基 础 知 识

对于大多数人来讲，学习一门编程语言常会经历某种"痛苦"的过程，人们往往对此望而却步。既然读者选择阅读本书，那么，我坚信读者对 Kotlin 程序设计抱有浓厚的兴趣，并立志于成为这一方面的专家。在此，我也祝贺读者在学习 Kotlin 语言这一方面迈出了勇敢的一步。

对于解决方案的问题领域，无论是应用程序开发、网络开发或者是分布式系统，Kotlin 语言均可视作一种较好的选择，并可提供所需的解决方案。也就是说，学习 Kotlin 是一种正确的方向，下面将对这门语言中的基础知识加以介绍。

Kotlin 是一门强类型、面向对象语言，可运行于 Java 虚拟机（JVM）上，并可用于开发各种问题领域中的应用程序。除了能够在虚拟机上运行之外，Kotlin 语言还可编译为 JavaScript，因而可用于开发客户端 Web 应用程序。另外，Kotlin 还可以直接编译成本地二进制文件，这些二进制文件在缺少虚拟机的情况下可通过 Kotlin/Native 运行于系统上。JetBrains 发布了 Kotlin 程序设计语言，这是一家位于俄罗斯圣彼得堡的软件公司，并负责对该语言加以维护。另外，Kotlin 这一名称取自于圣彼得堡附近的 Kotlin 岛。

Kotlin 语言旨在多个领域内提供工业强度级别的软件开发方案，但就目前来看，其主要应用体现在 Android 生态圈。在本书编写时，Kotlin 是谷歌作为 Android 官方语言所宣布的 3 种语言之一。Kotlin 与 Java 在语法上具有相似之处。事实上，其设计初衷是为了更好地替代 Java。因此，在软件开发中使用 Kotlin（而不是 Java）将会体现诸多优势。

本章主要涉及以下内容：
❑ Kotlin 的安装过程。
❑ Kotlin 程序设计语言的基础知识。
❑ Android Studio 的安装和设置。
❑ Gradle。
❑ Web 基础知识。

1.1 开始 Kotlin 之旅

当开发 Kotlin 应用程序时，首先需要在计算机上安装 Java 运行环境（JRE）。其中，JRE 可连同 Java 开发工具包（JDK）一起下载。当安装 Kotlin 时，需要使用到 JDK。

相应地，最为简单的 JDK 安装方式是使用 Oracle（Java 的拥有者）提供的某种 JDK 安装程序。对于所有的主流操作系统，Oracle 提供了不同的安装程序，读者可访问 http://www.oracle.com/technetwork/java/javase/downloads/index.html 下载相关的 JDK 版本，如图 1.1 所示。

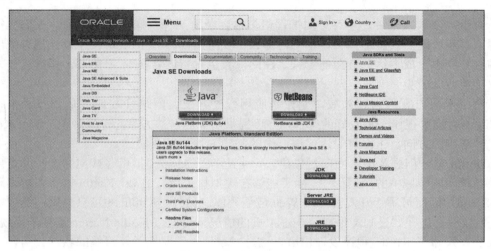

图 1.1　Java SE Web 页面

单击 JDK 下载按钮即可前往下载页面，如图 1.2 所示。其中，读者可针对相应的操作系统以及 CPU 架构选取相应的 JDK。此时，读者可选择合适的 JDK 并执行后续操作。

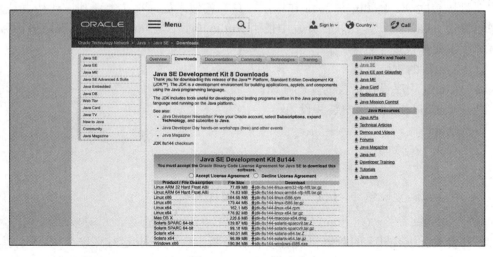

图 1.2　JDK 下载页面

1.1.1 安装 JDK

当在计算机设备上安装 JDK 时，根据当前操作系统，读者可检测一些必要的安装信息。

1. 在 Windows 设备上安装

通过下列 4 个步骤，可将 JDK 安装于 Windows 环境中。

（1）双击可下载的安装文件，并运行 JDK 安装程序。

（2）单击欢迎画面中的 Next 按钮，并选择需要安装的组件。其他内容可保持默认选项，随后单击 Next 按钮。

（3）接下来的窗口将提示用户选择安装的目标文件夹。当前，读者可选择默认文件夹（但需要留意该文件夹的位置，后续操作将会使用到该文件夹），随后单击 Next 按钮。

（4）遵循后续窗口中的各项说明，并单击 Next 按钮。其间，读者将被询问输入管理员密码。随后，Java 将被安装至用户的计算机中。

JDK 安装完毕后，还需要设置 JAVA_HOME 环境变量。对此，可执行下列步骤：

（1）打开 Control Panel。

（2）选择 Edit environment variable。

（3）在开启的窗口中，单击 New 按钮，此时将被提示添加新的环境变量。

（4）作为变量名输入 JAVA_HOME，并作为变量值输入 JDK 的安装路径。

（5）单击 OK 按钮以添加环境变量。

2. 在 macOS 设备上安装

当在 macOS 上安装 JDK 时，需要执行下列步骤：

（1）下载所需的 JDK .dmg 文件。

（2）找到下载后的.dmg 文件并双击该文件。

（3）包含 JDK 数据包图标的查找窗口将被打开。双击该图标并启动安装程序。

（4）单击简介窗口的 Continue 按钮。

（5）单击安装窗口的 Install 按钮。

（6）输入管理员账户和密码，并单击 Install Software 按钮。

至此，JDK 安装完毕，随后将显示配置窗口。

3. 在 Linux 上安装

通过 apt-get 命令，在 Linux 上安装 JDK 则较为简单和直观，对应步骤如下：

（1）更新计算机上的数据包索引，并在终端上运行下列命令：

```
sudo apt-get update
```

（2）运行下列命令并检测是否已经安装了 Java：

```
java -version
```

（3）如果输出了 Java 系统安装的版本信息，则说明 Java 已被安装完毕。如果未安装任何版本，则运行下列命令：

```
sudo apt-get install default-jdk
```

至此，JDK 已被安装在计算机上。

1.1.2 编译 Kotlin 程序

在安装了 JDK 后，还需要一种方式可编译和运行 Kotlin 程序。

Kotlin 程序可直接利用 Kotlin 命令行编译器进行编译；或者通过集成开发环境（IDE）构建和运行。

1. 与命令行编译器协同工作

命令行编译器可通过 Homebrew、SDKMAN!和 MacPorts 进行安装。另一种设置命令行编译器的方法是通过手动安装。

2. 在 macOS 上安装命令行编译器

Kotlin 命令行编译器可通过多种方式在 macOS 上安装。对此，较为常见的两种方式是 Homebrew 和 MacPorts。

（1）Homebrew

Homebrew 是一个 macOS 系统的数据包管理器，广泛地用于安装软件项目时数据包的安装。当安装 Homebrew 时，可查找 macOS 终端并运行下列命令：

```
/usr/bin/ruby -e "$(curl -fsSL
https://raw.githubusercontent.com/Homebrew/install/master/install)"
```

此时，用户需要等待数秒，以下载和安装 Homebrew。安装完毕后，可在终端上运行下列命令，以查看 Homebrew 是否可正常工作。

```
brew -v
```

如果系统终端上输出了 Homebrew 的当前版本，则表明 Homebrew 已经成功地安装在计算机上。

当安装了 Homebrew 后，可查找终端并执行下列命令：

```
brew install kotlin
```

稍作等待后即可完成安装过程。随后即可利用命令行编译器编译 Kotlin 程序。

（2）MacPorts

类似于 Homebrew，MacPorts 也是一个针对 macOS 的数据包管理器，并可通过下列步骤安装在系统上：

① 安装 Xcode 和 Xcode 命令行工具。
② 接受 Xcode 许可，即在终端上运行 xcodebuild -license 命令。
③ 安装所需的 MacPorts 版本。

读者可访问 https://www.macports.org/install.php 下载 MacPort。待下载完毕后，可查找终端并以超级用户身份运行 port install kotlin 命令，如下所示：

```
sudo port install kotlin
```

3. 在 Linux 上安装命令行编译器

利用 SDKMAN!，Linux 用户可方便地安装命令行编译器。

SDKMAN!可用于在基于 UNIX 的系统上安装数据包，例如 Linux 及其各种版本（Fedora 和 Solaris 版本）。SDKMAN!的安装过程包含以下 3 个步骤：

（1）利用 curl 命令将软件下载至系统上。定位系统并运行下列命令：

```
curl -s "https://get.sdkman.io" | bash
```

（2）随后将在终端中显示一系列的指令，用户可遵循这些指令并完成安装过程。安装完毕后，可运行下列命令：

```
source "$HOME/.sdkman/bin/sdkman-init.sh"
```

（3）运行下列命令：

```
sdk version
```

如果 SDKMAN!的版本号显示于终端窗口中，则说明安装成功。

在 SDKMAN!于系统中安装完毕后，可通过下列命令安装命令行编译器：

```
sdk install kotlin
```

4. 在 Windows 上安装命令行编译器

当在 Windows 环境下使用 Kotlin 命令行编译器时，需要执行以下步骤：

（1）访问 https://github.com/JetBrains/kotlin/releases/tag/v1.2.30，并下载软件的 GitHub 版本。

（2）定位并解压下载后的文件。

（3）打开解压后的 kotlinc\bin 文件夹。
（4）执行包含文件夹路径的命令提示符。
至此，读者可通过命令行使用 Kotlin 编译器。

1.1.3 运行第一个 Kotlin 程序

前述内容讨论了命令行编译器的设置过程，下面尝试运行一个简单的 Kotlin 程序。对此，可访问主目录并创建一个名为 Hello.kt 的新文件。注意，所有的 Kotlin 文件均包含了 .kt 后缀扩展名。

在文本编辑器中打开刚刚生成的文件，并输入下列内容：

```
// The following program prints Hello world to the standard system output.
fun main (args: Array<String>) {
  println("Hello world!")
}
```

保存文件，打开终端窗口并输入下列命令。

```
kotlinc hello.kt -include-runtime -d hello.jar
```

上述命令将程序编译为 hello.jar 可执行文件。这里，-include-runtime 标记用于指定需要编译为 JAR。在向命令中添加了该标记后，Kotlin 运行库将纳入 JAR 中。另外，-d 标记表明，需要调用编译器的输出结果。

在第一个 Kotlin 程序被编译后，下面将运行该程序。打开终端窗口（如果尚未开启，可访问保存 JAR 文件的所在目录），运行编译后的 JAR 文件，并执行下列命令：

```
java -jar hello.jar
```

在运行了上述命令后，"Hello world!"将被输出至显示窗口中。恭喜！读者已经成功地运行了第一个 Kotlin 程序。

1. 利用 Kotlin 编写脚本

如前所述，Kotlin 还可用于编写脚本。对于某些共同目标（自动执行任务），脚本是为特定的运行环境编写的程序。在 Kotlin 中，脚本文件包含了 .kts 后缀扩展名。

Kotlin 脚本的编写方式与 Kotlin 程序类似。实际上，采用 Kotlin 编写的脚本与 Kotlin 常规程序十分相似，二者间唯一的差别在于主函数。

在所选文件夹中生成一个文件，并将其命名为 NumberSum.kts。打开该文件并输入下列程序：

```
val x: Int = 1
val y: Int = 2
val z: Int = x + y
println(z)
```

相信读者已经猜到，上述脚本将输出两个数字（1 和 2）之和。保存该文件，并运行当前脚本，如下所示：

```
kotlinc -script NumberSum.kts
```

注意：
Kotlin 脚本无须进行编译。

2. 使用 REPL

REPL 表示 Read–Eval–Print Loop 的首字母缩写，REPL 是一个交互式的 shell 环境，在该环境中，程序可以在给定的即时结果下执行。通过运行 kotlinc 命令（无须任何参数），可调用交互式 shell 环境。

注意：
当在终端上运行 kotlinc 时，即可启动 Kotlin REPL。

如果成功地启动了 REPL，终端窗口中间显示一条欢迎消息，下一行中的"＞＞＞"则提示用户，REPL 正在等待输入内容。一如文本编辑器中所做的那样，可向终端窗口中输入代码，并从 REPL 中获取反馈结果，如图 1.3 所示。

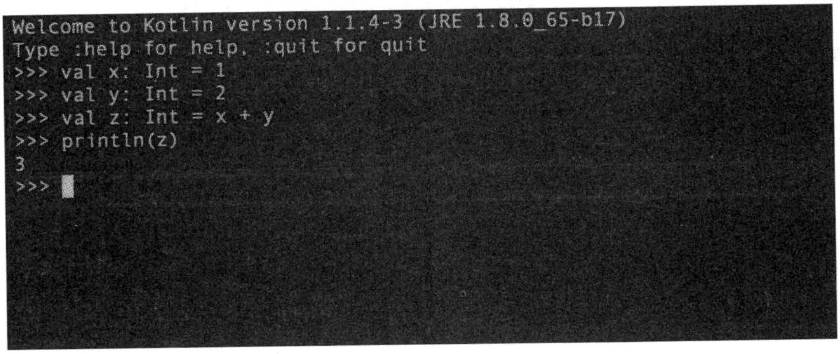

图 1.3 Kotlin REPL

其中，整数 1 和 2 分别被赋予 x 和 y 中。x 和 y 的求和结果存储于新变量 z 中。随后，z 值通过 print() 函数输出至显示窗口中。

1.1.4 在 IDE 中工作

利用命令行编写程序确实可行,但大多数场合下,最好使用专门为开发人员编写程序而构建的软件,尤其是开发大型项目时。

IDE 是一种计算机应用程序,对于软件开发来讲,其中面向程序员内置了相关的工具集。相应地,存在多种 IDE 可用于 Kotlin 开发。IntelliJ IDEA 中包含了较为全面的特性集,并可用于开发 Kotlin 应用程序。由于 IntelliJ IDEA 由 Kotlin 开发者们一手打造,与其他 IDE 相比,其优势更加明显,例如更好的工具集以及实时更新功能(包括 Kotlin 编程语言的最新特性)。

1. 安装 IntelliJ IDEA

读者可访问 JetBrains 网站,并针对 Windows、macOS 和 Linux 环境下载 IntelliJ IDEA,对应网址为 https://www.jetbrains.com/idea/download。在下载页面中,提供了两个下载版本:付费 Ultimate 版本和免费的 Community 版本,如图 1.4 所示。其中,Community 版本即可满足本章程序的需要。当然,读者可根据个人喜好下载相关版本。

图 1.4　IntelliJ IDEA 下载页面

2. 利用 IntelliJ 设置 Kotlin 项目

利用 IntelliJ 设置 Kotlin 项目主要包括以下几个步骤：

（1）启动 IntelliJ IDE 应用程序。
（2）单击 Create New Project 按钮。
（3）在左侧打开窗口中从现有项目选项中选择 Java。
（4）作为附加库向当前项目中添加 Kotlin/JVM。
（5）从窗口下拉菜单中选取 SDK 项目。
（6）单击 Next 按钮。
（7）选取模板（如有必要），并执行后续操作。
（8）在输入文本框中提供项目名称。当前项目命名为 HelloWorld。
（9）在输入文本框中设置项目位置。
（10）单击 Finish 按钮。

至此，项目创建完毕，并显示于 IDE 窗口中，如图 1.5 所示。

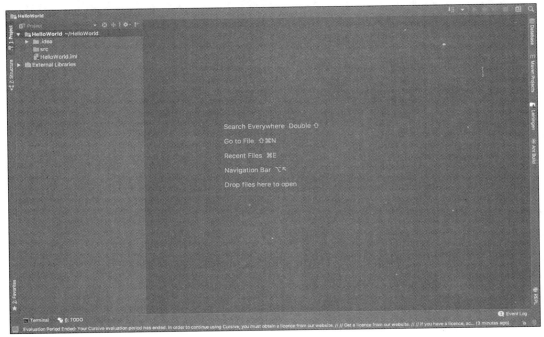

图 1.5　包含当前项目的 IDE 窗口

其中，窗口左侧显示了项目视图，包括项目文件的逻辑结构。

此外，还包含了以下两个文件夹。
- .idea 文件夹：包含了 IntelliJ 项目设置文件。
- src 文件夹：即项目的源文件夹，可于其中放置程序文件。

在项目构建完毕后，即可着手编写简单的程序。这里，可向源文件夹中添加名为 hello.kt 的文件（在 src 文件夹上单击鼠标右键，在弹出的快捷菜单中选择 New | Kotlin File/Class 命令，并将对应文件命名为 hello）。随后，可向文件中复制、粘贴下列代码：

```
fun main(args: Array<String>) {
  println("Hello world!")
}
```

当运行程序时，可单击主函数一侧的 Kotlin 图标，并选择 Run HelloKt，如图 1.6 所示。

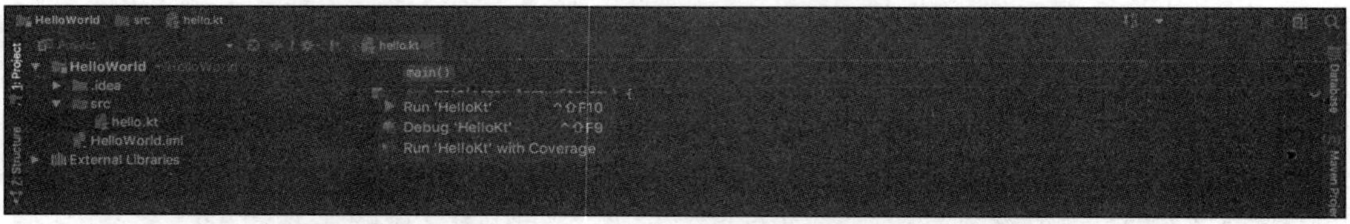

图 1.6　当前程序的运行结果

构建并运行当前项目，标准系统输出窗口中即会显示"Hello world!"。

1.2　Kotlin 编程语言基础知识

在开发环境和 IDE 设置完毕后，可对 Kotlin 语言本身稍作解释。我们将从语言的基本知识开始，并逐步深入更高级的主题，例如面向对象程序（OOP）设计。

1.2.1　Kotlin 知识

本节将讨论 Kotlin 语言的基础知识，并尝试构建代码块。下面首先考察变量。

1. 变量

变量表示为加载某个值的、内存位置的标识符。对此，一种简单的变量描述方式是：加载某个数值的标识符。考察下列程序：

```
fun main(args: Array<String>) {
  var x: Int = 1
}
```

在上述代码中，x 定义为一个变量，对应的加载值为 1。特别地，x 表示为一个整数变量。由于 x 定义为包含 Int 数据类型，因而将作为一个整数变量被引用。更准确地讲，x 表示为 Int 类实例。或许，读者会对术语"实例"和"类"感到迷惑，后续章节将对此加以讨论。当前，让我们把注意力主要集中于变量上。

当在 Kotlin 中定义一个变量时，可使用 var 关键字，进而表明可对该变量进行修改。而所声明变量的数据类型则位于冒号之后。需要注意的是，无须显式地定义变量的数据类型，Kotlin 支持类型推断机制，即根据定义推断对象的类型。因此，变量 x 的定义可记为：

```
var x = 1
```

另外，也可以在变量定义结尾处添加一个分号。通常情况下，分号无须添加，这一点与 JavaScript 十分类似，如下所示：

```
var x = 1 //I am a variable identified by x and I hold a value of 1
var y = 2 //I am a variable identified by y and I hold a value of 2
var z: Int = x + y //I am a variable identified by z and I hold a value
                   //of 3
```

相应地，在程序执行过程中，如果不希望修改变量值，可将其设置为不可变变量，并通过关键字 val 定义，如下所示。

（1）变量的作用域

变量作用域是指程序中变量的作用区域。也就是说，变量的作用域表示为变量可用的程序区域。Kotlin 变量包含了块作用域。因此，变量可以在对应块所定义的所有区域中使用，如下所示：

```
fun main(args: Array<String>) {
  // block A begins
  var a = 10
  var i = 1

  while (i < 10) {
    // block B begins
    val b = a / i
    print(b)
    i++
  }
  print(b) // Error occurs: variable b is out of scope
}
```

上述程序体现了块作用域的效果，其中包含了两个块。此处，函数定义开启了一个新块，在当前示例中将其标记为 A。在 A 中，声明了变量 a 和 i。因此，a 和 i 的作用域为 A。

A 中还创建了一个 while 循环,同时开启了 B 块。这里,循环声明标记了新块的开始位置。在 B 中,声明了一个变量 b。因此,b 值存在于 B 作用域中,且无法在其外部加以使用。当在 B 块之外试图输出 b 值时,将会产生错误。

需要注意的是,a 和 i 变量仍可在 B 块中加以使用——B 处于 A 的作用域范围内。

(2)局部变量

此类变量局部于某个作用域。在上述示例中,a、i 和 b 均为局部变量。

2. 运算数和运算符

运算符可视作指令中的部分内容,用于计算所操作的数值。另外,运算符对其运算数执行特定的运算,运算符示例包括+、-、*、/和%。运算符可根据所执行的操作类型,以及运算数的数量进行分类。

根据运算符所执行的操作类型,可将运算符划分为以下几种类型:

- 关系运算符。
- 赋值运算符。
- 逻辑运算符。
- 位运算符。

运算符类型的相关示例如表 1.1 所示。

表 1.1 运算符类型及其示例

运算符类型	示例
关系运算符	>, <, >=, <=, ==
赋值运算符	+=, -=, *=, /=, =
逻辑运算符	&&, ||, !
算术运算符	+, -, *, /
位运算符	and(bits), or(bits), xor(bits), inv(), shl(bits), shr(bits), ushr(bits)

根据运算数的数量,Kotlin 中包含了以下两种主要类型:

- 一元运算符。
- 二元运算符。

对应示例如表 1.2 所示。

表 1.2 一元运算符和二元运算符

运算符类型	描述	示例
一元运算符	仅包含一个运算数	!, ++, --
二元运算符	包含两个运算数	+, -, *, /, %, &&, ||

3. 数据类型

变量类型表示为变量可加载的数值集合。在大多数场合下，可显式地指定声明变量所加载的数值类型，并可通过具体的数据类型予以实现。

Kotlin 中包含了以下一些较为重要的类型：

- Int。
- Float。
- Double。
- Boolean。
- String。
- Char。
- Array。

（1）Int

Int 类型表示为 32 位有符号整数。当采用该类型声明一个变量时，变量的数值范围为整数集。也就是说，这一类变量仅可包含整数值。具体来说，Int 类型可加载 $-2147483648 \sim 2147483647$ 的整数值。

（2）Float

Float 类型表示为单精度 32 位浮点数。对于 Float 类型变量，该变量仅可加载浮点值，对应范围为 $\pm 3.40282347E+38F$（6~7 有效小数位）。例如：

```
var pi: Float = 3.142
```

（3）Double

Double 类型表示为双精度 64 位浮点数。类似于 Float 类型，Double 类型表明所声明的变量加载浮点值。Double 和 Float 类型间的主要差异在于：Double 类型可定义较大范围的数字，对应范围约为 $\pm 1.79769313486231570E+308$（15 个有效小数位）。例如：

```
var d: Double = 3.142
```

（4）布尔类型

布尔（Boolean）类型包含了 true 或 false 逻辑真值，如下所示：

```
var t: Boolean = true
var f: Boolean = false
```

布尔值可采用&&、||和!逻辑运算符进行操作，如表 1.3 所示。

表 1.3 布尔值的逻辑操作

运算符名称	表示	描述	运算符类型
逻辑与	&&	当两个运算数均为 true 时，结果为 true；否则为 false	二元运算符
逻辑或	||	至少一个运算数为 true 时，结果为 true；否则为 false	二元运算符
逻辑非	!	逆置布尔运算数	一元运算符

（5）字符串

字符串表示为字符序列。在 Kotlin 中，字符串通过字符串类表示。具体来说，字符串可通过双引号以及其中的字符序列加以定义，如下所示：

```
val martinLutherQuote: String = "Free at last, Free at last, Thank God
almighty we are free at last."
```

（6）字符

该类型用于表示字符。字符表示为一个信息单位，对应于字形或符号。在 Kotlin 中，字符定义为 Char 类型，并通过单引号表示，例如 '$'、'%' 和 '&'。

```
val c: Char = 'i' // I am a character
```

如前所述，字符串可表示为一个字符序列，如下所示：

```
var c: Char
val sentence: String = "I am made up of characters."

for (character in sentence) {
  c = character // Value of character assigned to c without error
  println(c)
}
```

（7）数组

数组表示为一种数据结构，并由元素或数值（包含对应的索引或键）集合构成。当存储数据元素集合时，数组十分有用，稍后即会在程序中使用到数组。

在 Kotlin 中，数组使用 arrayOf() 库方法加以构建。相应地，数组中的存储值可通过逗号分隔的序列予以传递，如下所示：

```
val names = arrayOf("Tobi", "Tonia", "Timi")
```

每个数组值均包含唯一的索引，用于指定数组中的位置，以及后续操作中的数值引用。另外，数组索引始于索引 0，并逐一递增（加 1）。

数组中既定索引位置处的数值,可通过调用 Array#get()方法或[]操作得到,如下所示:

```
val numbers = arrayOf(1, 2, 3, 4)
println(numbers[0]) // Prints 1
println(numbers.get(1)) // Prints 2
```

另外,还可对某一数组位置处的数值加以修改,如下所示:

```
val numbers = arrayOf(1, 2, 3, 4)
println(numbers[0]) // Prints 1
numbers[0] = 23
println(numbers[0]) // Prints 23
```

利用 length 属性,还可检测数组的尺寸,如下所示:

```
val numbers = arrayOf(1, 2, 3, 4)
println(numbers.length) // Prints 4
```

4. 函数

函数表示为一个代码块,经定义后即可重复使用。在编写程序时,较好的方法是将复杂的程序处理过程划分为较小的单元,并执行特定的任务。这一方式涵盖诸多优点,其中包括以下方面:

- 改善代码的可读性。在划分为较小的函数单元后,代码的可读性将显著提升,其理解范围也将有所降低。大多数时候,程序员只需要编写或调整一个大型代码库的部分内容。当采用函数时,需要读取程序并改进程序逻辑的上下文环境,仅限于编写对应逻辑的函数体中。
- 改进代码库的可维护性。使用代码库中的函数可便于程序的维护。如果某个程序功能需要修改,那么,仅须调整该功能所处的函数即可。

(1)函数声明

函数的声明可利用 fun 关键字加以实现。考察下列简单的函数定义:

```
fun printSum(a: Int, b: Int) {
  print(a + b)
}
```

该函数简单地输出两个数值(作为参数进行传递)之和。函数定义包含以下内容:

- 函数标识符。函数标识符表示为赋予函数的名称。如果希望在后续程序操作中调用该函数,则需要利用标识符引用该函数。在之前的函数声明中,printSum 表示为当前函数的标识符。
- 包含一对括号中以逗号分隔的参数列表。传递至函数中的数值称作函数的参数。

传递至函数中的所有参数需要定义为某种类型。具体类型位于冒号之后。
- 返回类型。函数的返回类型确定方式与变量和属性类似。也就是说,在最后一个括号后的冒号后指定相关类型。
- 函数体。

不难发现,上述函数并未设置返回类型。实际情况并非如此,函数包含了一个 Unit 返回类型。Unit 返回类型并不需要显式地予以指定。上述函数也可采用下列方式进行声明:

```
fun printSum(a: Int, b: Int): Unit {
  print(a + b)
}
```

注意:

函数并非总是需要标识符。不包含标识符的函数也称作匿名函数。在 Kotlin 中,匿名函数以 lambdas 这一形式体现。

(2)函数调用

函数在定义完毕后并未予以执行。对此,须对该函数加以调用并存在多种调用方式。例如,基于函数、方法的直接函数调用;利用 invoke()和 call()方法的间接调用。下列代码表示为基于函数自身的直接函数调用。

```
fun repeat(word: String, times: Int) {
  var i = 0

  while (i < times) {
    println(word)
    i++
  }
}

fun main(args: Array<String>) {
  repeat("Hello!", 5)
}
```

在编译并运行上述代码后,屏幕上将输出"Hello!"5 次。这里,"Hello!"是函数的第一个参数,5 则是第二个参数。随后,word 和 times 参数在 repeat 函数中被设置为"Hello!"和 5。只要 i 小于所指定的 times 值,while 循环即会运行并输出相关内容。同时,i++用于递增 i 值(加 1)。在每次循环迭代过程中,i 值将加 1。一旦 i 值等于 5,则循环终止。因此,"Hello!"将被输出 5 次。编译并运行当前程序将输出如图 1.7 所示的结果。

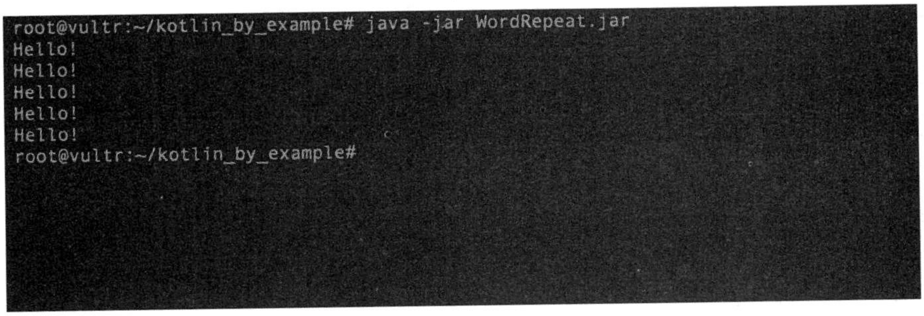

图 1.7 "Hello!" 将被输出 5 次

（3）返回值

顾名思义，返回值表示方法返回的数值。Kotlin 中的函数可在执行时返回数值。函数返回值的类型则通过函数的返回类型加以定义，如下所示：

```kotlin
fun returnFullName(firstName: String, surname: String): String {
  return "${firstName} ${surname}"
}

fun main(args: Array<String>) {
  val fullName: String = returnFullName("James", "Cameron")
  println(fullName) // prints: James Cameron
}
```

在上述代码中，returnFullName 函数作为输入参数接收两个字符串，并在被调用时返回一个字符串。相应地，翻译类型在函数头中加以定义；而返回的字符串则通过字符串模板生成，如下所示：

```
"${firstName} ${surname}"
```

第一个名称值和最后一个名称值将内插至字符串中。

（4）函数命名规则

在 Kotlin 中，函数的命名规则与 Java 类似。当命名方法时，可采用驼峰规则——首个字母采用大写形式，且不包含空格和标点符号，如下所示：

```kotlin
//Good function name
fun sayHello() {
  println("Hello")
}

//Bad function name
fun say_hello() {
```

```
    println("Hello")
}
```

5. 注释

当编写代码时，可能需要记录与代码相关的一些信息，这可通过注释予以实现。在 Kotlin 中，存在以下 3 种注释方式：

- ❏ 单行注释。
- ❏ 多行注释。
- ❏ 文档注释。

（1）单行注释

顾名思义，这一类注释只占据一行。单行注释始于//。在编译过程中，//之后的全部字符均被忽略。考察下列代码：

```
val b: Int = 957 // This is a single line comment
// println(b)
```

由于执行输出操作的函数已被注释掉，因而 b 值不会输出至控制台中。

（2）多行注释

多行注释将占据多行，并以/*开始，以*/结束，如下所示：

```
/*
 * I am a multiline comment.
 * Everything within me is commented out.
 */
```

（3）文档注释

这种注释类型与多行注释类似，主要的差别在于：文档注释将在程序中记录代码。文档注释始于/**并以*/结束，如下所示：

```
/**
 * Adds an [item] to the queue.
 * @return the new size of the queue.
 */
fun enqueue(item: Object): Int { ... }
```

6. 控制程序流

在编写程序时，常见情形是控制程序的执行方式。当根据相关条件或程序状态制定角色时，处理控制问题将十分必要。对此，Kotlin 包含了多种结构，例如为人们所熟悉的 if、while 和 for 结构；而一些结构则是 Kotlin 特有的内容，例如 when 结构。本节将讨论与控制流相关的各种程序结构。

（1）条件表达式

条件表达式常用于程序流的分支结构，并根据条件测试结构执行或忽略程序语句。这里，条件语句可视作程序的决策点。

Kotlin 包含两种分支处理结构，即 if 表达式和 when 表达式。

（2）if 表达式

根据条件判断结果，if 表达式用于制定逻辑决策。对此，可利用 if 关键字编写 if 表达式，如下所示：

```
val a = 1

if (a == 1) {
 print("a is one")
}
```

上述 if 表达式测试 a == 1（读作 a 等于 1）条件是否为 true。如果该条件为 true，将在屏幕上输出字符串"a is one"；否则不输出任何内容。

if 表达式通常会包含一个或多个 else 或 else if 关键字，并以此进一步控制程序流。考察下列 if 表达式：

```
val a = 4
if (a == 1) {
 print("a is equal to one.")
} else if (a == 2) {
   print("a is equal to two.")
} else {
   print("a is neither one nor two.")
}
```

上述表达式首先测试 a 是否等于 1，该测试结果为 false，因而将继续测试后续条件。第二个条件表达式的结果也为 false，因而将执行最后一条语句。最终，屏幕上将输出"a is neither one nor two"。

（3）when 表达式

when 表达式则是另一种程序控制方式。考察下列代码以查看其工作方式：

```
fun printEvenNumbers(numbers: Array<Int>) {
  numbers.forEach {
    when (it % 2) {
      0 -> println(it)
    }
```

```
  }
}
fun main (args: Array<String>) {
  val numberList: Array<Int> = arrayOf(1, 2, 3, 4, 5, 6)
  printEvenNumbers(numberList)
}
```

上述 printEvenSum 函数接收一个整型数组，并将其作为唯一参数。本章稍后将对数组予以介绍，当前可将其视为数值序列集合。在该示例中，所传递的数组包含了整数空间内的数值。每个数组元素通过 forEach 方法进行遍历，且每个数字在 when 表达式中被测试。

此处，it 引用了 forEach 方法所遍历的当前值。另外，% 运算符表示为二元运算符，并对两个运算数进行计算，即第一个运算数除以第二个运算符，并返回除法运算后的余数。因此，when 表达式测试当前遍历值是否（何时）被 2 除且余数为 0。若是，则该值为偶数且输出至屏幕上。

为了进一步查看程序的工作方式，可将上述代码复制、粘贴至文件中，随后编译并运行该程序，对应结果如图 1.8 所示。

图 1.8　when 表达式测试

（4）Elvis 运算符

Elvis 运算符则是 Kotlin 中一种较为简洁的结构，如下所示：

```
(expression) ?: value2
```

下列代码显示了 Elvis 运算符在 Kotlin 中的应用：

```
val nullName: String? = null
val firstName = nullName ?: "John"
```

如果 nullName 中的值不为 null，Elvis 运算符将返回该值；否则返回"John"字符串。因此，firstName 被赋予了 Elvis 运算符的返回值。

7．循环

循环语句确保代码块中的语句集合可被重复执行。也就是说，循环操作可保证程序中的多条语句被多次执行。Kotlin 中的循环结构包含 for 循环、while 循环和 do…while 循环。

（1）for 循环

Kotlin 中的 for 循环遍历提供了迭代器的任意对象，并与 Ruby 语言中的 for…in 循环类似。for 循环包含下列语法形式：

```
for (obj in collection) { … }
```

如果仅包含一条语句，那么，for 循环中的块结构将不再必需。这里，集合是一种提供迭代器的结构类型。考察下列程序：

```
val numSet = arrayOf(1, 563, 23)

for (number in numSet) {
  println(number)
}
```

numSet 数组中的每个值将被当前循环所遍历，并被赋予变量 number 中。随后，number 将被输出至标准系统输出窗口中。

> **注意：**
> 数组中的每个数据元素均包含一个索引。索引表示为数组中数据元素的位置。在 Kotlin 中，数组的索引以 0 开始。

如果希望输出每个数值的索引（而非数值本身），则可使用下列代码：

```
for (index in numSet.indices) {
  println(index)
}
```

另外，还可指定迭代器变量的类型，如下所示：

```
for (number: Int in numSet) {
  println(number)
}
```

（2）while 循环

只要满足特定的条件，while 循环将执行块结构中的指令。while 循环可通过关键字 while 标记，对应形式如下所示：

```
while (condition) { … }
```

与 for 循环的情况一样，如果循环结构中仅包含一条语句，那么，块结构是可选的。在 while 循环中，当满足条件时，块结构中的语句将被重复执行。考察下列代码：

```
val names = arrayOf("Jeffrey", "William", "Golding", "Segun", "Bob")
var i = 0

while (!names[i].equals("Segun")) {
  println("I am not Segun.")
  i++
}
```

在上述程序中，while 循环中的代码块将被执行，并输出"I am not Segun"，直至对应名称为 Segun。当遇到 Segun 时，循环将终止，且不输出任何内容，如图 1.9 所示。

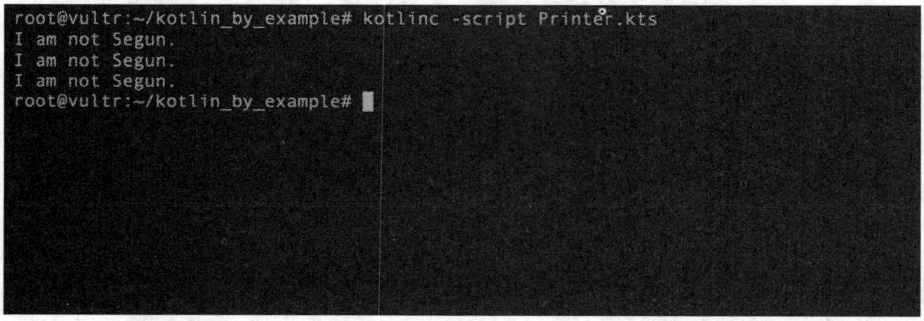

```
root@vultr:~/kotlin_by_example# kotlinc -script Printer.kts
I am not Segun.
I am not Segun.
I am not Segun.
root@vultr:~/kotlin_by_example#
```

图 1.9 while 循环

（3）break 和 continue 关键字

通常，当声明循环时，如果满足相关条件，则需要退出循环，或者开始下一次迭代过程，这可通过 break 和 continue 关键字实现。下面考察一些相关示例。打开 Kotlin 脚本文件，并复制下列代码：

```
data class Student(val name: String, val age: Int, val school: String)

val prospectiveStudents: ArrayList<Student> = ArrayList()
val admittedStudents: ArrayList<Student> = ArrayList()

prospectiveStudents.add(Student("Daniel Martinez", 12, "Hogwarts"))
prospectiveStudents.add(Student("Jane Systrom", 22, "Harvard"))
prospectiveStudents.add(Student("Matthew Johnson", 22, "University of Maryland"))
prospectiveStudents.add(Student("Jide Sowade", 18, "University of Ibadan"))
prospectiveStudents.add(Student("Tom Hanks", 25, "Howard University"))
```

```kotlin
for (student in prospectiveStudents) {
  if (student.age < 16) {
    continue
  }
  admittedStudents.add(student)

  if (admittedStudents.size >= 3) {
    break
  }
}

println(admittedStudents)
```

上述程序从学生名单中选取被录取的学生。程序开始处定义了一个数据类，并对每名学生的数据建模；随后创建两个数组列表。其中，一个数组列表加载学生信息，以供录取使用；另一个列表中则用于加载已被录取的学生名单。

随后的 5 行代码负责将学生添加至列表中。接下来，将声明一个循环用于遍历列表中的全部学生。如果学生年龄小于 16 岁，循环将跳至下一次操作。也就是说，不符合年龄的学生不予录取（因而也不会添加至录取学生列表中）。

如果学生的年龄大于或等于 16 岁，则被添加至录取列表中。随后，利用 if 表达式判断录取学生的数量是否大于或等于 3。若条件为 true，则程序跳出当前循环且迭代过程结束。程序的最后一行代码将输出列表中的学生名单。

运行该程序，将得到如图 1.10 所示的结果。

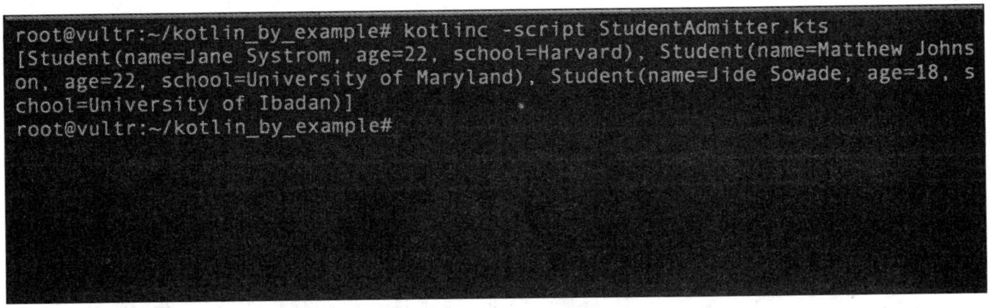

```
root@vultr:~/kotlin_by_example# kotlinc -script StudentAdmitter.kts
[Student(name=Jane Systrom, age=22, school=Harvard), Student(name=Matthew Johnson, age=22, school=University of Maryland), Student(name=Jide Sowade, age=18, school=University of Ibadan)]
root@vultr:~/kotlin_by_example#
```

图 1.10 输出列表中的学生名单

（4）do...while 循环

do...while 循环类似于 while 循环，唯一的不同之处在于，循环的条件测试在第一次循环迭代之后进行，对应形式如下所示：

```
do {
...
} while (condition)
```

当条件测试为 true 时，代码块中的语句再次被执行，如下所示：

```
var i = 0

do {
 println("I'm in here!")
 i++
} while (i < 10)

println("I'm out here!")
```

NullPointerException 是每一名 Java 程序员都会遇到的情况。Kotlin 类型系统具备空安全特性，并试图在代码中消除空引用的出现。相应地，Kotlin 支持可空类型以及非空类型（即是否可加载 null 值相关类型）。

为了清晰地解释 NullPointerException，考察下列 Java 程序：

```
class NullPointerExample {

 public static void main(String[] args) {
  String name = "James Gates";
  System.out.println(name.length()); // Prints 11

  name = null; // assigning a value of null to name
  System.out.println(name.length()); // throws NullPointerException
 }
}
```

上述程序向标准系统输出窗口输出字符串变量的长度。当编译并运行该程序时，将抛出空指针异常，执行过程将中途停止，如图 1.11 所示。

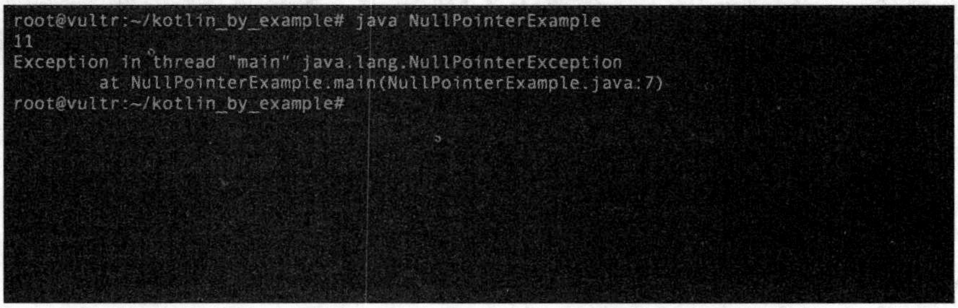

图 1.11　空指针异常

读者是否可尝试解释 NullPointerException 出现的原因？这里，String#length 方法包含了空引用，因此，程序将终止执行并抛出异常。显然，这并不是我们所期望的结果。

在 Kotlin 中，可防止向 name 对象赋予 null 值，如下所示：

```
var name: String = "James Gates"
println(name.length)

name = null // null value assignment not permitted
println(name.length)
```

在图 1.12 中，Kotlin 类型系统检测到将 null 值赋予 name 对象，并提示程序员予以修正。

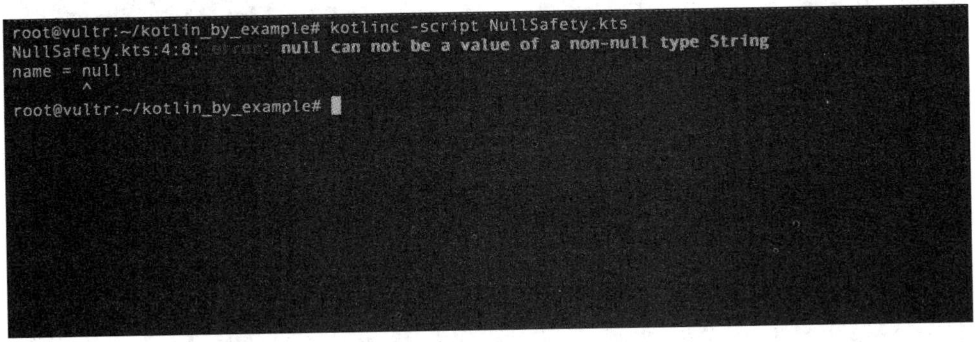

```
root@vultr:~/kotlin_by_example# kotlinc -script NullSafety.kts
NullSafety.kts:4:8: error: null can not be a value of a non-null type String
name = null
       ^
root@vultr:~/kotlin_by_example#
```

图 1.12　Kotlin 类型系统检测到将 null 值赋予 name 对象

此时，读者可能会想，如果程序员打算传递 null 值，情况又当如何？相应地，程序员可向变量类型添加一个 "?"，并将该变量值声明为可空类型，如下所示：

```
var name: String? = "James"
println(name.length)

name = null // null value assignment permitted
println(name.length)
```

尽管变量 name 声明为可空类型，在运行上述程序时依然会出现错误，其原因在于：需要以一种安全的方式访问变量的属性 length，这可通过 "?." 加以实现，如下所示：

```
var name: String? = "James"
println(name?.length)

name = null // null value assignment permitted
println(name?.length)
```

在使用了"?."安全操作符后，程序则按照所期望的要求运行，且不会再抛出空指针异常。类型系统意识到，此时引用了空指针，并禁止调用空对象上的 length()方法。类型安全输出内容如图 1.13 所示。

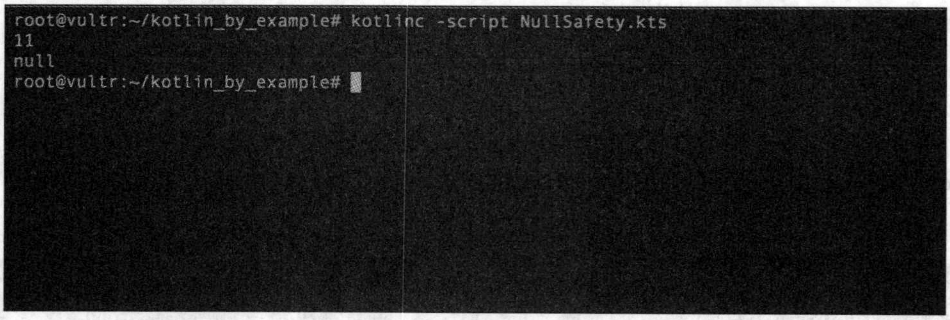

图 1.13　类型安全输出

"?."安全操作符的替代方案则是使用"!!"操作符。"!!"操作符使得程序员可继续执行程序，一旦函数调用在空引用上进行，则会抛出 KotlinNullPointerException。

我们可以进一步考察替换后的效果。图 1.14 显示了程序的输出结果。当使用"!!"操作符时，将抛出 KotlinNullPointerException。

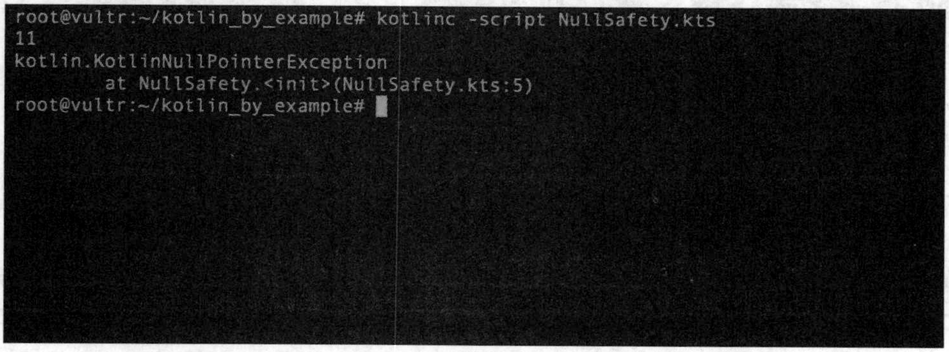

图 1.14　抛出 KotlinNullPointerException

8. 数据包

数据包表示为相关类、接口、枚举、注解和函数的逻辑分组。随着源文件不断增加，有必要将这些文件整合至相关集合中，从而可提升应用程序的可维护性，防止命名冲突，并实现更好的访问控制。

数据包可利用 package 关键字以及随后的包名称加以创建，如下所示：

```
package foo
```

每个程序文件只能有一个数据包语句。如果没有指定程序文件的数据包，则将文件的内容放入默认包中。

通常情况下，程序员需要使用到所声明数据包外部的其他类和类型，这可通过导入数据包资源予以实现。如果两个类属于同一个数据包，则不需要进行额外的导入工作，如下所示：

```
package animals
data class Buffalo(val mass: Int, val maxSpeed: Int, var isDead: Boolean = false)
```

在下列代码片段中，类并不需要被导入当前程序中，该类与 Lion 类处于同一个数据包（animals），如下所示：

```
package animals
class Lion(val mass: Int, val maxSpeed: Int) {

 fun kill(animal: Buffalo) { // Buffalo type used with our import
   if (!animal.isDead) {
     println("Lion attacking animal.")
     animal.isDead = true
     println("Lion kill successful.")
   }
 }
}
```

当导入独立包中的类、函数、接口以及类型时，须使用 import 关键字，随后是相应的包名。例如，下列 main 函数存在于默认包中。因此，如果需要使用到 main 函数中的 Lion 类和 Buffalo 类，需要通过 import 关键字对其进行导入，如下所示：

```
import animals.Buffalo
import animals.Lion

fun main(args: Array<String>) {
 val lion = Lion(190, 80)
 val buffalo = Buffalo(620, 60)
 println("Buffalo is dead: ${buffalo.isDead}")
 lion.kill(buffalo)
 println("Buffalo is dead: ${buffalo.isDead}")
}
```

1.2.2 面向对象程序设计

截至目前，前述内容已在多个示例中使用到了类这一概念，但尚未对此予以详细讨论。本节将介绍类以及 Kotlin 中其他的面向对象结构。

1. 简介

在高级程序设计语言面世之初，程序均采用面向过程方式加以编写，其中采用了一系列的定义良好的结构化步骤编写程序。

随着软件产业的不断发展，计算机程序也越加庞大，因而有必要设计一种更优的方案以解决软件设计问题。因此，面向对象程序设计语言应运而生。

面向对象语言围绕对象和数据组织模块，而不再是动作以及序列逻辑。在面向对象程序设计语言中，实现了对象、类和接口的编写、扩展和继承机制，以打造工业级别的软件产品。

类是一个可修改和可扩展的程序模板，用于创建对象，并通过变量、常量和属性来维护状态。

类中一般定义了特征和行为。这里，特征体现为变量，而行为则以方法的形式予以实现。其中，方法表示为特定于某个类或者类集合的函数；同时，类具有从其他类继承特征和行为的能力，这种能力称作继承。

Kotlin 是一种完全面向对象的程序设计语言，并支持面向对象语言的全部特性。在 Kotlin 中，仅支持单继承机制，这一点与 Java 和 Ruby 保持一致。某些语言，例如 C++，则支持多重继承。多重继承的一个副作用是管理问题，例如名称冲突问题。另外，继承自其他类的类称作子类，而所继承的类则称作超类。

作为一种结构，接口则需要强制执行类中的特征和行为。基于接口的行为强制操作其实现可描述为：在某个类中实现接口。另外，接口也可扩展为另一个接口，这一点与类结构类似。

最后，对象则表示为类实例，其中包含了自身的唯一状态。

2. 与类协同工作

类可通过 class 关键字加以声明，并于随后附以类名，如下所示：

```
class Person
```

与之前示例程序类似，类也需要相应的代码体。其中设置了相关特征和行为。类体可通过花括号实现，如下所示：

```
class HelloPrinter {
  fun printHello() {
    println("Hello!")
  }
}
```

在上述代码片段中，类命名为 HelloPrinter，其中声明了一个函数。相应地，声明于类中的函数称作方法，也称作行为。当方法声明完毕后，即可被所有的类实例加以使用。

3．生成对象

声明类实例（或者对象实例）与变量的声明十分类似。下列代码生成了一个 HelloPrinter 类实例：

```
val printer = HelloPrinter()
```

这里，printer 表示为 HelloPrinter 类实例。HelloPrinter 类名后的括号用于调用 HelloPrinter 类的主构造方法。其中，构造方法类似于一个函数，但用于初始化某个类型的对象。

HelloPrinter 类中声明的函数可直接通过 printer 对象调用，如下所示：

```
printer.printHello() // Prints hello
```

少数情况下，可能需要函数直接从类中调用，且不需要创建一个对象。对此，可使用伴生对象。

4．伴生对象

通过关键字 companion 和 object，可在类中声明伴生对象。随后，即可使用伴生对象中的静态函数，如下所示：

```
class Printer {
  companion object DocumentPrinter {
    fun printDocument() = println("Document printing successful.")
  }
}

fun main(args: Array<String>) {
  Printer.printDocument() // printDocument() invoked via companion object
  Printer.Companion.printDocument() // also invokes printDocument()
                                    //via a companion object
}
```

某些时候，需要对伴生对象设置一个标识符。对此，可将具体名称放置于 object 关键字之后。考察下列示例：

```
class Printer {
  companion object DocumentPrinter { // Companion object identified by DocumentPrinter
    fun printDocument() = println("Document printing successful.")
  }
}

fun main(args: Array<String>) {
  Printer.DocumentPrinter.printDocument() // printDocument() invoked via
                                          // a named companion object
}
```

5．属性

类中可定义属性，并通过 var 和 val 关键字加以声明。例如，在下列代码片段中，Person 类包含了 3 个属性，即 age、firstName 和 surname。

```
class Person { var age = 0
  var firstName = ""
  var surname = ""
}
```

属性可通过类实例加以访问，对应操作方式为：实例标识符+"."+属性名。例如，在下列代码片段中，创建了 Person 类实例（名为 person），并通过访问相关属性，向 firstName、surname 和 age 属性赋值。

```
val person = Person()
person.firstName = "Raven"
person.surname = "Spacey"
person.age = 35
```

1.3 Kotlin 的优点

如前所述，Kotlin 旨在设计成一个更好的 Java，因而与 Java 相比，Kotlin 的优势主要体现在以下几方面：

❑ 空安全性。NullPointerException 常出现于 Java 中，通过空安全类型系统，Kotlin

在一定程度上缓解了这一问题。
- 扩展函数。函数可更方便地添加至类中,并可通过多种方式扩展各项功能,即 Kotlin 中的扩展函数。
- 单例模式。在 Kotlin 程序中,可更加方便地实现单例模式,而 Java 在这一方面则表现得较为烦琐。
- 数据类。当编写程序时,一种常见的情形是定义一个类,其唯一功能是加载数据,而这一枯燥的工作常会涉及大量代码。Kotlin 的数据类则简化了这项操作——创建加载数据的类仅须编写一行代码。
- 函数类型。与 Java 不同,Kotlin 包含函数类型,这也使得函数可作为参数接收另一个函数,同时也可返回一个函数。

1.4 利用 Kotlin 开发 Android 应用程序

前述内容对 Kotlin 所体现的一些强大功能进行了简要的分析。在接下来的章节中,我们将探讨如何在 Android 应用程序开发中使用这些特性——这也是 Kotlin 的一个亮点。

下面首先对当前任务设置相关系统。开发 Android 应用程序的一个主要的需求是获得一个合适的 IDE——虽然这并非必需,但会简化开发过程。对此存在多种选择方案,其中最受欢迎的 IDE 包括以下几款:
- Android Studio。
- Eclipse。
- IntelliJ IDE。

对于 Android 开发来讲,Android Studio 则是一款功能最为强大的 IDE。本书所涉及的与 Android 相关的内容均采用 Android Studio。

1.4.1 设置 Android Studio

在本书编写时,Android Studio 3.0 为当前最新版本,并附带完整的 Kotlin 支持。

读者可访问 https://developer.android.com/studio/preview/index.html,以下载 Android Studio 3.0 的 Canary 版本。在下载完毕后,可打开下载包或可执行文件,并遵循相关的安装指令。其间,安装向导可指导用户完成安装步骤,如图 1.15 所示。

随后将提示用户选择 Android Studio 配置类型,如图 1.16 所示。

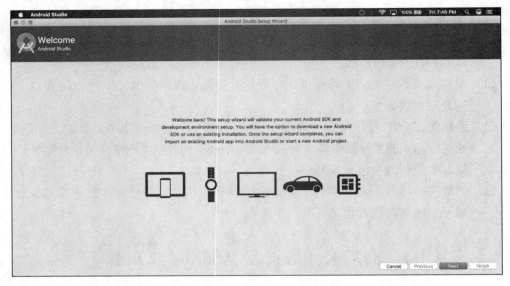

图 1.15　Android Studio 安装向导

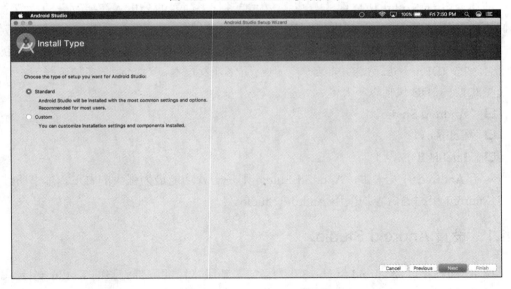

图 1.16　Android Studio 配置类型

接下来，选择 Standard 配置项，在随后的操作中单击 Verify Settings 窗口上的 Finish 按钮。此时，Android Studio 将下载配置所需的组件，在下载过程中，读者须稍作等待，如图 1.17 所示。

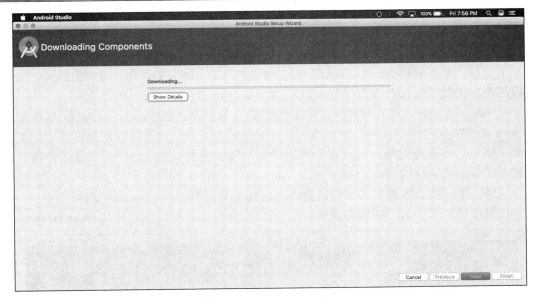

图 1.17　下载配置所需的组件

待组件下载完毕后，单击 Finish 按钮。接下来将显示开始界面，并可使用 Android Studio，如图 1.18 所示。

图 1.18　登录界面

1.4.2 构建第一个 Android 应用程序

下面讨论如何利用 Android Studio 创建简单的 Android 应用程序，即 HelloApp。该应用程序在单击相关按钮后将在屏幕上显示"Hello World!"。

在 Android Studio 的开始界面中，单击 Start a new Android Studio project，进而配置与 App 相关的某些细节内容，例如应用程序名称、公司域名以及项目所处位置。如果尚无公司域名，可在公司域名输入框中填写任何有效的域名。当前项目仅供展示使用，因而暂不需要使用合法的域名。

另外，还需要设置项目的保存位置，同时选中复选框以包含 Kotlin 支持。

在参数填写完毕后，随后将显示如图 1.19 所示的窗口，即 Target Android Devices。

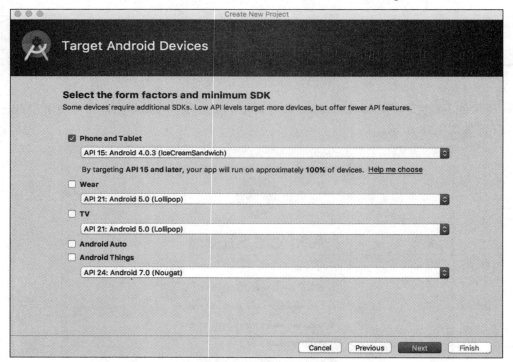

图 1.19　Target Android Devices 窗口

这里需要指定目标设备。考虑到当前应用程序运行于智能手机设备上，因而可选中 Phone and Tablet 复选框。Phone and Tablet 下方是一个选项下拉菜单，用于指定所建项目的目标 API 级别（level）。此处，API 级别表示为一个整数，用于唯一标识 Android 平台所提供的框架 API 版本。此处选择 API level 15 并进入下一个窗口，即 Add an Activity to Mobile 窗口，如图 1.20 所示。

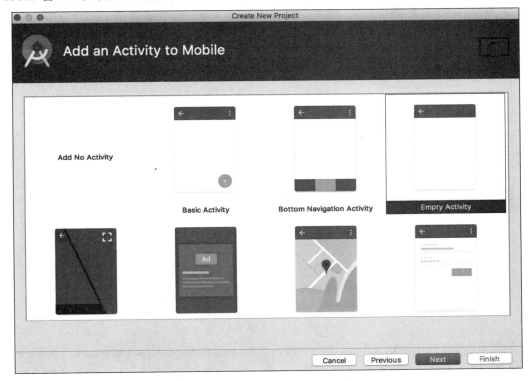

图 1.20 Add an Activity to Mobile 窗口

在下一个窗口中，需要选择一个活动（activity）并添加至当前应用程序中。这里，活动是包含了唯一用户界面的单一画面——类似于一个窗口。第 2 章将对此予以详细介绍。当前，可选择空活动并进入下一个窗口。

接下来需要配置所指定的活动，并将其命名为 HelloActivity，同时确保选中 Generate Layout File 和 Backwards Compatibility 复选框，如图 1.21 所示。

单击 Finish 按钮后，稍作等待后即完成项目的设置。

当设置结束后，即可看到包含了项目文件的 IDE 窗口。

💡 提示：

在项目开发期间，常会遇到与缺少必要的项目组件相关的错误。对此，可通过 SDK 管理器进行下载。

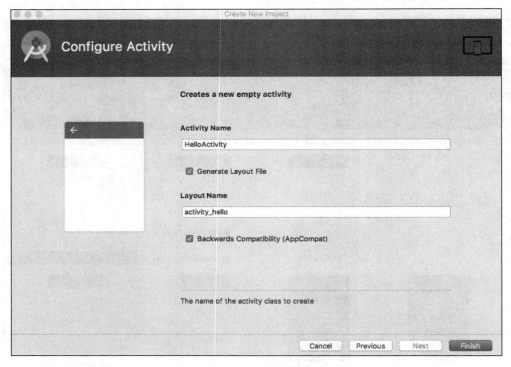

图 1.21　配置活动

确保已经打开了 IDE 的项目窗口（在导航栏上选择 View | Tool Windows | Project 命令）；同时，Android 视图当前已从下拉列表（位于 Project 窗口上方）中被选取。随后将会看到窗口左侧中的相关文件，如下所示。

- app | java | com.mydomain.helloapp | HelloActivity.java：即应用程序的主活动。当构建并运行应用程序时，系统将生成该活动实例。
- app | res | layout | activity_hello.xml：HelloActivity 用户界面定义于该 XML 文件中，其中包含了置于 ConstraintLayout 的 ViewGroup 中的 TextView 元素。TextView 的文本内容被设置为"Hello World!"。
- app | manifests | AndroidManifest.xml：AndroidManifest 文件用于描述应用程序的基本特征。除此之外，该文件中还定义了应用程序的组件。
- Gradle Scripts | build.gradle：当前项目中显示了两个 build.gradle 文件。其中，第一个 build.gradle 文件用于当前项目；第二个 build.gradle 文件则用于应用程序组件。对于 Gradle 工具的编译过程配置，以及应用程序的构建，一般常会用到组件的 build.gradle 文件。

> **注意：**
> Gradle 是一个开源的构建自动化系统，用于项目配置的声明。在 Android 中，Gradle 用作构建工具，旨在构建数据包以及管理应用程序间的依赖关系。

1. 构建用户界面

用户界面（UI）是用户与应用程序交互的主要方式。Android 应用程序的用户界面可通过创建、操控布局文件予以实现。这里，布局文件是位于 app | res | layout 中的 XML 文件。

当对 HelloApp 构建布局文件时，需要执行以下 3 个步骤：

（1）向布局文件中添加 LinearLayout。

（2）将 TextView 置于 LinearLayout 中，并移出所包含的 android:text 属性。

（3）向 LinearLayout 中添加一个按钮。

打开 activity_hello.xml 文件，随后将显示布局编辑器。如果该编辑器位于 Design 视图中，可将其调整至 Text 视图中，即切换布局编辑器下方的选项。当前，布局编辑器如图 1.22 所示。

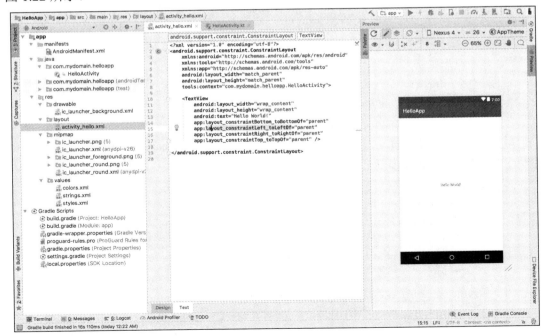

图 1.22　布局编辑器

LinearLayout 表示为一个 ViewGroup，并以单列形式在水平或垂直方向上排列子视图。读者可从下列代码块中复制所需的 LinearLayout 代码片段，并将其粘贴到 TextView 之前的 ConstraintLayout 中。

```
<LinearLayout
        android:id="@+id/ll_component_container"
        android:layout_width="match_parent"
        android:layout_height="match_parent"
        android:orientation="vertical"
        android:gravity="center">
</LinearLayout>
```

此处，将 activity_hello.xml 文件中的 TextView 粘贴至 LinearLayout 元素中，并移除 android:text 属性，如下所示：

```
<LinearLayout
        android:id="@+id/ll_component_container"
        android:layout_width="match_parent"
        android:layout_height="match_parent"
        android:orientation="vertical"
        android:gravity="center">
<TextView
        android:id="@+id/tv_greeting"
        android:layout_width="wrap_content"
        android:layout_height="wrap_content"
        android:textSize="50sp" />
</LinearLayout>
```

最后，还需要向布局文件中添加按钮元素，该元素为 LinearLayout 的子元素。当创建一个按钮时，可使用 Button 元素，如下所示：

```
<LinearLayout
        android:id="@+id/ll_component_container"
        android:layout_width="match_parent"
        android:layout_height="match_parent"
        android:orientation="vertical"
        android:gravity="center">
<TextView
        android:id="@+id/tv_greeting"
        android:layout_width="wrap_content"
        android:layout_height="wrap_content"
        android:textSize="50sp" />
```

```
<Button
    android:id="@+id/btn_click_me"
    android:layout_width="wrap_content"
    android:layout_height="wrap_content"
    android:layout_marginTop="16dp"
    android:text="Click me!"/>
</LinearLayout>
```

切换到布局编辑器的设计视图,当显示用户界面时,即可查看到内容的变化方式,如图1.23所示。

图1.23 显示用户界面,并显示内容的变化方式

当前布局仍存在一个问题,当单击"CLICK ME!"按钮时,该按钮并未执行任何操作。对此,可向该按钮添加单击事件监听器。打开 HelloActivity.java 文件,编辑函数,针对"CLICK ME!"按钮的单击事件添加相关逻辑,并导入所需的数据包。对应代码如下所示:

```
package com.mydomain.helloapp
import android.support.v7.app.AppCompatActivity
import android.os.Bundle
import android.text.TextUtils
import android.widget.Button
```

```kotlin
import android.widget.TextView
import android.widget.Toast

class HelloActivity : AppCompatActivity() {

  override fun onCreate(savedInstanceState: Bundle?) {
    super.onCreate(savedInstanceState)
    setContentView(R.layout.activity_hello)
    val tvGreeting = findViewById<TextView>(R.id.tv_greeting)
    val btnClickMe = findViewById<Button>(R.id.btn_click_me)
    btnClickMe.setOnClickListener {
      if (TextUtils.isEmpty(tvGreeting.text)) {
        tvGreeting.text = "Hello World!"
      } else {
        Toast.makeText(this, "I have been clicked!",
                    Toast.LENGTH_LONG).show()
      }
    }
  }
}
```

在上述代码片段中，通过使用 findViewById 函数，我们在 activity_hello 布局文件中添加了对 TextView 和 Button 元素的引用。findViewById 函数可用于获取指向布局元素的引用，此类元素位于当前所设置的内容视图中。onCreate 函数的第二行代码将 HelloActivity 的内容视图设置为 activity_hello.xml 布局。

在 findViewById 函数右侧，是位于尖括号中的 TextView 类型，这称作函数泛型，并强制传递至 findViewById 的资源 ID 隶属于一个 TextView 元素。

在添加了引用对象后，可将 onClickListener 设置为 btnClickMe。监听器用于应用程序中所出现的事件。为了在元素的单击操作上执行某个动作，可将包含所执行动作的 lambda 传递至元素的 setOnClickListener 方法中。

当单击 btnClickMe 后，将检测 tvGreeting 并查看是否已设置为包含相关的文本内容。如果 TextView 未被设置文本内容，那么，对应的文本将设置为"Hello World!"；否则，将显示包含"I have been clicked!"文本的提示框。

2．运行应用程序

当运行应用程序时，可单击 IDE 窗口右上方的 Run 'app' (^R)按钮，并选择一个部署目标。随后，HelloApp 将在部署目标上被构建、安装和运行，如图 1.24 所示。

图 1.24　构建、安装和运行 HelloApp

读者可使用某种预置虚拟设备或创建自定义虚拟设备，进而将其用作部署目标。除此之外，还可将物理 Android 设备通过 USB 连接至计算机上，并将其选作部署目标。具体选择方案取决于读者。在选取了部署设备后，单击 OK 按钮即可构建、运行当前应用程序。

运行应用程序后，读者即可看到所创建的布局效果，如图 1.25 所示。

单击"CLICK ME!"后，将显示"Hello World!"，如图 1.26 所示。

图 1.25　所创建的布局效果　　　　图 1.26　显示"Hello World!"

连续单击"CLICK ME!"按钮将显示"I have been clicked!"文本提示消息，如图 1.27 所示。

图1.27 单击"CLICK ME!"按钮将显示"I have been clicked!"文本提示消息

1.5 Web 基础知识

大多数应用程序都会以某种方式与服务器通信。在阅读后续章节之前,读者需要了解一些与 Web 相关的概念,本节将对此予以简要介绍。

1.5.1 Web 的含义

Web 是一个复杂的系统,通过一个或多个协议,可与公共网络上的其他系统进行通信。相应地,协议是管理设备间信息交换的正式的、定义良好的规则系统。

1.5.2 超文本传输协议

Web 上的所有通信都是按照协议进行的。对此,超文本传输协议(HTTP)是一个特别重要的、用于实现系统间通信的协议。

数十亿的图像、视频、文本文件、文档和其他文件每天都在互联网上传输，这些文件都是通过 HTTP 传输的。HTTP 是分布式和超媒体信息系统的应用协议。可以说，它是跨互联网通信的一个基本组成部分。可靠性是使用 HTTP 进行跨系统数据传输的一个主要优点，其原因在于使用了可靠的协议，如传输控制协议（TCP）和互联网协议（IP）。

1.5.3 客户端和服务器

Web 客户端是使用 HTTP 与 Web 服务器通信的应用程序；而 Web 服务器则是为 Web 客户端提供（或服务于）Web 资源的计算机设备。任何可提供 Web 内容的数据形式均可视为 Web 资源。Web 资源可以是媒体文件、HTML 文档、网关等。客户端须通过 Web 内容实现各种功能，如信息呈现和数据操作。

客户端和服务器通过 HTTP 进行通信。使用 HTTP 的一个主要原因是它在数据传输中非常可靠。HTTP 的使用可确保在请求-响应周期中不会发生数据丢失现象。

1.5.4 HTTP 请求和响应

顾名思义，HTTP 请求是 Web 客户机通过 HTTP 发送给服务器的 Web 资源请求；HTTP 响应由服务器发送，并响应于 HTTP 事务中的请求。请求和响应的关系如图 1.28 所示。

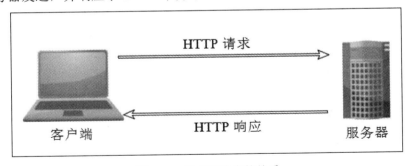

图 1.28 请求和响应的关系

1.5.5 HTTP 方法

HTTP 支持许多请求方法，这些方法也可以称为命令。HTTP 方法指定由服务器执行的操作类型。一些常用的 HTTP 方法如表 1.4 所示。

表 1.4 常用的 HTTP 方法

HTTP 方法	描 述
GET	检索客户端中已命名的资源
POST	在客户端和服务器之间传送数据
DELETE	删除服务器端的某项命名资源
PUT	将客户端收集的数据存储在服务器上的命名资源中
OPTIONS	返回服务器支持的 HTTP 方法
HEAD	检索不包含任何内容的 HTTP 头

1.6 本 章 小 结

　　本章介绍了 Kotlin 语言及其基本知识。在这个过程中，我们学习了如何在计算机上安装 Kotlin、IDE 的含义、如何在 IDE 中编写 Kotlin 程序、如何编写和运行 Kotlin 脚本，以及如何使用 REPL。此外，本章还讲解了如何使用 IntelliJ IDEA 和 Android Studio，之后实现了一个简单的 Android 应用程序。最后，本章还阐述了与网络相关的基本概念。

　　在第 2 章中，我们将通过创建一个 Android 应用程序，以深入了解 Kotlin 程序。此外，还将讨论 Android 应用程序体系结构、Android 应用程序的重要组成部分，以及更多的主题。

第 2 章 构建 Android 应用程序——俄罗斯方块游戏

在第 1 章中，我们简要地介绍了与核心 Kotlin 语言相关的关键主题，相信读者已经了解了 Kotlin 的基本原理，以及强大的面向对象编程方法，并将其用于我们的软件开发中。在此基础上，本章将开发一个 Android 应用程序，以使我们学到的知识得以很好地应用。

本章主要涉及以下内容：
- Android 应用程序组件。
- 视图。
- 视图组。
- 利用 XML 实现布局。
- 字符串和尺寸。
- 处理输入事件。

本章将通过实践方法来学习这些主题，通过 Android 应用程序这一形式，实现一个经典游戏的布局和组件，即俄罗斯方块游戏。在利用 Android 应用程序开发游戏之前，下面首先对 Android 操作系统做一个简短的概述。

2.1 Android 概述

Android 是一个基于 Linux 的移动操作系统，由谷歌开发和维护，主要面向移动电话和平板电脑等智能移动设备。另外，与 Android 操作系统交互的主要界面是图形用户界面（GUI）。Android 设备用户与操作系统环境之间的操控和交互，主要是通过可视化的触摸界面加以实现的，例如单击或滑动等手势操作。

软件可通过 App 的形式安装于 Android OS 中。App 表示为一个应用程序，可在某种环境下运行，并执行一项或多项任务，以实现特定的目标或目标集合。在移动设备上安装应用程序的能力为用户和应用程序开发人员提供了巨大的机会。其中，用户利用应用程序提供的功能实现日常目标，开发人员根据软件应用程序的需求条件，开发满足用户需求的应用程序，并可能获得利润。

对于开发人员来说，Android 为开发高性能应用程序提供了大量的实用工具。这些应用程序可以针对不同的市场，如娱乐、企业和电子商务。另外，应用程序也可以游戏的形式出现。

本章将详细介绍 Android 应用程序框架提供的一些实用工具。

Android 应用程序框架为用户提供了一些组件，从而可以利用这些组件为俄罗斯方块应用程序构建用户界面。Android 中的组件是可重用的程序模板或对象，可用于定义应用程序的各项功能。Android 应用程序框架提供的一些重要组件包括以下几种：

- 活动（Activity）。
- 意图（Intent）。
- 意图过滤器。
- 片段（Fragment）。
- 服务。
- 加载器。
- 内容提供商。

2.1.1 活动

活动代表了一个 Android 组件，它是实现应用程序流和组件到组件交互的核心。活动以类的形式实现。一个活动的实例被 Android 系统用于代码初始化操作。

在创建应用程序的用户界面时，活动是一个非常重要的组件。它提供了一个窗口，可以绘制用户界面元素。简单地说，应用程序屏幕是在活动的基础上创建的。

2.1.2 意图

意图可促进活动间的通信。在 Android 应用程序中，意图可以被视为消息器，用于从应用程序组件请求操作的消息对象。意图可针对动作加以使用，比如请求启动一个活动并在 Android 系统环境中传送广播。

相应地，存在以下两种类型的意图：

- 隐式意图。
- 显式意图。

隐式意图表示为一类消息器对象，且并未特意标识一个应用程序组件以执行某个动作，而是指定了一个执行动作，并允许存在于另一个应用程序中的组件执行该动作。

显式意图通过显式方式指定应执行某个动作的应用程序组件。这一类意图可在应程序内执行相关动作，例如启动某项活动，如图 2.1 所示。

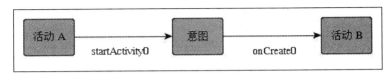

图 2.1　显式意图

2.1.3　意图过滤器

意图过滤器表示为应用程序清单中的一个声明，用于指定组件接收的意图类型。这在许多场合下均十分有用，例如应用程序中的某项活动处理另一个应用程序中的组件所请求的特定动作。针对处理外部请求的活动，意图过滤器可声明于应用程序清单文件中。如果不希望某项活动处理隐式意图，则无须对此声明意图过滤器。

2.1.4　片段

片段（fragment）表示为一个应用程序组件，体现了活动中用户界面的部分内容。与活动类似，片段也包含了可被修改的布局，并在活动窗口中被绘制。

2.1.5　服务

与大多数其他组件不同，服务并不提供用户界面，且用于执行应用程序中的后台处理。另外，服务不需要创建它的应用程序在前台运行。

2.1.6　加载器

作为一种组件，加载器可实现数据源的数据加载，例如内容提供商，以供后续活动或片段中的显示操作使用。

2.1.7　内容提供商

这一类组件有助于控制数据资源的访问行为，相关资源可能存储于应用程序内部或外部。除此之外，内容提供商还可通过公开的应用程序编程接口促进与另一个应用程序的数据共享，如图 2.2 所示。

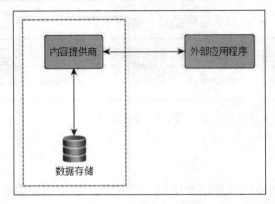

图 2.2　内容提供商

2.2　理解俄罗斯方块游戏

在开发俄罗斯方块游戏之前，需要理解游戏的规则及其约束条件。

俄罗斯方块是一类配对型视频游戏，其中使用到了贴图块。这里，贴图块由 4 个相互连接的方块图案组成，如图 2.3 所示。

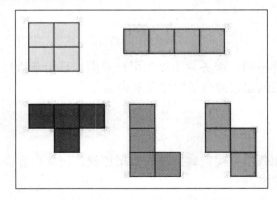

图 2.3　俄罗斯方块游戏中的贴图块

在俄罗斯方块游戏中，随机图块序列由屏幕上方下落，并通过玩家进行控制。其间，可对每个图块执行多种运动操作，包括左移、右移以及旋转操作。另外，还可增加图块的下落速度。游戏的目标是通过下降的图块生成水平方向的一行图块。当产生一行图块时，整个一行图块将被消去。

在理解了游戏规则后，下面首先讨论应用程序的用户界面。

2.3 创建用户界面

如前所述,用户界面是指应用程序用户与 App 之间的一种交互方式。在介绍相关代码之前,需要先期勾勒出 UI 的图形表达结果,并可通过多种工具予以实现,例如 Photoshop。目前,我们仅须描绘出 UI 的大致模样,如图 2.4 所示。

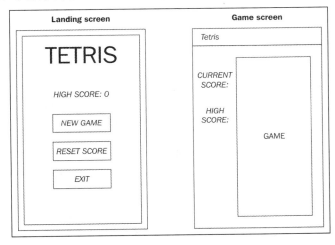

图 2.4 UI 示意图

在上述示意图中可以看出,当前应用程序需要两个不同的区域,即配置区域和体验游戏的游戏区域,且分别需要使用到两个独立的活动,即 MainActivity 和 GameActivity。

MainActivity 表示为应用程序的入口点,包含了用户界面以及与配置窗口相关的全部逻辑。其中,配置窗口的 UI 包含了应用程序标题,显示用户积分榜的视图,以及执行不同动作的 3 个按钮。这里,NEW GAME 按钮将引领用户体验游戏,RESET SCORE 按钮将用户的积分榜重置为 0,EXIT 按钮将关闭当前应用程序。

GameActivity 则是游戏体验区域的程序化模板。其中,将创建视图以及用户和游戏间的逻辑交互。该活动的 UI 包含了一个操作栏(应用程序的标题将显示于其上),两个文本视图(用于显示用户的当前分值以及高分积分榜),以及游戏体验过程中的布局元素。

当前,我们已经了解了应用程序中所需的主要活动,以及用户界面的大致状态。下面讨论用户界面的实际实现。

在 Android Studio 中创建新的 Android 项目,并将其命名为"Tetris no activity"。当

打开 IDE 窗口时，将会看到项目的对应结构。

首先需要向当前项目中添加 MainActivity，此处 MainActivity 应为空活动。右击资源数据包，在弹出的快捷菜单中选择 New | Activity | Empty Activity 命令，即可将 MainActivity 添加至项目中，如图 2.5 所示。

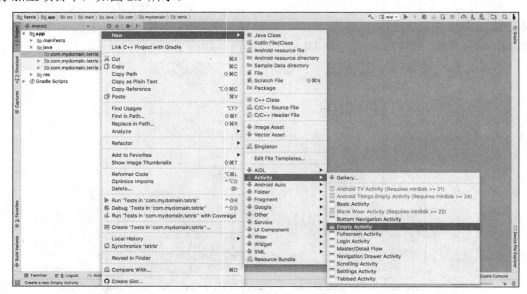

图 2.5　将 MainActivity 添加至项目中

随后，将当前活动命名为 MainActivity，并确保选中 Generate Layout File、Launcher Activity 以及 Backwards Compatibility (AppCompat) 复选框。

一旦将当前活动添加至项目中，即可访问其布局资源文件，如下所示：

```
<?xml version="1.0"
encoding="utf-8"?><android.support.constraint.ConstraintLayout
xmlns:android="http://schemas.android.com/apk/res/android"
    xmlns:app="http://schemas.android.com/apk/res-auto"
    xmlns:tools="http://schemas.android.com/tools"
    android:layout_width="match_parent"
    android:layout_height="match_parent"
    tools:context="com.mydomain.tetris.MainActivity">
</android.support.constraint.ConstraintLayout>
```

资源文件的第一行代码用于指定文件版本以及字符编码方案。在该文件中，采用了 utf-8 字符编码方案。此处，UTF 是指 Unicode 转换格式。作为一种编码格式，UTF 与美国信息交换标准码（ASII 码）具有相同的紧凑格式，且包含了任意的 Unicode 字符，而后者是文本文件中常用的字符格式。随后的 8 行代码定义了 MainActivity UI 中所显示的 ConstraintLayout。

在进一步讨论之前，下面首先详细考察 ConstraintLayout。

2.3.1 ConstraintLayout

ConstraintLayout 表示为视图组的类型，并支持灵活的定位以及应用程序微件的尺寸重置。相应地，各种约束类型均可应用于 ConstraintLayout 上。

1. 边距

边距是指两个布局元素之间的空白区域。当在某个元素上设置侧边距时，则适用于对应的布局约束条件。也就是说，在目标端和源端（添加了边距的元素一侧）之间作为空白区域添加边距，如图 2.6 所示。

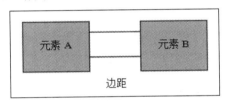

图 2.6　边距

2. 链

链表示为约束条件，并在单一轴向上提供了分组行为。相应地，对应轴向可以是水平轴或垂直轴，如图 2.7 所示。

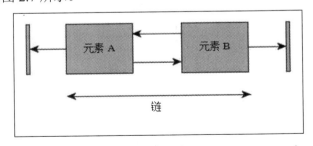

图 2.7　链

如果元素集合均为双向连接，那么，该集合称作链。

3. 尺寸约束

此类约束条件关系到置于某个布局中的微件尺寸。尺寸约束可在微件上进行设置，同时使用到 ConstraintLayout，如图 2.8 所示。

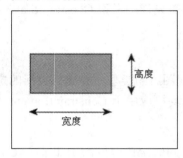

图 2.8　尺寸约束

这里，微件的尺寸可通过 android:layout_width 和 android:layout_height 加以确定，如下所示：

```
<TextView
 android:layout_height="16dp"
 android:layout_width="32dp"/>
```

大多数时候，可能需要某个微件与其父视图分组具有相同的尺寸，对此，可将 match_parent 值赋予尺寸属性中，如下所示：

```
<LinearLayout
 android:layout_width="120dp"
 android:layout_height="100dp">
<TextView
 android:layout_width="match_parent"
 android:layout_height="match_parent"/>
</LinearLayout>
```

另外，如果希望微件的尺寸不固定，同时封装其中的元素，那么，应将 wrap_content 值赋予尺寸属性中，如下所示：

```
<TextView
 android:layout_width="wrap_content"
 android:layout_height="wrap_content"
 android:text="I wrap around the content within me"
 android:textSize="15sp"/>
```

上述内容详细讨论了 ConstraintLayout，以及微件约束条件。下面查看 activity_

main.xml 文件，如下所示：

```xml
<android.support.constraint.ConstraintLayout
xmlns:android="http://schemas.android.com/apk/res/android"
    xmlns:app="http://schemas.android.com/apk/res-auto"
    xmlns:tools="http://schemas.android.com/tools"
    android:layout_width="match_parent"
    android:layout_height="match_parent"
    tools:context="com.mydomain.tetris.MainActivity">
</android.support.constraint.ConstraintLayout>
```

对于 ConstraintLayout 元素，不难发现，其宽度和高度尺寸被设置为 match_parent。这意味着，ConstraintLayout 尺寸被设置为与当前窗口中的尺寸相匹配。包含 xmlns: prefix 的属性用于定义 XML 命名空间。相应地，针对所有 XML 命名空间属性设置的值表示为命名空间 URI。这里，URI 是统一资源标识符的缩写。顾名思义，URI 用于标识命名空间所需的资源。

tools:context 通常被设置为 XML 布局文件中的根元素，并指定布局所关联的活动，当前为 MainActivity。

前述内容探讨了 activity_main.xml 布局，下面向其中添加某些布局元素。在前述游戏界面示意图中，全部布局元素均以垂直排列方式设置。对此，可使用 LinearLayout，如下所示：

```xml
<android.support.constraint.ConstraintLayout
xmlns:android="http://schemas.android.com/apk/res/android"
    xmlns:app="http://schemas.android.com/apk/res-auto"
    xmlns:tools="http://schemas.android.com/tools"
    android:layout_width="match_parent"
    android:layout_height="match_parent"
    tools:context="com.mydomain.tetris.MainActivity">
    <LinearLayout
        android:layout_width="match_parent"
        android:layout_height="match_parent"
        app:layout_constraintBottom_toBottomOf="parent"
        app:layout_constraintLeft_toLeftOf="parent"
        app:layout_constraintRight_toRightOf="parent"
        app:layout_constraintTop_toTopOf="parent"
        android:layout_marginVertical="16dp"
        android:orientation="vertical">
    </LinearLayout>
</android.support.constraint.ConstraintLayout>
```

考虑到 LinearLayout 应与其父元素包含相同的尺寸，因而可将 android:layout_width 和 android:layout_height 设置为 match_parent。随后，可使用 app:layout_constraintBottom_toBottomOf、app:layout_constraintLeft_toLeftOf、app:layout_constraintRight_toRightOf 以及 app:layout_constraintTop_toTopOf 属性确定 LinearLayout 的边界约束条件。

- app:layout_constraintBottom_toBottomOf：将元素的底边对齐到另一个元素的底部。
- app:layout_constraintLeft_toLeftOf：将元素的左边对齐到另一个元素的左侧。
- app:layout_constraintRight_toRightOf：将元素的右边对齐到另一个元素的右侧。
- app:layout_constraintTop_toTopOf：将元素的上方对齐到另一个元素的上方。

此时，LinearLayout 的各边均与父元素的边缘对齐——ConstraintLayout。android:layout_marginVertical 将 16dp 的边距添加至当前元素的上方和下方。

2.3.2 定义尺寸资源

通常情况下，在一个布局文件中，可包含多个元素，并向属性指定同一约束条件值，这一类值应添加至某个尺寸资源文件中。下面将对此创建一个尺寸源文件。在应用程序项目视图中，访问 res | values，并在 dimens 目录中创建新值资源文件，如图 2.9 所示。

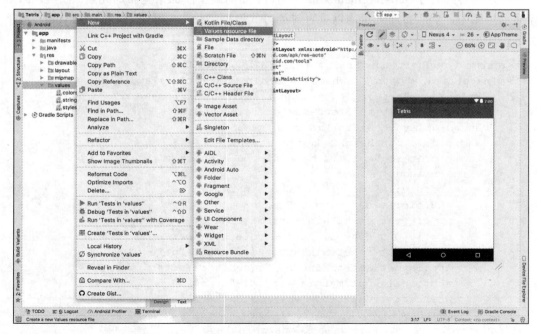

图 2.9 创建新值资源文件

其他文件属性则保留默认状态，如图2.10所示。

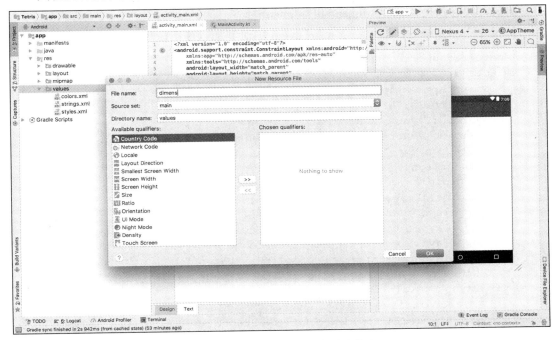

图2.10　其他文件属性保留默认状态

待创建完毕后，打开对应文件，相关内容如下所示：

```
<?xml version="1.0" encoding="utf-8"?>
<resources></resources>
```

dimens.xml文件中的第一行代码表示当前文件中所使用的XML版本，以及字符编码方案。第二行代码包含了一个<resources>资源标签，当前尺寸将在该标签内声明。随后可添加一些尺寸值，如下所示：

```
<?xml version="1.0" encoding="utf-8"?>
<resources>
    <dimen name="layout_margin_top">16dp</dimen>
    <dimen name="layout_margin_bottom">16dp</dimen>
    <dimen name="layout_margin_start">16dp</dimen>
    <dimen name="layout_margin_end">16dp</dimen>
    <dimen name="layout_margin_vertical">16dp</dimen>
</resources>
```

新尺寸通过<dimen>标签加以声明。一般情况下，尺寸名称应采用 Snake Case 方式书写。另外，尺寸值应添加至<dimen>和</dimen>标签之间。

待尺寸添加完毕后，即可在 LinearLayout 中对其加以使用，如下所示：

```
<android.support.constraint.ConstraintLayout
xmlns:android="http://schemas.android.com/apk/res/android"
    xmlns:app="http://schemas.android.com/apk/res-auto"
    xmlns:tools="http://schemas.android.com/tools"
    android:layout_width="match_parent"
    android:layout_height="match_parent"
    tools:context="com.mydomain.tetris.MainActivity">
    <LinearLayout
      android:layout_width="match_parent"
      android:layout_height="match_parent"
      app:layout_constraintBottom_toBottomOf="parent"
      app:layout_constraintLeft_toLeftOf="parent"
      app:layout_constraintRight_toRightOf="parent"
      app:layout_constraintTop_toTopOf="parent"
      android:layout_marginTop="@dimen/layout_margin_top"
 <!— layout_margin_top dimension reference —>
      android:layout_marginBottom="@dimen/layout_margin_bottom"
 <!— layout_margin_top dimension reference —>
      android:orientation="vertical"
      android:gravity="center_horizontal">
    </LinearLayout>
</android.support.constraint.ConstraintLayout>
```

我们已经设置了 LinearLayout 视图组，现在需要向它添加所需的布局视图。在此之前，首先需要理解视图以及视图组这两个概念。

2.3.3 视图

视图表示为一种布局元素，占据屏幕的一个设定区域，并负责绘制和事件处理。视图是 UI 元素和微件（例如文本框、输入框和按钮）的基类，所有的视图均扩展了 View 类。

视图可在某个资源文件的 XML 布局中创建。考察下列代码：

```
<TextView
android:layout_width="wrap_content"
android:layout_height="wrap_content"
android:text="Roll the dice!"/>
```

除了直接在布局文件中生成视图,还可在程序文件中以编程方式实现。例如,可创建一个 TextView 类实例,并向其构造方法中传递 context,进而生成文本视图。对应代码片段如下所示:

```
class MainActivity : AppCompatActivity() {
  override fun onCreate(savedInstanceState: Bundle?) {
    super.onCreate(savedInstanceState)
    setContentView(R.layout.activity_main)
    val textView: TextView = TextView(this)
  }
}
```

2.3.4 视图组

视图组则是一种较为特殊的视图类型,其中包含了多个视图。包含一个或多个视图的视图组一般称作父视图;而所包含的视图则称作子视图。视图组表示为多个其他视图容器的父类。视图组的例子包括 LinearLayout、CoordinatorLayout、ConstraintLayout、RelativeLayout、AbsoluteLayout、GridLayout 以及 FrameLayout。

视图组可在源文件的 XML 布局中创建,如下所示:

```
<LinearLayout
 android:layout_width="wrap_content"
 android:layout_height="wrap_content"
 android:layout_marginTop="16dp"
 android:layout_marginBottom="16dp"/>
```

类似于视图,视图组也可在组件类中以编程方式创建。下列代码片段创建了一个 LinearLayout 类实例,并将 MainActivity 的 context 传递至其构造方法中,进而设置线性布局。

```
class MainActivity : AppCompatActivity() {
  override fun onCreate(savedInstanceState: Bundle?) {
    super.onCreate(savedInstanceState)
    setContentView(R.layout.activity_main)
    val linearLayout: LinearLayout = LinearLayout(this)
  }
}
```

在理解了视图和视图组这两个概念后,可向当前布局中添加多个视图。相应地,可利用<TextView>元素向布局中添加文本视图，并通过<Button>元素添加按钮，如下所示：

```xml
<?xml version="1.0" encoding="utf-8"?>
<android.support.constraint.ConstraintLayout
xmlns:android="http://schemas.android.com/apk/res/android"
    xmlns:app="http://schemas.android.com/apk/res-auto"
    xmlns:tools="http://schemas.android.com/tools"
    android:layout_width="match_parent"
    android:layout_height="match_parent"
    tools:context="com.mydomain.tetris.MainActivity">
  <LinearLayout
        android:layout_width="match_parent"
        android:layout_height="match_parent"
        app:layout_constraintBottom_toBottomOf="parent"
        app:layout_constraintLeft_toLeftOf="parent"
        app:layout_constraintRight_toRightOf="parent"
        app:layout_constraintTop_toTopOf="parent"
        android:layout_marginTop="@dimen/layout_margin_top"
        android:layout_marginBottom="@dimen/layout_margin_bottom"
        android:orientation="vertical">
    <TextView
            android:layout_width="wrap_content"
            android:layout_height="wrap_content"
            android:text="TETRIS"
            android:textSize="80sp"/>
    <TextView
            android:id="@+id/tv_high_score"
            android:layout_width="wrap_content"
            android:layout_height="wrap_content"
            android:text="High score: 0"
            android:textSize="20sp"
            android:layout_marginTop="@dimen/layout_margin_top"/>
    <LinearLayout
            android:layout_width="match_parent"
            android:layout_height="0dp"
            android:layout_weight="1"
            android:orientation="vertical">
      <Button
            android:id="@+id/btn_new_game"
            android:layout_width="wrap_content"
            android:layout_height="wrap_content"
```

```
            android:text="New game"/>
    <Button
            android:id="@+id/btn_reset_score"
            android:layout_width="wrap_content"
            android:layout_height="wrap_content"
            android:text="Reset score"/>
    <Button
            android:id="@+id/btn_exit"
            android:layout_width="wrap_content"
            android:layout_height="wrap_content"
            android:text="exit"/>
    </LinearLayout>
  </LinearLayout>
</android.support.constraint.ConstraintLayout>
```

在游戏界面示意图中，已经添加了两个文本视图以加载应用程序标题、积分榜以及 3 个按钮，进而可执行所需动作。此处使用了两个新属性，即 android:id 和 android:layout_weight。其中，android:id 用于设置布局元素的唯一标识符。在同一个布局中，不可存在两个相同的 ID。对于视图在其父容器中应该占用多少空间，android:layout_weight 属性用于设置优先值，如下所示：

```
<LinearLayout
    android:layout_width="match_parent"
    android:layout_height="match_parent"
    android:orientation="vertical">
    <Button
        android:layout_width="70dp"
        android:layout_height="40dp"
        android:text="Click me"/>
    <View
        android:layout_width="70dp"
        android:layout_height="0dp"
        android:layout_weight="1"/>
</LinearLayout>
```

在上述代码片段中，线性布局中包含了两个子视图。其中，按钮的尺寸约束分别为 70dp 和 40dp。另外一方面，视图的宽度显式地设置为 70dp，其高度设置为 0dp。在 android:layout_weight 属性设置为 1 后，视图的高度被设置为覆盖父视图中的所有剩余空间。

图 2.11 显示了当前布局设计的预览图。

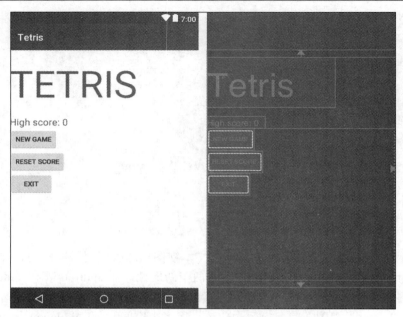

图 2.11　当前布局设计的预览图

与之前的示意图相比，图 2.11 中的某些元素似乎消失了。另外，当前布局内容实现了右对齐，而非前述居中排列。对此，可在线性布局视图组中使用 android:gravity 属性。在下列代码片段中，通过 android:gravity 属性，在两个线性布局中均实现了布局微件的居中设置。

```xml
<?xml version="1.0" encoding="utf-8"?>
<android.support.constraint.ConstraintLayout
xmlns:android="http://schemas.android.com/apk/res/android"
    xmlns:app="http://schemas.android.com/apk/res-auto"
    xmlns:tools="http://schemas.android.com/tools"
    android:layout_width="match_parent"
    android:layout_height="match_parent"
    tools:context="com.mydomain.tetris.MainActivity">
 <LinearLayout
     android:layout_width="match_parent"
     android:layout_height="match_parent"
     app:layout_constraintBottom_toBottomOf="parent"
     app:layout_constraintLeft_toLeftOf="parent"
     app:layout_constraintRight_toRightOf="parent"
     app:layout_constraintTop_toTopOf="parent"
```

```xml
        android:layout_marginTop="@dimen/layout_margin_top"
        android:layout_marginBottom="@dimen/layout_margin_bottom"
        android:orientation="vertical"
        android:gravity="center">
    <!-- Aligns child elements to the centre of view group -->
    <TextView
        android:layout_width="wrap_content"
        android:layout_height="wrap_content"
        android:text="TETRIS"
        android:textSize="80sp"/>
    <TextView
        android:id="@+id/tv_high_score"
        android:layout_width="wrap_content"
        android:layout_height="wrap_content"
        android:text="High score: 0"
        android:textSize="20sp"
        android:layout_marginTop="@dimen/layout_margin_top"/>
    <LinearLayout
        android:layout_width="match_parent"
        android:layout_height="0dp"
        android:layout_weight="1"
        android:orientation="vertical"
        android:gravity="center">
    <!-- Aligns child elements to the centre of view group -->
    <Button
        android:id="@+id/btn_new_game"
        android:layout_width="wrap_content"
        android:layout_height="wrap_content"
        android:text="New game"/>
    <Button
        android:id="@+id/btn_reset_score"
        android:layout_width="wrap_content"
        android:layout_height="wrap_content"
        android:text="Reset score"/>
    <Button
        android:id="@+id/btn_exit"
        android:layout_width="wrap_content"
        android:layout_height="wrap_content"
        android:text="exit"/>
    </LinearLayout>
  </LinearLayout>
</android.support.constraint.ConstraintLayout>
```

其中，android:gravity 设置为 center，微件最终以期望的方式排列。图 2.12 显示了当前布局视图组中添加了 android:gravity 视图组后的效果。

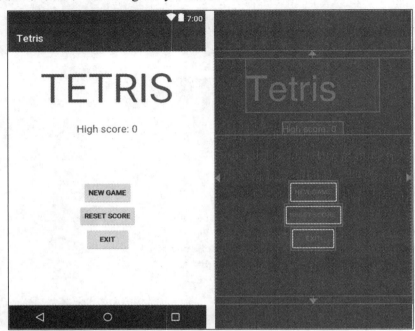

图 2.12　添加了 android:gravity 视图组后的效果

2.3.5　定义字符串资源

前述内容讨论了如何将硬编码字符串作为数值传递至元素属性中（需要设置文本）。总体来说，这并非是最佳方法且一般应尽量避免。相应地，字符串值应添加至字符串资源文件中。

strings.xml 是字符串资源的默认文件，且位于 res | values 目录中，如图 2.13 所示。

通过<string> XML 标签，字符串值可作为字符串资源予以添加。针对目前所用的所有字符串值，需要添加字符串资源。对此，可向字符串资源文件中加入下列代码。

```xml
<resources>
 <string name="app_name">Tetris</string>
 <string name="high_score_default">High score: 0</string>
 <string name="new_game">New game</string>
 <string name="reset_score">Reset score</string>
```

```
<string name="exit">exit</string>
</resources>
```

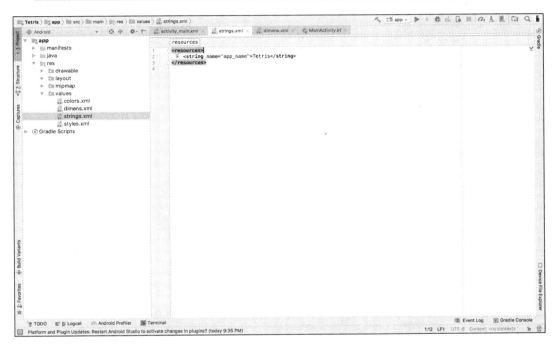

图 2.13　字符串资源的默认文件

另外，有必要编辑 MainActivity 布局文件，进而使用所创建的资源。字符串资源可以用@strings/前缀字符串资源名来引用，如下所示：

```
<?xml version="1.0" encoding="utf-8"?>
<android.support.constraint.ConstraintLayout
xmlns:android="http://schemas.android.com/apk/res/android"
    xmlns:app="http://schemas.android.com/apk/res-auto"
    xmlns:tools="http://schemas.android.com/tools"
    android:layout_width="match_parent"
    android:layout_height="match_parent"
    tools:context="com.mydomain.tetris.MainActivity">
  <LinearLayout
        android:layout_width="match_parent"
        android:layout_height="match_parent"
        app:layout_constraintBottom_toBottomOf="parent"
        app:layout_constraintLeft_toLeftOf="parent"
```

```xml
        app:layout_constraintRight_toRightOf="parent"
        app:layout_constraintTop_toTopOf="parent"
        android:layout_marginTop="@dimen/layout_margin_top"
        android:layout_marginBottom="@dimen/layout_margin_bottom"
        android:orientation="vertical"
        android:gravity="center">
<!-- Aligns child elements to the centre of view group -->
<TextView
        android:layout_width="wrap_content"
        android:layout_height="wrap_content"
        android:text="@string/app_name"
        android:textAllCaps="true"
        android:textSize="80sp"/>
<TextView
        android:id="@+id/tv_high_score"
        android:layout_width="wrap_content"
        android:layout_height="wrap_content"
        android:text="@string/high_score_default"
        android:textSize="20sp"
        android:layout_marginTop="@dimen/layout_margin_top"/>
<LinearLayout
        android:layout_width="match_parent"
        android:layout_height="0dp"
        android:layout_weight="1"
        android:orientation="vertical"
        android:gravity="center">
    <!-- Aligns child elements to the centre of view group -->
    <Button
        android:id="@+id/btn_new_game"
        android:layout_width="wrap_content"
        android:layout_height="wrap_content"
        android:text="@string/new_game"/>
    <Button
        android:id="@+id/btn_reset_score"
        android:layout_width="wrap_content"
        android:layout_height="wrap_content"
        android:text="@string/reset_score"/>
    <Button
        android:id="@+id/btn_exit"
        android:layout_width="wrap_content"
        android:layout_height="wrap_content"
        android:text="@string/exit"/>
</LinearLayout>
```

```
    </LinearLayout>
</android.support.constraint.ConstraintLayout>
```

2.3.6 处理输入事件

在用户与应用程序交互的循环过程中,通过与微件之间的交互,用户可以为流程的执行提供某种形式的输入,这一类输入可通过事件予以捕捉。在 Android 应用程序中,事件可从用户所交互的、特定的视图对象中被捕捉。对于输入事件处理,View 类提供了所需的结构和处理过程。

事件监听器是应用程序中的一个处理程序,并等待 UI 事件的出现。应用程序中可设置多种类型的事件,一些较为常见的事件包括单击事件、触摸事件、长时间按键事件以及文本变化事件。

为了捕捉微件事件并执行某个动作,事件监听器需要在视图中被设置,这可以通过调用一个视图的 set.Listener()方法来实现,并向方法中传递一个 lambda 或指向某个函数的引用。

下列示例展示了按钮上的单击事件的捕捉操作。其中,Lambda 传递至视图类的 **setOnClickListener** 方法中。

```
val button: Button = findViewById<Button>(R.id.btn_send)
button.setOnClickListener {
  // actions to perform on click event
}
```

相应地,也可采用指向函数的引用替代 lambda,如下所示:

```
class MainActivity : AppCompatActivity() {
  override fun onCreate(savedInstanceState: Bundle?) {
    super.onCreate(savedInstanceState)
    setContentView(R.layout.activity_main)
    val btnExit: Button = findViewById<Button>(R.id.btn_exit)
    btnExit.setOnClickListener(this::handleExitEvent)
  }
  fun handleExitEvent(view: View) {
    finish()
  }
}
```

视图类中定义了大量的监听器 setter 方法,如下所示。

❑ setOnClickListener():设置一个函数,并在视图单击操作中被调用。

❑ setOnContextClickListener()：设置一个函数，并在视图的上下文单击操作中被调用。
❑ setOnCreateContextMenuListener()：设置一个函数，并在创建视图快捷菜单时被调用。
❑ setOnDragListener()：设置一个函数，并在视图中出现拖曳事件时被调用。
❑ setOnFocusChangeListener()：设置一个函数，并在视图焦点变化时被调用。
❑ setOnHoverChangeListener()：设置一个函数，并在视图上出现悬停事件时被调用。
❑ setOnLongClickListener()：设置一个函数，并在视图上出现长按操作时被调用。
❑ setOnScrollChangeListener()：设置一个函数，并在视图的滚动位置（X 或 Y）发生变化时被调用。

注意：

事件监听器是应用程序中的一个处理程序，并等待 UI 事件的出现。

在理解了如何处理出入事件后，下面讨论 MainActivity 逻辑实现。

当前主要的活动界面是 App 操作栏，鉴于当前视图并不需要使用到这一元素，因而可暂且将其隐藏，如图 2.14 所示。

图 2.14　App 操作栏

App 操作栏也称作动作栏，且定义为 ActionBar 类的实例。布局中的动作栏微件实例可通过 supportActionBar 访问器变量获得。下列代码将得到一个动作栏，如果未返回空引用，则对其加以隐藏。

```
package com.mydomain.tetris
import android.support.v7.app.AppCompatActivity
import android.os.Bundle
import android.support.v7.app.ActionBar
import android.view.View
import android.widget.Button

class MainActivity : AppCompatActivity() {
  override fun onCreate(savedInstanceState: Bundle?) {
    super.onCreate(savedInstanceState)
    setContentView(R.layout.activity_main)
```

```
    val appBar: ActionBar? = supportActionBar

    if (appBar != null) {
      appBar.hide()
    }
  }
}
```

虽然上述代码实现了所需任务，但通过 Kotlin 的类型安全系统，代码量还可大幅降低，如下所示：

```
package com.mydomain.tetris
import android.support.v7.app.AppCompatActivity
import android.os.Bundle
import android.view.View
import android.widget.Button

class MainActivity : AppCompatActivity() {
  override fun onCreate(savedInstanceState: Bundle?) {
    super.onCreate(savedInstanceState)
    setContentView(R.layout.activity_main)
    supportActionBar?.hide()
  }
}
```

如果 supportActionBar 并非是空对象引用，同时未执行其他操作，则可调用 hide()方法，这可有效地防止出现空指针异常。

对于当前布局中的微件，需要创建对应的对象引用。这对于许多场合均十分有用，例如监听器注册。视图的对象引用可通过下列方式得到：将视图的资源 ID 传递至 findViewById()中。下列代码片段将对象引用传递至 MainActivity（位于 MainActivity.kt 文件内）中。

```
package com.mydomain.tetris
import android.support.v7.app.AppCompatActivity
import android.os.Bundle
import android.view.View
import android.widget.Button
import android.widget.TextView

class MainActivity : AppCompatActivity() {

  var tvHighScore: TextView? = null
```

```
    override fun onCreate(savedInstanceState: Bundle?) {
        super.onCreate(savedInstanceState)
        setContentView(R.layout.activity_main)
        supportActionBar?.hide()

        val btnNewGame = findViewById<Button>(R.id.btn_new_game)
        val btnResetScore = findViewById<Button>(R.id.btn_reset_score)
        val btnExit = findViewById<Button>(R.id.btn_exit)
        tvHighScore = findViewById<TextView>(R.id.tv_high_score)
    }
}
```

当前，用户界面元素的对象引用已设置完毕，下面需要处理其中的事件。对于布局内的所有按钮，需要定义单击监听器（毕竟，单击按钮后需要执行相关操作）。

如前所述，New Game 按钮的唯一功能是令用户进入游戏中（即体验游戏）。对此，需要使用到显式意图。相应地，可向 MainActivity（位于 MainActivity.kt 文件中）添加一个私有函数，其中包含了 New Game 按钮单击操作所包含的执行逻辑，并通过 setOnClickListener() 调用设置指向该函数的引用，如下所示：

```
package com.mydomain.tetris
import android.support.v7.app.AppCompatActivity
import android.os.Bundle
import android.view.View
import android.widget.Button
import android.widget.TextView

class MainActivity : AppCompatActivity() {

    var tvHighScore: TextView? = null

    override fun onCreate(savedInstanceState: Bundle?) {
        super.onCreate(savedInstanceState)
        setContentView(R.layout.activity_main)
        supportActionBar?.hide()
        val btnNewGame = findViewById<Button>(R.id.btn_new_game)
        val btnResetScore = findViewById<Button>(R.id.btn_reset_score)
        val btnExit = findViewById<Button>(R.id.btn_exit)
        tvHighScore = findViewById<TextView>(R.id.tv_high_score)
```

```
    btnNewGame.setOnClickListener(this::onBtnNewGameClick)
}

private fun onBtnNewGameClick(view: View) { }
}
```

随后，可创建新的空活动，并将其命名为 GameActivity。在该活动构建完毕后，可利用相关意图启动 New Game 按钮单击操作上的活动，如下所示：

```
private fun onBtnNewGameClick(view: View) {
  val intent = Intent(this, GameActivity::class.java)
  startActivity(intent)
}
```

函数中的第一行代码将生成 Intent 类的新实例，并将当前上下文和所需的活动类传递至构造方法中。注意，此处传递了 this 作为构造方法中的第一个参数。其中，调用 this 关键字将引用当前实例。因此，实际上将当前活动（MainActivity）作为第一个参数传递至构造方法中。这里，读者可能会稍感疑惑：当需要上下文作为第一个参数时，为何传递一个活动，并作为 Intent 构造方法中的第一个参数？其原因在于：所有活动均为 Context 抽象类的扩展。因此，全部活动均处于自身正确的上下文环境中。

startActivity()方法被调用来启动一个没有预期结果的活动。当某个意图作为其唯一参数被传递时，将启动一个无预期结果的活动。读者可尝试运行该应用程序，并查看单击按钮后的效果。

注意：

Context 表示为 Android 应用程序框架中的一个抽象类，某个上下文的实现由 Android 系统提供。Context 允许访问特定的应用程序资源，访问并调用应用程序级别的操作，例如启动某项活动、发送广播以及接收意图。

下面实现 EXIT 和 RESET SCORE 按钮的单击操作，如下所示：

```
package com.mydomain.tetris
import android.content.Intent
import android.support.v7.app.AppCompatActivity
import android.os.Bundle
import android.view.View
import android.widget.Button
import android.widget.TextView

class MainActivity : AppCompatActivity() {
```

```kotlin
var tvHighScore: TextView? = null

override fun onCreate(savedInstanceState: Bundle?) {
  super.onCreate(savedInstanceState)
  setContentView(R.layout.activity_main)
  supportActionBar?.hide()

  val btnNewGame = findViewById<Button>(R.id.btn_new_game)
  val btnResetScore = findViewById<Button>(R.id.btn_reset_score)
  val btnExit = findViewById<Button>(R.id.btn_exit)
  tvHighScore = findViewById<TextView>(R.id.tv_high_score)
  btnNewGame.setOnClickListener(this::onBtnNewGameClick)
  btnResetScore.setOnClickListener(this::onBtnResetScoreClick)
  btnExit.setOnClickListener(this::onBtnExitClick)
}

private fun onBtnNewGameClick(view: View) {
  val intent = Intent(this, GameActivity::class.java)
  startActivity(intent)
}

private fun onBtnResetScoreClick(view: View) {}

private fun onBtnExitClick(view: View) {
  System.exit(0)
}
}
```

当整数 0 作为参数被传递时，onBtnExitClick 函数中的 System.exit()被调用，将终止程序的进一步执行并退出。最后一项需要完成的任务是处理单击事件，进而实现重置积分榜时的操作逻辑。对此，首先需要实现某些数据存储逻辑，以存储高分值。此处将使用到 SharedPreferences。

2.3.7 与 SharedPreferences 协同工作

SharedPreferences 定义为一个接口，用于存储、访问和修改数据。SharedPreferences API 支持键-值对集合的数据存储操作。

下面将利用 SharedPreferences 接口并针对当前 App 处理数据存储问题。首先，可在项目的源目录 storage 中创建一个数据包（右击源目录，在弹出的快捷菜单中选择 New | Package 命令），如图 2.15 所示。

第 2 章 构建 Android 应用程序——俄罗斯方块游戏

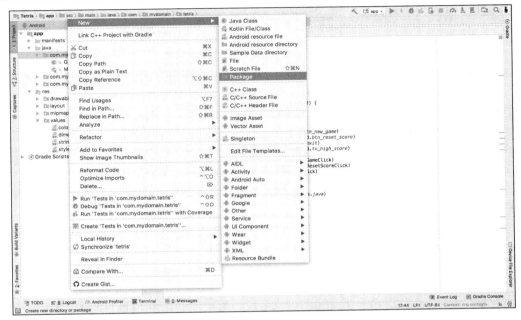

图 2.15 创建一个数据包

接下来，在 storage 数据包中定义一个新的 Kotlin 类 AppPreferences，并向该类文件中输入下列代码：

```
package com.mydomain.tetris.storage
import android.content.Context
import android.content.SharedPreferences

class AppPreferences(ctx: Context) {

  var data: SharedPreferences = ctx.getSharedPreferences
                      ("APP_PREFERENCES", Context.MODE_PRIVATE)

  fun saveHighScore(highScore: Int) {
    data.edit().putInt("HIGH_SCORE", highScore).apply()
  }

  fun getHighScore(): Int {
    return data.getInt("HIGH_SCORE", 0)
  }

  fun clearHighScore() {
```

```
    data.edit().putInt("HIGH_SCORE", 0).apply()
  }
}
```

在上述代码片段中，当生成类实例时，需要向类的构造方法中传递 Context，进而可访问 getSharedPreferences()方法，该方法将获取特定的预置文件。另外，该文件可通过字符串名称（即传递至 getSharedPreferences()方法中的第一个参数）被识别。

saveHighScore()函数接收一个整数参数，即所保存的最高积分值，并作为该方法的唯一参数。data.edit()将返回一个 Editor 对象，进而可修改预置文件。当存储预置文件中的整数时，可调用编辑器的 putInt()方法。其中，传递至 putInt()方法中的第一个参数表示为键字符串，用于访问相应的存储值。该方法的第二个参数为整数值，当前表示为所存储的最高积分值。

通过调用 data.getInt()、getHighScore()方法将返回高积分值。这里，getInt()函数由 SharedPreferences 实现，并提供了整数存储值的读取访问操作。HIGH_SCORE 则表示为检索值的唯一标识符。其中，传递至该函数中的第二个参数 0 定义了所返回的默认值——此时不存在与特定键对应的数值。

clearHighScore()将分值重置为 0。也就是说，简单地利用 0 值覆写与 HIGH_SCORE 键对应的数值。

目前，AppPreferences 工具类定义完毕，下面继续完成 MainActivity 中的 onBtnResetScoreClick()函数，如下所示：

```
private fun onBtnResetScoreClick(view: View) {
  val preferences = AppPreferences(this)
  preferences.clearHighScore()
}
```

当单击高分值重置按钮时，对应值将被重置为 0。当执行此类动作时，读者可能希望得到用户的某些反馈。对此，可使用 Snackbar 以实现这一功能。

当在 Android 应用程序中使用 Snackbar 类时，需要将 Android 设计支持库依赖项添加到模块级的构建脚本中。针对于此，可在 build.gradle 的依赖项闭包中添加下列代码：

```
implementation 'com.android.support:design:26.1.0'
```

接下来，模块级别的 build.gradle 脚本如下所示：

```
apply plugin: 'com.android.application'
apply plugin: 'kotlin-android'
apply plugin: 'kotlin-android-extensions'

android {
  compileSdkVersion 26
```

```
buildToolsVersion "26.0.1"
defaultConfig {
  applicationId "com.mydomain.tetris"
  minSdkVersion 15
  targetSdkVersion 26
  versionCode 1
  versionName "1.0"
  testInstrumentationRunner "android.support.test.runner
  .AndroidJUnitRunner"

}
buildTypes {
  release {
    minifyEnabled false
    proguardFiles getDefaultProguardFile('proguard-android.txt'),
                'proguard-rules.pro'
  }
 }
}

dependencies {
  implementation fileTree(dir: 'libs', include: ['*.jar'])
  implementation "org.jetbrains.kotlin:
                kotlin-stdlib-jre7:$kotlin_version"
  implementation 'com.android.support:appcompat-v7:26.1.0'
  implementation 'com.android.support.constraint:
                constraint-layout:1.0.2'
  testImplementation 'junit:junit:4.12'
  androidTestImplementation 'com.android.support.test:runner:1.0.1'
  androidTestImplementation 'com.android.support.test.espresso:espressocore:
3.0.1'
  implementation 'com.android.support:design:26.1.0'
  // adding android design support library
}
```

在修改完毕之后,单击编辑器窗口显示消息中的 Sync Now 按钮,即可同步当前项目,如图 2.16 所示。

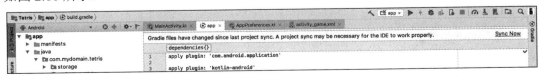

图 2.16 同步当前项目

下面修改 onBtnResetClick()方法，在执行了积分榜重置后，将以 Snackbar 的形式提供用户反馈信息，如下所示：

```
private fun onBtnResetScoreClick(view: View) {
  val preferences = AppPreferences(this)
  preferences.clearHighScore()
  Snackbar.make(view, "Score successfully reset",
           Snackbar.LENGTH_SHORT).show()
}
```

单击 RESET SCORE 按钮即可成功地重置玩家的高分积分榜，如图 2.17 所示。

图 2.17　重置玩家的高分积分榜

在进一步讨论之前，还需要更新显示于 MainActivity 中的、高积分文本视图中的文本内容，并体现相应的重置分数。这可通过修改文本视图中的文本内容加以实现，如下所示：

```
private fun onBtnResetScoreClick(view: View) {
  val preferences = AppPreferences(this)
  preferences.clearHighScore()
  Snackbar.make(view, "Score successfully reset",
           Snackbar.LENGTH_SHORT).show()
  tvHighScore?.text = "High score: ${preferences.getHighScore()}"
}
```

2.3.8 实现游戏活动布局

当前,我们已经创建了主活动布局。在结束本章之前,还有必要针对 GameActivity 构建布局。对此,可打开 activity_game.xml 文件,并添加下列代码:

```xml
<?xml version="1.0" encoding="utf-8"?>
<android.support.constraint.ConstraintLayout
    xmlns:android="http://schemas.android.com/apk/res/android"
    xmlns:app="http://schemas.android.com/apk/res-auto"
    xmlns:tools="http://schemas.android.com/tools"
    android:layout_width="match_parent"
    android:layout_height="match_parent"
    tools:context="com.mydomain.tetris.GameActivity">
   <LinearLayout
        android:layout_width="match_parent"
        android:layout_height="match_parent"
        android:orientation="horizontal"
        android:weightSum="10"
        android:background="#e8e8e8">
     <LinearLayout
         android:layout_width="wrap_content"
         android:layout_height="match_parent"
         android:orientation="vertical"
         android:gravity="center"
         android:paddingTop="32dp"
         android:paddingBottom="32dp"
         android:layout_weight="1">
       <LinearLayout
           android:layout_width="wrap_content"
           android:layout_height="0dp"
           android:layout_weight="1"
           android:orientation="vertical"
           android:gravity="center">
         <TextView
             android:layout_width="wrap_content"
             android:layout_height="wrap_content"
             android:text="@string/current_score"
             android:textAllCaps="true"
             android:textStyle="bold"
             android:textSize="14sp"/>
         <TextView
```

```xml
            android:id="@+id/tv_current_score"
            android:layout_width="wrap_content"
            android:layout_height="wrap_content"
            android:textSize="18sp"/>
    <TextView
            android:layout_width="wrap_content"
            android:layout_height="wrap_content"
            android:layout_marginTop="@dimen/layout_margin_top"
            android:text="@string/high_score"
            android:textAllCaps="true"
            android:textStyle="bold"
            android:textSize="14sp"/>
    <TextView
            android:id="@+id/tv_high_score"
            android:layout_width="wrap_content"
            android:layout_height="wrap_content"
            android:textSize="18sp"/>
    </LinearLayout>
    <Button
            android:id="@+id/btn_restart"
            android:layout_width="wrap_content"
            android:layout_height="wrap_content"
            android:text="@string/btn_restart"/>
    </LinearLayout>
    <View
        android:layout_width="1dp"
        android:layout_height="match_parent"
        android:background="#000"/>
    <LinearLayout
        android:layout_width="0dp"
        android:layout_height="match_parent"
        android:layout_weight="9">
    </LinearLayout>
  </LinearLayout>
</android.support.constraint.ConstraintLayout>
```

该布局内的大多数视图属性之前均有所应用，因而此处不再赘述。例外情况是 android:background 和 android:layout_weightSum 属性。

android:background 属性用于设置视图或视图组的背景颜色。相应地，#e8e8e8 和#000 作为值传递至两个实例中，android:background 则用于当前布局中。这里，#e8e8e8 表示为灰色的十六进制颜色代码，而#000 则表示黑色代码。

android:layout_weightSum 定义了视图组中的最大权值和，其计算方式为：视图组中所有子视图的 layout_weight 值之和。activity_game.xml 中的第一个线性布局将全部子视图的权值和声明为 10。因此，线性布局的直接子元素分别包含了 1 和 9 的布局权值。

此处使用了 3 个字符串资源，且之前尚未添加至字符串资源文件中。下面将下列字符串资源添加至 strings.xml 中，如下所示：

```xml
<string name="high_score">High score</string>
<string name="current_score">Current score</string>
<string name="btn_restart">Restart</string>
```

最后，对于高分积分榜以及当前积分文本视图，还需要向游戏活动中加入一些简单的逻辑，如下所示：

```kotlin
package com.mydomain.tetris

import android.os.Bundle
import android.support.v7.app.AppCompatActivity
import android.widget.Button
import android.widget.TextView
import com.mydomain.tetris.storage.AppPreferences

class GameActivity: AppCompatActivity() {

 var tvHighScore: TextView? = null
 var tvCurrentScore: TextView? = null
 var appPreferences: AppPreferences? = null

 public override fun onCreate(savedInstanceState: Bundle?) {
   super.onCreate(savedInstanceState)
   setContentView(R.layout.activity_game)
   appPreferences = AppPreferences(this)

   val btnRestart = findViewById<Button>(R.id.btn_restart)

   tvHighScore = findViewById<TextView>(R.id.tv_high_score)
   tvCurrentScore = findViewById<TextView>(R.id.tv_current_score)

   updateHighScore()
   updateCurrentScore()
 }

 private fun updateHighScore() {
   tvHighScore?.text = "${appPreferences?.getHighScore()}"
 }
```

```
 private fun updateCurrentScore() {
   tvCurrentScore?.text = "0"
 }
}
```

在上述代码片段中,生成了指向布局视图元素的对象引用。除此之外,还声明了 **updateHighScore()** 和 **updateCurrentScore()** 函数,这两个函数在生成视图时被调用,并将默认的分值显示于当前分值中,同时将布局文件中的高分值设置于文本视图中。

保存修改后的项目,构建并运行当前应用程序。当应用程序显示所创建的布局时,单击 **NEW GAME** 按钮,如图 2.18 所示。

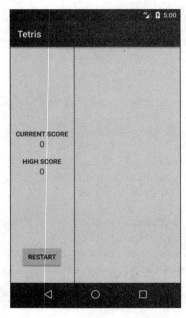

图 2.18　所创建的布局

布局右侧当前未包含任何内容,同时,这也是游戏的体验区域。第 3 章将实现这一部分内容。最后一项需要了解的内容是 App 清单文件,下面将对此加以讨论。

2.4　App 清单文件

App 清单文件是每个 Android 应用程序均包含的 XML 文件,并位于应用程序根文件夹的清单文件中。清单文件加载了与 Android 操作系统的应用程序相关的重要信息。在应用程序运行之前，Android 系统需要读取应用程序的 androidManifest.xml 文件中的信息。其中，需要在 App 清单文件中注册的一些信息包括以下方面：

- 应用程序的 Java 包名。
- 应用程序中的活动。
- 用于应用程序中的服务。
- 将隐式意图转发至某个活动的意图过滤器。
- 用于应用程序中的广播接收者的描述。
- 应用程序中与内容提供商相关的数据。
- 实现了各种应用程序组件的类。
- 应用程序所需的授权。

下列代码片段展示了 androidManifest.xml 文件的一般结构，其中包含了清单文件中所有的元素和声明。

```xml
<?xml version="1.0" encoding="utf-8"?>
<manifest>
  <uses-permission />
  <permission />
  <permission-tree />
  <permission-group />
  <instrumentation />
  <uses-sdk />
  <uses-configuration />
  <uses-feature />
  <supports-screens />
  <compatible-screens />
  <supports-gl-texture />

  <application>
    <activity>
      <intent-filter>
        <action />
        <category />
        <data />
      </intent-filter>
      <meta-data />
    </activity>
    <activity-alias>
      <intent-filter>
```

```xml
        . . .
      </intent-filter>
      <meta-data />
    </activity-alias>
    <service>
      <intent-filter>
        . . .
      </intent-filter>
      <meta-data/>
    </service>
    <receiver>
      <intent-filter> . . . </intent-filter>
      <meta-data />
    </receiver>
    <provider>
      <grant-uri-permission />
      <meta-data />
      <path-permission />
    </provider>
    <uses-library />
  </application>
</manifest>
```

不难发现，清单文件中涵盖了大量的元素，本书将对其中的大部分内容予以介绍。实际上，当前俄罗斯方块游戏中已经涉及了部分清单元素。读者可打开该游戏的 androidManifest.xml 文件，对应内容如下所示：

```xml
<?xml version="1.0" encoding="utf-8"?>
<manifest xmlns:android="http://schemas.android.com/apk/res/android"
    package="com.mydomain.tetris">
  <application
      android:allowBackup="true"
      android:icon="@mipmap/ic_launcher"
      android:label="@string/app_name"
      android:roundIcon="@mipmap/ic_launcher_round"
      android:supportsRtl="true"
      android:theme="@style/AppTheme">
    <activity android:name=".MainActivity">
      <intent-filter>
        <action android:name="android.intent.action.MAIN" />
        <category android:name="android.intent.category.LAUNCHER" />
      </intent-filter>
```

```
        </activity>
        <activity android:name=".GameActivity" />
    </application>
</manifest>
```

清单文件中的元素采用字母顺序排列,其中包括以下方面:
- <action>。
- <activity>。
- <application>。
- <category>。
- <intent-filter>。
- <manifest>。

2.4.1 <action>

<action>用于向意图过滤器中添加某个动作,该元素通常是<intent-filter>元素的子元素。意图过滤器应包含一个或多个此类元素。若未针对意图过滤器声明 action 元素,该过滤器将不会接收 Intent 对象,对应语法格式如下所示:

```
<action name=""/>
```

上述 name 属性指定了所处理的 action 的名称。

2.4.2 <activity>

该元素声明了应用程序中的一个活动。全部活动都需要在 App 清单文件中加以声明,以使 Android 系统对此有所了解。<activity>通常置于父<application>元素中。下列代码片段通过<activity>元素在清单文件中显示了活动声明。

```
<activity android:name=".GameActivity" />
```

其中,name 属性用于指定实现了所声明活动的类名。

2.4.3 <application>

该元素表示为应用程序声明,并包含了子元素以声明应用程序中的组件。下列代码显示了<application>的使用方式。

```xml
<application
    android:allowBackup="true"
    android:icon="@mipmap/ic_launcher"
    android:label="@string/app_name"
    android:roundIcon="@mipmap/ic_launcher_round"
    android:supportsRtl="true"
    android:theme="@style/AppTheme">
  <activity android:name=".MainActivity">
    <intent-filter>
      <action android:name="android.intent.action.MAIN" />
      <category android:name="android.intent.category.LAUNCHER" />
    </intent-filter>
  </activity>
  <activity android:name=".GameActivity" />
</application>
```

上述代码片段中的<application>元素使用了6个属性，如下所示。

- android:allowBackup：用于确定应用程序是否参与备份和恢复基础结构。当该属性设置为 true 时，当前应用程序将通过 Android 系统备份；否则，Android 系统将不会生成应用系统备份。
- android:icon：用于确定应用程序的图标资源；除此之外，还用于确定应用程序组件的图标资源。
- android:label：整体上用于确定应用程序的默认标记。此外，还用于确定应用程序组件的默认标记。
- android:roundIcon：当需要使用到圆形图标资源时，该属性用于确定所用的图标。当启动程序请求使用 App 图标时，Android 框架将返回 android:icon 或 android:roundIcon，具体返回内容取决于设备构建配置。由于可返回二者中的一个图标，因而须针对两个属性指定一个资源。
- android:supportsRtl：该属性用于确定应用程序是否支持自右向左（RTL）布局。当该属性设置为 true 时，应用程序将支持此类布局方式；否则，应用程序将不支持 RTL 布局。
- android:theme：该属性针对应用程序中的所有活动，确定了定义默认主题的样式资源。

2.4.4 <category>

该元素表示为<intent-filter>的子元素，用于确定其父意图过滤器组件的分类名称。

2.4.5 <intent-filter>

该元素用于确定活动、服务、广播接收者组件所响应的意图类型。意图过滤器通常在包含<intent-filter>元素的父组件中加以声明。

2.4.6 <manifest>

<manifest>表示为 App 清单文件的根元素,其中包含了单一的<application>元素,并确定了 xmlns:android 和数据包属性。

2.5 本章小结

本章考察了 Android 应用程序框架,其中涉及 7 种基本的 Android App 组件:活动、意图、意图过滤器、片段、服务、加载器以及内容提供商。

除此之外,本章还介绍了布局的构建过程、约束布局、布局约束类型、字符串、尺寸资源、视图、视图组以及与 SharedPreferences 的协同工作方式。第 3 章将深入讨论俄罗斯方块游戏的场景,并实现游戏操作以及应用程序逻辑。

第 3 章　俄罗斯方块游戏的逻辑和功能

第 2 章讨论了与经典游戏俄罗斯方块相关的一些内容，制定了应用程序的布局，并实现了所设置的布局元素。其中，我们针对应用程序创建了两个活动，即 MainActivity 和 GameActivity。除此之外，还实现了视图的基本特征和行为，但并未涉及 App 的核心体验过程，本章将完成这一功能，主要包括以下内容：
- 异常处理。
- MVP 模式。

3.1　实现游戏体验过程

关于游戏的体验过程，本章主要考察 GameActivity。图 3.1 显示了与此对应的效果。

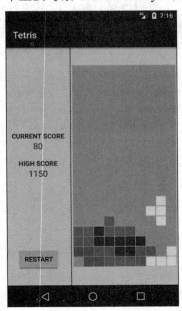

图 3.1　GameActivity 的最终效果

待了解了游戏体验过程的最终效果后，下面开始介绍具体的开发过程。

在第 2 章中，曾讨论了基于贴图的图块匹配游戏，此类贴图经组合后可形成较大的图块。这里，图块由 4 个正方形图案组成，且彼此垂直连接。

3.1.1 图块建模

图块对于俄罗斯方块游戏来说十分重要，需要通过编程方式对此类元素进行建模。对此，可将每个图块视为构造块。构造块包含了一组特性，并可归类至不同的特征和行为中。

1．构造块特征

下列内容体现了构造块所包含的某些特征。

- 形状：图块均包含固定的形状且无法被修改。
- 尺寸：图块包含尺寸特征，即高度和宽度。
- 颜色：图块通常会涵盖某种颜色，对应元素不会发生改变，并在图块的存在过程中进行维护。
- 位置特征：在任意时刻，图块均包含-X 和 Y 轴上的位置。

2．图块的行为

图块的主要行为体现在其独特的运动上，包括平移和旋转操作。其中，平移运动指的是空间两点间的直线运动。在俄罗斯方块中，图块可实现左移、右移以及下移操作。另外，旋转也是刚体运动中的一种运动类型。也就是说，旋转运动涉及自由空间内的对象的旋转行为。在俄罗斯方块游戏中，所有的图块均可实现旋转运动。

在了解了基本的图块特征和行为后，读者可能会产生疑问：如何构成此类块状图案？注意，全部图块特征均应用于一个图块上，其中涉及以下两项内容：

- 图块由 4 个贴图构成。
- 图块中的全部贴图彼此垂直排列。

下面将上述特征转换为程序模型，且仍然从形状建模开始讨论。

3．图块形状建模

形状建模方法取决于多种情况，例如须测量的形状类型，以及在空间维度中建模的特定形状。

与二维形状建模相比，三维形状建模则较为困难。在俄罗斯方块中，图块建模主要在二维环境下进行。在开始采用编程方式建模之前，需要了解即将建模的实际形状。俄罗斯方块游戏存在 7 种基本的图块，即图 3.2 中的 O、I、T、L、J、S 和 Z。

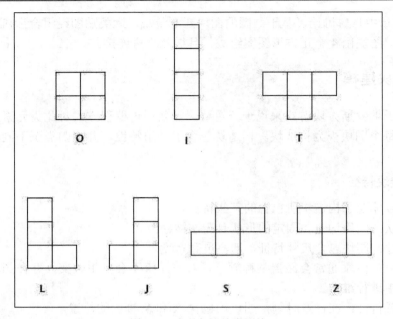

图 3.2 游戏中的基本图块

上述形状占据了边缘构成的边界，对应形状所涉及的空间区域可视为一个轮廓线，这与相框中的相片有几分类似。相应地，我们需要对此进行建模，进而包含独立的形状。鉴于图块形状的二维特征，此处可采用二维字节数组加载图框信息。其中，字节表示为信息的数位单元，一般由 8 位组成，且每一位表示为二进制位。在计算机中，这也是最小的数据单元，对应值为 0 或 1。

对应思想可描述为：使用一个二维数组对形状图框建模，图框所覆盖的区域，其字节值为 1；而那些没有被它覆盖的区域，对应值为 0。考察图 3.3 中的图块。

这里，可以将该形状视为字节的二维数组，其中包含了两行 3 列，如图 3.4 所示。

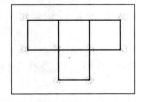

图 3.3 图块

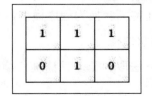

图 3.4 图块的二维数组

字节值 1 被赋予数组单元中，并以此构成了图框的形状。另外一方面，字节值 0 则不属于图框形状。当采用类时，建模过程变得十分简单。首先，需要定义一个函数以生成所需的字节数组结构,并用于存储图框字节。另外,还可在资源数据包中创建一个新的数据包,并将其命名为 helpers。在该数据包中,生成一个文件 HelperFunctions.kt,其中包含开发该 App 过程中的全部帮助函数。随后,打开 HelperFunctions.kt 文件,并向该文件中输入下列代码:

```
package com.mydomain.tetris.helpers

fun array2dOfByte(sizeOuter: Int, sizeInner: Int): Array<ByteArray>
        = Array(sizeOuter) { ByteArray(sizeInner) }
```

上述代码定义了 array2dOfByte()函数,该函数接收两个参数。其中,第一个参数表示数组的行数;第二个参数则表示为字节数组的列数。array2dOFByte()方法生成并返回包含特定属性的新数组。下面将尝试定义 Frame 类,并在资源数据包中创建一个新的数据包,同时将其命名为 models。全部对象模型均在该数据包内打包。在 models 数据包中,在文件 Frame.kt 中定义 Frame 类,并输入下列代码:

```
package com.mydomain.tetris.models

import com.mydomain.tetris.helpers.array2dOfByte

class Frame(private val width: Int) {
  val data: ArrayList<ByteArray> = ArrayList()

  fun addRow(byteStr: String): Frame {
    val row = ByteArray(byteStr.length)

    for (index in byteStr.indices) {
      row[index] = "${byteStr[index]}".toByte()
    }
    data.add(row)
    return this
  }

  fun as2dByteArray(): Array<ByteArray> {
    val bytes = array2dOfByte(data.size, width)
    return data.toArray(bytes)
  }
}
```

Frame 类包含了两个属性,即 width 和 data。其中,width 表示为一个整数属性,并加载图框的宽度值(即图框字节数组中的列数)。data 属性则加载 ByteArray 值空间中的元素数组列表。另外，函数 addRow()和 get(). addRow()接收一个字符串参数，并将每个字符串字符转换为字节表达，随后将字节表达结构添加至字节数组中。接下来，该字节数组将添加至数据列表中。get()方法将数据数组列表转换为字节数组，并返回该数组。

一旦图框建模完毕并加载了图块后，即可对游戏中的各种形状建模。对此，可使用 enum 类。在此之前，可在模块数据包中创建 a Shape.kt 文件，并首先对图 3.5 所示的形状建模。

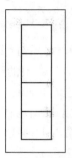

图 3.5　对图块形状建模

如前所述，可将图框视作二维字节数组。具体来说，该字节数组包含了 4 行 1 列数据，且每个单元格中利用字节值 1 进行填充。据此，即可对当前形状建模。在 Shape.kt 文件中，可定义一个 Shape 枚举类，如下所示：

```
enum class Shape(val frameCount: Int, val startPosition: Int) {
  Tetromino(2, 2) {
    override fun getFrame(frameNumber: Int): Frame {
      return when (frameNumber) {
        0 -> Frame(4).addRow("1111")
        1 -> Frame(1)
                .addRow("1")
                .addRow("1")
                .addRow("1")
                .addRow("1")
        else -> throw IllegalArgumentException("$frameNumber is an invalid
                                                 frame number.")
      }
    }
  };
  abstract fun getFrame(frameNumber: Int): Frame
}
```

enum 类的声明方式可描述为：在 class 关键字之前放置 enum 关键字。Shape 枚举类的主构造方法接收两个参数。其中，第一个参数为 frameCount，该整数值指定了形状中的图框数量。第二个参数为 startPosition，用于确定对应形状在 X 轴上的起始位置。在 enum 类文件中，还声明了一个 getFrame()函数。注意，该函数使用了 abstract 关键字。抽象函数不包含任何实现（因而不包含函数体），并用于抽象某种行为，随后的扩展类须对此予以实现。enum 类中的对应代码如下所示：

```
Tetromino(2, 2) {
 override fun getFrame(frameNumber: Int): Frame {
  return when (frameNumber) {
   0 -> Frame(4).addRow("1111")
   1 -> Frame(1)
               .addRow("1")
               .addRow("1")
               .addRow("1")
               .addRow("1")
   else -> throw IllegalArgumentException("$frameNumber is an invalid
                                           frame number.")
  }
 }
};
```

在上述代码块中，enum 实例提供了抽象函数的具体实现。该实例的标识符为 Tetromino。作为参数，可将整数值 2 传递至 Tetromino 构造方法中的 frameCount 和 startPosition 属性。除此之外，Tetromino 还提供了 getFrame()函数实现，即覆写 Shape 所声明的 getFrame()函数。相应地，通过 override 关键字，可实现函数的覆写操作。Tetromino 中的 getFrame()函数接收一个 frameNumber 整数，进而指定了所返回的 Tetromino 图框。读者可能会对此感到疑问：为何 Tetromino 可包含多个图框？这仅仅是图块旋转效果所致。之前看到的单列图块可实现左旋或右旋，进而可得到如图 3.6 所示的效果。

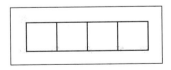

图 3.6　图块的旋转

若传递至 getFrame()函数中的 frameNumber 为 0，getFrame()函数将返回一个 Frame 对象，该对象针对水平状态的 Tetromino 图框建模。若 frameNumber 为 1，则返回垂直状态下的图框对象。

如果 frameNumber 既不为 0，也不为 1，那么，函数将抛出 IllegalArgumentException。

> **注意**：

除了作为一个对象之外，Tetromino 也可表示为一个常量。一般情况下，enum 类常用于定义常量。由于需要实现固定的形状集合，因而 enum 类非常适合对图块形状建模。

当理解了 Shape 类的工作方式后，即可对图块形状建模，如下所示：

```
enum class Shape(val frameCount: Int, val startPosition: Int) {
```

下面通过单一图框且起始位置为 1 创建图块形状。这里建模的图块表示为正方形或者 "O" 形状的图块，如下所示：

```
Tetromino1(1, 1) {
  override fun getFrame(frameNumber: Int): Frame {
    return Frame(2)
            .addRow("11")
            .addRow("11")
  }
},
```

下面利用两个图框构建图块形状，且起始位置为 1。此处建模的图块为 "Z" 形状的图块，如下所示：

```
Tetromino2(2, 1) {
  override fun getFrame(frameNumber: Int): Frame {
    return when (frameNumber) {
      0 -> Frame(3)
              .addRow("110")
              .addRow("011")
      1 -> Frame(2)
              .addRow("01")
              .addRow("11")
              .addRow("10")
      else -> throw IllegalArgumentException("$frameNumber is an invalid
                                        frame number.")
    }
  }
},
```

下面利用两个图框，且起始位置为 1 生成图块形状。这里，建模图块表示为 "S" 形状图块，如下所示：

```
Tetromino3(2, 1) {
  override fun getFrame(frameNumber: Int): Frame {
```

```
        return when (frameNumber) {
          0 -> Frame(3)
                    .addRow("011")
                    .addRow("110")
          1 -> Frame(2)
                    .addRow("10")
                    .addRow("11")
                    .addRow("01")
          else -> throw IllegalArgumentException("$frameNumber is
                                      an invalid frame number.")
        }
      }
    },
```

下面利用两个图框,且起始位置为 2 创建图块。这里,建模图块表示为 "T" 形图块,如下所示:

```
Tetromino4(2, 2) {
  override fun getFrame(frameNumber: Int): Frame {
    return when (frameNumber) {
      0 -> Frame(4).addRow("1111")
      1 -> Frame(1)
                .addRow("1")
                .addRow("1")
                .addRow("1")
                .addRow("1")
      else -> throw IllegalArgumentException("$frameNumber is an
                                    invalid frame number.")
    }
  }
},
```

下面利用 4 个图框,且起始位置为 1 创建图块。这里,建模图块表示为 "T" 形图块,如下所示:

```
Tetromino5(4, 1) {
  override fun getFrame(frameNumber: Int): Frame {
    return when (frameNumber) {
      0 -> Frame(3)
                .addRow("010")
                .addRow("111")
      1 -> Frame(2)
                .addRow("10")
```

```
                .addRow("11")
                .addRow("10")
      2 -> Frame(3)
                .addRow("111")
                .addRow("010")
      3 -> Frame(2)
                .addRow("01")
                .addRow("11")
                .addRow("01")
      else -> throw IllegalArgumentException("$frameNumber is an
                                        invalid frame number.")
    }
  }
},
```

下面利用 4 个图框,且起始位置为 1 创建图块。这里,建模图块表示为 "J" 形状的图块,如下所示:

```
Tetromino6(4, 1) {
  override fun getFrame(frameNumber: Int): Frame {
    return when (frameNumber) {
      0 -> Frame(3)
                .addRow("100")
                .addRow("111")
      1 -> Frame(2)
                .addRow("11")
                .addRow("10")
                .addRow("10")
      2 -> Frame(3)
                .addRow("111")
                .addRow("001")
      3 -> Frame(2)
                .addRow("01")
                .addRow("01")
                .addRow("11")
      else -> throw IllegalArgumentException("$frameNumber is
                                  an invalid frame number.")
    }
  }
},
```

下面利用 4 个图框,且起始位置为 1 创建图块。这里,建模图块表示为 "L" 形状的图块,如下所示:

```
    Tetromino7(4, 1) {
      override fun getFrame(frameNumber: Int): Frame {
        return when (frameNumber) {
          0 -> Frame(3)
                  .addRow("001")
                  .addRow("111")
          1 -> Frame(2)
                  .addRow("10")
                  .addRow("10")
                  .addRow("11")
          2 -> Frame(3)
                  .addRow("111")
                  .addRow("100")
          3 -> Frame(2)
                  .addRow("11")
                  .addRow("01")
                  .addRow("01")
          else -> throw IllegalArgumentException("$frameNumber is
                                        an invalid frame number.")
        }
      }
    };

    abstract fun getFrame(frameNumber: Int): Frame
}
```

在对方块和形状建模后,下一步采用编程方式建模的内容是方块自身。此处,将展示 Kotlin 与 Java 之间的无缝衔接操作,也就是说,通过 Java 实现建模过程,并在 models 目录中定义名为 Block 的新 Java 类(models | New | Java Class)。下面将添加相应的实例变量,并以此展现方块的特征。考察下列代码:

```
package com.mydomain.tetris.models;
import android.graphics.Color;
import android.graphics.Point;

public class Block {
  private int shapeIndex;
  private int frameNumber;
  private BlockColor color;
  private Point position;

  public enum BlockColor {
```

```
    PINK(Color.rgb(255, 105, 180), (byte) 2),
    GREEN(Color.rgb(0, 128, 0), (byte) 3),
    ORANGE(Color.rgb(255, 140, 0), (byte) 4),
    YELLOW(Color.rgb(255, 255, 0), (byte) 5),
    CYAN(Color.rgb(0, 255, 255), (byte) 6);
    BlockColor(int rgbValue, byte value) {
      this.rgbValue = rgbValue;
      this.byteValue = value;
    }

    private final int rgbValue;
    private final byte byteValue;
  }
}
```

在上述代码中，我们加入了 4 个实例变量，即 shapeIndex、frameNumber、color 和 position。其中，shapeIndex 加载图块形状的索引；frameNumber 将记录图块形状包含的图框数量；color 则加载图块的颜色特征；而 position 用于记录图块当前的空间位置。

同时，enum 模板 BlockColor 也将添加至 Block 类中。enum 创建了一个 BlockColor 实例常量集，每个实例均包含 rgbValue 和 byteValue 属性。其中，rgbValue 表示一个整数，并唯一标识 Color.rgb()方法所指定的 RGB 颜色。Color 则是 Android 应用程序框架提供的一个类，而 rgb()则是定义于 Color 类中的类方法。相应地，调用 Colour.rgb()方法 5 次将分别定义粉色、绿色、橘黄色、黄色和青色。

在 Block 类中，分别使用了 private 和 public 关键字。这些关键字都不是为了吸引眼球而添加的，它们都有各自的用处。private、public 连同 protected 关键字统称为访问修饰符。

ⓘ 注意：

作为一类关键字，访问修饰符用于指定类、方法、函数、变量和结构的访问限制。Java 中设置了 3 种访问修饰符，即 private、public 和 protected。在 Kotlin 中，访问修饰符也称作可见性修饰符，包括 public、protected、private 和 internal。

（1）私有访问修饰符（private）

声明为 private 的方法、变量、构造方法以及结构仅可在声明类中被访问。例外情况是私有顶级（top-level）函数和属性，它们针对同一文件中的所有成员均为可见。类中的私有变量可通过声明了 getter 和 setter 方法（并授权访问）的外部类进行访问。在 Java

中，setter 和 getter 的定义方式如下所示：

```
public class Person {
  Person(String fullName, int age) {
    this.fullName = fullName;
    this.age = age;
  }

  private String fullName;
  private int age;

  public String getFullName() {
    return fullName;
  }

  public int getAge() {
    return age;
  }
}
```

在 Kotlin 中，setter 和 getter 的定义如下所示：

```
public class Person(private var fullName: String) {
  var name: String
  get() = fullName
  set(value) {
    fullName = value
  }
}
```

使用 private 访问修饰符是程序中数据隐藏的主要方式。信息隐藏也称作封装。

（2）公有访问修饰符（public）

声明为 public 的方法、变量、构造方法和结构可在声明类的外部自由地被访问。不同数据包中的 public 类在使用之前必须被导入，下列类使用了 public 访问修饰符：

```
public class Person { .. }
```

（3）保护访问修饰符（protected）

声明为受保护的变量、方法、函数和结构只能由相同包中的类访问，或者由子类访问，这些类存在于单独的包中，如下所示：

```
public class Person(private var fullName: String) {
  protected name: String
```

```
    get() = fullName
    set(value) {
        fullName = value
    }
}
```

（4）内部可见性修饰符

内部可见性修饰符用于声明同一个模块（module）内可见的成员。这里，模块表示为整体编译的 Kotlin 文件集合，例如一个 Maven 项目、Gradle 资源集合、IntelliJ IDEA 模块，或者是利用 Ant 任务调用编译的文件集合。internal 修饰符的使用方式与其他可见性修饰符类似，如下所示：

```
internal class Person { }
```

在读者理解了访问修饰符和可见性修饰符后，即可继续实现 Block 类。下面将针对该类定义一个构造方法，并将所创建的实例变量设置为初始状态。从句法角度来看，Java 中的构造方法定义不同于 Kotlin，如下所示：

```
public class Block {
    private int shapeIndex;
    private int frameNumber;
    private BlockColor color;
    private Point position;
```

构造方法定义如下所示：

```
    private Block(int shapeIndex, BlockColor blockColor) {
        this.frameNumber = 0;
        this.shapeIndex = shapeIndex;
        this.color = blockColor;
        this.position = new Point(AppModel.FieldConstants
                            .COLUMN_COUNT.getValue() / 2, 0);
    }

    public enum BlockColor {
        PINK(Color.rgb(255, 105, 180), (byte) 2),
        GREEN(Color.rgb(0, 128, 0), (byte) 3),
        ORANGE(Color.rgb(255, 140, 0), (byte) 4),
        YELLOW(Color.rgb(255, 255, 0), (byte) 5),
        CYAN(Color.rgb(0, 255, 255), (byte) 6);
        BlockColor(int rgbValue, byte value) {
            this.rgbValue = rgbValue;
            this.byteValue = value;
```

```
    }
    private final int rgbValue;
    private final byte byteValue;
  }
}
```

需要注意的是,上述构造方法定义设置了 private 访问权限,其原因在于,此处并不希望该构造方法在 Block 类外部被访问。由于其他类仍然需要通过某种方式创建图块实例,因而须对此定义一个静态方法,即 createBlock 方法,如下所示:

```
public class Block {
  private int shapeIndex;
  private int frameNumber;
  private BlockColor color;
  private Point position;
```

构造方法定义如下所示:

```
  private Block(int shapeIndex, BlockColor blockColor) {
    this.frameNumber = 0;
    this.shapeIndex = shapeIndex;
    </span> this.color = blockColor;
    this.position = new Point( FieldConstants.COLUMN_COUNT
                        .getValue()/2, 0);
  }

  public static Block createBlock() {
    Random random = new Random();
    int shapeIndex = random.nextInt(Shape.values().length);
    BlockColor blockColor = BlockColor.values()
        [random.nextInt(BlockColor.values().length)];

    Block block = new Block(shapeIndex, blockColor);
    block.position.x = block.position.x - Shape.values()
        [shapeIndex].getStartPosition();
    return block;
  }
  public enum BlockColor {
      PINK(Color.rgb(255, 105, 180), (byte) 2),
      GREEN(Color.rgb(0, 128, 0), (byte) 3),
      ORANGE(Color.rgb(255, 140, 0), (byte) 4),
```

```
    YELLOW(Color.rgb(255, 255, 0), (byte) 5),
    CYAN(Color.rgb(0, 255, 255), (byte) 6);
    BlockColor(int rgbValue, byte value) {
      this.rgbValue = rgbValue;
      this.byteValue = value;
    }

    private final int rgbValue;
    private final byte byteValue;
  }
}
```

createBlock()随机选取 Shape 枚举类中图块形状的索引以及 BlockColor，并将两个随机选取值赋予 shapeIndex 和 blockColor 中。新的 Block 实例利用这两个值（作为参数被传递）被创建，同时设置了 X 轴向上的图块位置。最后，createBlock()方法返回经创建和初始化后的图块。

这里，还需要向 Block 类中添加相应的 getter 和 setter，用以访问图块实例中较为重要的属性，对应方法如下所示：

```
public static int getColor(byte value) {
  for (BlockColor colour : BlockColor.values()) {
    if (value == colour.byteValue) {
      return colour.rgbValue;
    }
  }
  return -1;
}

public final void setState(int frame, Point position) {
  this.frameNumber = frame;
  this.position = position;
}

@NonNull
public final byte[][] getShape(int frameNumber) {
  return Shape.values()[shapeIndex].getFrame(frameNumber).as2dByteArray();
}

public Point getPosition() {
  return this.position;
}
```

```
public final int getFrameCount() {
  return Shape.values()[shapeIndex].getFrameCount();
}

public int getFrameNumber() {
  return frameNumber;
}

public int getColor() {
  return color.rgbValue;
}

public byte getStaticValue() {
  return color.byteValue;
}
```

其中，@NonNull 表示为 Android 应用程序框架提供的注解，并以此说明某个字段、参数或方法返回结果不可为 null。在上述代码片段中，@NonNull 设置于 getShape()方法之前，表明该方法不可返回 null 值。

> **注意：**
> 在 Java 中，注解是一种元数据形式，并可添加至 Java 源代码中。另外，注解可用于类、方法、变量、参数和数据包上。同样，注解也可在 Kotlin 中声明和使用。

@NotNull 注解位于 android.support.annotation 包中，在 Block.java 文件中，可在开始处予以导入，如下所示：

```
import android.support.annotation.NonNull;
```

需要注意的是，在 Block 类的构造方法中，当前图块实例的位置实例变量其设置方式如下所示：

```
this.position = new Point(FieldConstants.COLUMN_COUNT.getValue()/2, 0);
```

其中，图块生成的列数为 10，这一常量值将在应用程序代码中使用多次，因而可将其声明为常量。对此，可在应用程序源数据包中创建一个包，并将名为 FieldConstants 的 Kotlin 文件添加至该数据包中。随后，可添加游戏区域的行、列常量。当前，该字段包含了 10 列、20 行，如下所示：

```
enum class FieldConstants(val value: Int) {
  COLUMN_COUNT(10), ROW_COUNT(20);
}
```

接下来，可将包含 FieldConstants 枚举类的数据包导入 Block.java 中，并利用常量值 COLUMN_COUNT 替换整数 10，如下所示：

```
this.position = new Point( FieldConstants.COLUMN_COUNT.getValue()/2, 0);
```

至此，Block 类的编程建模暂告一段落。

3.1.2　构建应用程序模型

前述内容讨论了构成俄罗斯方块的特定组件的建模过程，下面开始着手定义应用程序逻辑，并创建应用程序模型以实现必要的俄罗斯方块游戏的逻辑内容。这里，程序逻辑可视作视图和图块组件之间的中间接口，如图 3.7 所示。

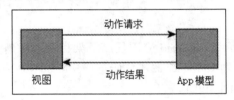

图 3.7　程序逻辑

其中，视图将向 App 模型发送动作请求；相应地，模型将执行该动作，并向视图发送反馈信息。类似于之前创建的模型，此处也需要针对 App 模型定义一个独立的类。对此，可创建一个名为 AppModel.kt 的 Kotlin 文件，并向该文件中添加 AppModel 类，同时导入 Point、FieldConstants、array2dOfByte 和 AppPreferences，如下所示：

```
package com.mydomain.tetris.models

import android.graphics.Point
import com.mydomain.tetris.constants.FieldConstants
import com.mydomain.tetris.helpers.array2dOfByte
import com.mydomain.tetris.storage.AppPreferences

class AppModel
```

某些 AppModel 函数分别负责记录当前积分值、tetris 字段状态、当前图块、当前游戏状态以及图块的运行行为。另外，AppModel 应可直接访问应用程序的 SharedPreferences 文件中存储的数值，即通过所定义的 AppPreferences 类。初看之下，此类需求较为复杂；实际上，其实现过程十分简单。

首先需要添加 AppModel 所使用的一些常量，其中包括游戏体验过程中的游戏状态，

以及运动行为。针对于此,可使用下列枚举类:

```
class AppModel {
  enum class Statuses {
    AWAITING_START, ACTIVE, INACTIVE, OVER
  }

  enum class Motions {
    LEFT, RIGHT, DOWN, ROTATE
  }
}
```

其中定义了 4 个状态常量。具体来讲,AWAITING_START 表示为游戏启动之前的状态;ACTIVE 表示游戏进程中的状态;OVER 表示为游戏结束时的状态。

本章前述内容曾讨论到,图块包含了 4 个独立的运动状态。也就是说,图块可实现右移、左移、上移和下移。对此,Motions 枚举类中分别定义了 LEFT、RIGHT、UP、DOWN 和 ROTATE,以表示此类独特的运动行为。

在添加了所需的常量后,即可加入相应的 AppModel 类属性,如下所示:

```
package com.mydomain.tetris.models

import android.graphics.Point
import com.mydomain.tetris.constants.FieldConstants
import com.mydomain.tetris.helpers.array2dOfByte
import com.mydomain.tetris.storage.AppPreferences

class AppModel {
  var score: Int = 0
  private var preferences: AppPreferences? = null

  var currentBlock: Block? = null
  var currentState: String = Statuses.AWAITING_START.name

  private var field: Array<ByteArray> = array2dOfByte(
    FieldConstants.ROW_COUNT.value,
    FieldConstants.COLUMN_COUNT.value
  )

  enum class Statuses {
    AWAITING_START, ACTIVE, INACTIVE, OVER
  }
```

```
enum class Motions {
  LEFT, RIGHT, DOWN, ROTATE
  }
}
```

　　score 表示为一个整数属性，用于加载游戏会话过程中玩家的当前分值。Preferences 定义为一个 private 属性，加载 AppPreferences 对象，并提供了应用程序 SharedPreferences 文件的直接访问能力。属性 currentBlock 用于加载当前图块的平移状态。currentState 将加载游戏状态。Statuses.AWAITING_START.name 将以 AWAITING_START 字符串形式返回 Statuses.AWAITING_START 的名称。另外，游戏的当前状态将被初始化为 AWAITING_START，其原因在于，这将是 GameActivity 启动后需要转换的第一个状态。最后，field 定义为一个二维数组，并用作游戏体验区域。

　　随后，还需要添加一些 setter 和 getter 函数，即 setPreferences()、setCellStatus()和 getCellStatus()，并将其加入 AppModel 中，如下所示：

```
fun setPreferences(preferences: AppPreferences?) {
  this.preferences = preferences
}

fun getCellStatus(row: Int, column: Int): Byte? {
  return field[row][column]
}

private fun setCellStatus(row: Int, column: Int, status: Byte?) {
  if (status != null) {
    field[row][column] = status
  }
}
```

　　setPreferences()方法将 AppModel 的 preferences 属性设置为 AppPreferences（作为参数传递至函数中）。getCellStatus()方法返回 field 二维数组中特定行、列位置处的单元格状态。setCellStatus()方法将 field 中的单元格状态设置为特定的字节。

　　同时，模型中还需定义相应的状态检测函数，并以此确定游戏的当前状态。鉴于存在的 3 种游戏状态，因而须针对每种状态定义共计 3 个函数，即 isGameAwaitingStart()、isGameActive()和 isGameOver()，如下所示：

```
class AppModel {

  var score: Int = 0
```

```kotlin
private var preferences: AppPreferences? = null

var currentBlock: Block? = null
var currentState: String = Statuses.AWAITING_START.name

private var field: Array<ByteArray> = array2dOfByte(
  FieldConstants.ROW_COUNT.value,
  FieldConstants.COLUMN_COUNT.value
)

fun setPreferences(preferences: AppPreferences?) {
  this.preferences = preferences
}

fun getCellStatus(row: Int, column: Int): Byte? {
  return field[row][column]
}

private fun setCellStatus(row: Int, column: Int, status: Byte?) {
  if (status != null) {
    field[row][column] = status
  }
}

fun isGameOver(): Boolean {
  return currentState == Statuses.OVER.name
}

fun isGameActive(): Boolean {
  return currentState == Statuses.ACTIVE.name
}

fun isGameAwaitingStart(): Boolean {
  return currentState == Statuses.AWAITING_START.name
}

enum class Statuses {
  AWAITING_START, ACTIVE, INACTIVE, OVER
}

enum class Motions {
```

```
    LEFT, RIGHT, DOWN, ROTATE
  }
}
```

取决于游戏的各自状态，上述 3 个方法将返回相应的布尔值。截至目前，我们尚未使用到 AppModel 中的 score。下面将定义一个函数，用于增加 score 所持有的分值，即 boostScore()函数，如下所示：

```
private fun boostScore() {
  score += 10
  if (score > preferences?.getHighScore() as Int)
    preferences?.saveHighScore(score)
}
```

当调用 boostScore()函数时，该函数将玩家分值加 10，并于随后检测当前分值是否大于预置文件中所记录的最高分值。若是，最高分值将被改写为当前分值。

在介绍了上述较为基本的函数和字段之后，下面开始讨论一些较为复杂的函数。其中，第一个函数是 generateNextBlock()，如下所示：

```
private fun generateNextBlock() {
  currentBlock = Block.createBlock()
}
```

generateNextBlock()函数创建新的图块实例，同时将 currentBlock 设置为新创建的实例。

在进一步讨论方法定义之前，下面首先创建一个 enum 类，并加载单元格的常量值。对此，可在常量数据包中生成 CellConstants.kt 文件，并添加下列源代码：

```
package com.mydomain.tetris.constants

enum class CellConstants(val value: Byte) {
  EMPTY(0), EPHEMERAL(1)
}
```

读者可能会对这一类常量的具体含义有所疑惑。回忆一下，当创建 Frame 类并对图块进行建模时，曾定义了 addRow()函数，并接收 1 或 0 字符串作为参数。其中，1 表示图框构成的单元格，0 表示图框之外的单元格；随后，将 1 或 0 值转换为字节表达方式。在后续函数中，将会操控此类字节，并对其定义对应的常量值。

相应地，可向 AppModel 中导入新创建的 enum 类，并在接下来的函数中对其加以使用，如下所示：

```
private fun validTranslation(position: Point, shape: Array<ByteArray>):
Boolean {
```

```
        return if (position.y < 0 || position.x < 0) {
            false
        } else if (position.y + shape.size > FieldConstants.ROW_COUNT.value) {
            false
        } else if (position.x + shape[0].size > FieldConstants
                    .COLUMN_COUNT.value) {
            false
        } else {
            for (i in 0 until shape.size) {
                for (j in 0 until shape[i].size) {
                    val y = position.y + i
                    val x = position.x + j

                    if (CellConstants.EMPTY.value != shape[i][j] &&
                        CellConstants.EMPTY.value != field[y][x]) {
                        return false
                    }
                }
            }
            true
        }
    }
```

相应地，可将 validTranslation()方法添加至 AppModel 中。顾名思义，该函数根据条件集检测游戏区域内图块的平移运动是否有效。若平移行为可正常工作，该函数返回 true，否则返回 false。其中，前 3 个条件测试俄罗斯方块的平移位置是否有效；else 代码块检测俄罗斯方块行将移至的单元格是否为空。若不为空，则返回 false。

对于 validTranslation()函数，需要定义一个调用函数。对此，可声明一个 moveValid()方法实现这一操作，并将其添加至 AppModel，如下所示：

```
private fun moveValid(position: Point, frameNumber: Int?): Boolean {
    val shape: Array<ByteArray>? = currentBlock?
                                    .getShape(frameNumber as Int)
    return validTranslation(position, shape as Array<ByteArray>)
}
```

moveValid()调用了 validTranslation()检测玩家所执行的运动行为是否被允许。若是，则返回 true，否则返回 false。除此之外，还需要定义其他一些重要的方法，其中包括 generateField()、resetField()、persistCellData()、assessField()、translateBlock()、blockAdditionPossible()、shiftRows()、startGame()、restartGame()、endGame()以及 resetModel()。

下面首先讨论 generateField()，并将下列代码添加至 AppModel。

```kotlin
fun generateField(action: String) {
  if (isGameActive()) {
    resetField()
    var frameNumber: Int? = currentBlock?.frameNumber
    val coordinate: Point? = Point()
    coordinate?.x = currentBlock?.position?.x
    coordinate?.y = currentBlock?.position?.y

    when (action) {
      Motions.LEFT.name -> {
        coordinate?.x = currentBlock?.position?.x?.minus(1)
      }
      Motions.RIGHT.name -> {
        coordinate?.x = currentBlock?.position?.x?.plus(1)
      }
      Motions.DOWN.name -> {
        coordinate?.y = currentBlock?.position?.y?.plus(1)
      }
      Motions.ROTATE.name -> {
        frameNumber = frameNumber?.plus(1)

        if (frameNumber != null) {
          if (frameNumber >= currentBlock?.frameCount as Int) {
            frameNumber = 0
          }
        }
      }
    }

    if (!moveValid(coordinate as Point, frameNumber)) {
      translateBlock(currentBlock?.position as Point,
              currentBlock?.frameNumber as Int)
      if (Motions.DOWN.name == action) {
        boostScore()
        persistCellData()
        assessField()
        generateNextBlock()

        if (!blockAdditionPossible()) {
          currentState = Statuses.OVER.name;
          currentBlock = null;
          resetField(false);
```

```
      }
    }
  } else {
    if (frameNumber != null) {
      translateBlock(coordinate, frameNumber)
      currentBlock?.setState(frameNumber, coordinate)
    }
  }
}
```

generateField()将生成区域的刷新结果,并由作为参数传递至 generateField()的相关动作所决定。

generateField()被调用后,首先检测游戏是否处于活动状态。如果游戏处于活动状态,将获取图框号和坐标。随后,所请求的动作将通过 when 表达式确定。一旦确定了所请求的动作后,对于左移、右移以及下移动作,将适当调整图块坐标。当请求旋转运动时,frameNumber 将被调整为相应的图框号,以显示俄罗斯方块对应的旋转状态。

接下来,generateField()方法通过 moveValid()判断所请求运行是否为有效运动。对于无效运动行为,图块通过 translateBlock()方法在游戏区域内保持当前位置不变。

generateField()方法分别调用了 resetField()、persistCellData()和 assessField()方法,下面将其添加至 AppModel 中。

```
private fun resetField(ephemeralCellsOnly: Boolean = true) {
  for (i in 0 until FieldConstants.ROW_COUNT.value) {
    (0 until FieldConstants.COLUMN_COUNT.value)
    .filter { !ephemeralCellsOnly || field[i][it] ==
              CellConstants.EPHEMERAL.value }
    .forEach { field[i][it] = CellConstants.EMPTY.value }
  }
}

private fun persistCellData() {
  for (i in 0 until field.size) {
    for (j in 0 until field[i].size) {
      var status = getCellStatus(i, j)

      if (status == CellConstants.EPHEMERAL.value) {
        status = currentBlock?.staticValue
        setCellStatus(i, j, status)
      }
```

```kotlin
      }
    }
  }

  private fun assessField() {
    for (i in 0 until field.size) {
      var emptyCells = 0;

      for (j in 0 until field[i].size) {
        val status = getCellStatus(i, j)
        val isEmpty = CellConstants.EMPTY.value == status
        if (isEmpty)
          emptyCells++
      }
      if (emptyCells == 0)
        shiftRows(i)
    }
  }
```

可能读者已经意识到,这里并未实现 translateBlock()。下面将该方法连同 blockAdditionPossible()、shiftRows()、startGame()、restartGame()、endGame()和resetModel() 方法添加至 AppModel,如下所示:

```kotlin
  private fun translateBlock(position: Point, frameNumber: Int) {
    synchronized(field) {
      val shape: Array<ByteArray>? = currentBlock?.getShape(frameNumber)

      if (shape != null) {
        for (i in shape.indices) {
          for (j in 0 until shape[i].size) {
            val y = position.y + i
            val x = position.x + j

            if (CellConstants.EMPTY.value != shape[i][j]) {
              field[y][x] = shape[i][j]
            }
          }
        }
      }
    }
  }

  private fun blockAdditionPossible(): Boolean {
```

```kotlin
    if (!moveValid(currentBlock?.position as Point,
        currentBlock?.frameNumber)) {
      return false
    }
    return true
}

private fun shiftRows(nToRow: Int) {
  if (nToRow > 0) {
    for (j in nToRow - 1 downTo 0) {
      for (m in 0 until field[j].size) {
        setCellStatus(j + 1, m, getCellStatus(j, m))
      }
    }
  }

  for (j in 0 until field[0].size) {
    setCellStatus(0, j, CellConstants.EMPTY.value)
  }
}

fun startGame() {
  if (!isGameActive()) {
    currentState = Statuses.ACTIVE.name
    generateNextBlock()
  }
}

fun restartGame() {
  resetModel()
  startGame()
}

fun endGame() {
  score = 0
  currentState = AppModel.Statuses.OVER.name
}

private fun resetModel() {
  resetField(false)
  currentState = Statuses.AWAITING_START.name
  score = 0
}
```

如果所请求的运动为下移,且运动处于无效状态,则表明图块已经到达游戏区域底部。此时,玩家的分值通过 boostScore()被调整,且游戏区域中的所有单元格通过 persistCellData()方法保持其状态。随后,assessField()方法将被调用,并逐行遍历游戏区域,并检测所填充行中的所有单元格,如下所示:

```
private fun assessField() {
  for (i in 0 until field.size) {
    var emptyCells = 0;

    for (j in 0 until field[i].size) {
      val status = getCellStatus(i, j)
      val isEmpty = CellConstants.EMPTY.value == status

      if (isEmpty)
        emptyCells++
    }

    if (emptyCells == 0)
      shiftRows(i)
  }
}
```

当一行中的全部单元格均被填充,该行将被消除并通过 shiftRow()。在游戏区域计算完毕后,新图块将通过 generateNextBlock()方法生成,如下所示:

```
private fun generateNextBlock() {
  currentBlock = Block.createBlock()
}
```

在新图块置于游戏区域之前,AppModel 须确保当前游戏区域未被填充,同时,图块可通过 blockAdditionPossible()方法移至游戏区域内,如下所示:

```
private fun blockAdditionPossible(): Boolean {
  if (!moveValid(currentBlock?.position as Point,
      currentBlock?.frameNumber)) {
    return false
  }
  return true
}
```

如果无法实现图块的添加信息,这将表明,全部图块累积于游戏区域的上方,从而导致游戏结束。最终,游戏的当前状态将被设置为 Statuses.OVER,currentBlock 被设置为 null，同时清空游戏区域。

另外一方面，如果运动行为从开始即为有效，对应图块将通过 translateBlock()移至其新坐标处，当前图块状态将被设置为新坐标以及 frameNumber。

待设置完毕后，即可成功地创建应用程序模型，并处理游戏逻辑。下面将构建视图以利用 AppModel。

3.1.3 创建 TetrisView

截至目前，前述内容已经实现了相关类，并对俄罗斯方块的图块、图框以及形状建模；除此之外，还实现了 AppModel 类，以整合视图以及编程组件之间的交互行为。如果缺少此类视图，那么，用户将无法与 AppModel 进行交互。若用户缺少与游戏之间的交流方式，该游戏将难以称之为一部完整的作品。本节将实现 TetrisView，即玩家体验俄罗斯方块游戏的用户界面。

接下来，可创建名为 view 的数据包，并将 TetrisView.kt 文件加入其中。考虑到 TestrisView 表示为一个 View，因而需要扩展 View 类。对此，可将下列代码添加至 TetrisView.kt 中：

```kotlin
package com.mydomain.tetris.views

import android.content.Context
import android.graphics.Canvas
import android.graphics.Color
import android.graphics.Paint
import android.graphics.RectF
import android.os.Handler
import android.os.Message
import android.util.AttributeSet
import android.view.View
import android.widget.Toast
import com.mydomain.tetris.constants.CellConstants
import com.mydomain.tetris.GameActivity
import com.mydomain.tetris.constants.FieldConstants
import com.mydomain.tetris.models.AppModel
import com.mydomain.tetris.models.Block

class TetrisView : View {

    private val paint = Paint()
    private var lastMove: Long = 0
    private var model: AppModel? = null
```

```kotlin
    private var activity: GameActivity? = null
    private val viewHandler = ViewHandler(this)
    private var cellSize: Dimension = Dimension(0, 0)
    private var frameOffset: Dimension = Dimension(0, 0)

    constructor(context: Context, attrs: AttributeSet) :
            super(context, attrs)

    constructor(context: Context, attrs: AttributeSet, defStyle: Int) :
            super(context, attrs, defStyle)

    companion object {
      private val DELAY = 500
      private val BLOCK_OFFSET = 2
      private val FRAME_OFFSET_BASE = 10
    }
}
```

TetrisView 类扩展了 View 类，应用程序视图元素均需要对该类进行扩展。由于 View 类型中包含了一个初始化构造方法，因而对于 TetrisView 需要声明两个次级构造方法，进而根据调用哪一个次级构造方法初始化视图类的相关构造方法。

paint 属性表示为 android.graphics.Paint 实例。Paint 类包含了与绘制文本、位图和几何形状相关的样式和色彩信息。lastMove 用于记录上一次移动的时间（以毫秒计）。Model 实例用于加载 AppModel 实例，该实例与 TetrisView 交互并控制游戏体验。Activity 表示为所创建的 GameActivity 类实例。cellSize 和 frameOffset 属性分别表示游戏中单元格的尺寸以及图框偏移。

Android 应用程序框架并未提供 ViewHandler 和 Dimension，因而需要手动对其予以实现。

1. 实现 ViewHandler

由于图块在游戏区域中以固定的时间间隔移动，因而需要某种方式设置一个线程，以处理图块的运动，进而休眠或唤醒该线程以使图块在一定的时间量后处于运动状态。一种方法是使用句柄处理消息延迟请求，并在延迟后继续处理消息。根据 Android 文档中的描述，句柄可发送或处理与线程 MessageQueue 关联的 Meaasge 对象。也就是说，每个句柄实例均与某个线程和该线程的消息队列有所关联。

ViewHandler 则是须针对 TetrisView 实现的自定义句柄，以满足视图的消息发送和处理需求。由于 ViewHandler 定义为 Handler 的子类，因而需要扩展 Handler，并向 ViewHandler

类中加入必需的行为。

对此，可作为 private 类，向 TetrisView 中添加 VieHandler 类，如下所示：

```kotlin
private class ViewHandler(private val owner: TetrisView) : Handler() {

  override fun handleMessage(message: Message) {
    if (message.what == 0) {
      if (owner.model != null) {
        if (owner.model!!.isGameOver()) {
          owner.model?.endGame()
          Toast.makeText(owner.activity, "Game over",
                     Toast.LENGTH_LONG).show();
        }
        if (owner.model!!.isGameActive()) {
          owner.setGameCommandWithDelay(AppModel.Motions.DOWN)
        }
      }
    }
  }

  fun sleep(delay: Long) {
    this.removeMessages(0)
    sendMessageDelayed(obtainMessage(0), delay)
  }
}
```

作为参数，ViewHandler 类在其构造方法中接收一个 TetrisView 实例，并覆写了其超类中的 handleMessage() 函数。handleMessage() 函数负责检测所发送的消息内容。其中，what 定义为一个整数值，表明所发送的消息。如果 what 等于 0，所传递的 TetrisView 的实例（拥有者）则包含一个不等于 0 的模型，此时，将对游戏状态进行检测。如果游戏结束，则调用 AppModel 的 endGame() 函数，并弹出消息对话框以提示玩家游戏处于结束状态。如果游戏处于活动状态，那么，将触发下移运动。

sleep() 方法简单地移除之前所发送的消息，并发送包含延迟（由 delay 参数指定）的新消息。

2. 实现 Dimension

Dimension 仅需加载两个属性，即 width 和 height。因此，这可视作数据类型的最佳候选者。相应地，可将下列 private 类添加至 TetrisView 类中，如下所示：

```kotlin
private data class Dimension(val width: Int, val height: Int)
```

其中包含了当前属性，以及所需的 setter 和 getter。

3. 实现 TetrisView

相信读者已经猜测到，当前，与 TetrisView 相关的工作还远未结束。首先，需要针对视图的 model 和 activity 属性实现相应的 setter 和 getter，并将其添加至 TetrisView 类中，如下所示：

```kotlin
fun setModel(model: AppModel) {
    this.model = model
}

fun setActivity(gameActivity: GameActivity) {
    this.activity = gameActivity
}
```

setModel()和 setActivity()定义为 model 和 activity 属性的 setter 函数。顾名思义，setModel()函数设置视图当前所使用的模型；setActivity()函数则负责设置所使用的活动。下面添加 3 个辅助方法，即 setGameCommand()、setGameCommandWithDelay()和 updateScore()，如下所示：

```kotlin
fun setGameCommand(move: AppModel.Motions) {
  if (null != model && (model?.currentState ==
AppModel.Statuses.ACTIVE.name)) {
    if (AppModel.Motions.DOWN == move) {
      model?.generateField(move.name)
      invalidate()
      return
    }
    setGameCommandWithDelay(move)
  }
}

fun setGameCommandWithDelay(move: AppModel.Motions) {
  val now = System.currentTimeMillis()

  if (now - lastMove > DELAY) {
    model?.generateField(move.name)
    invalidate()
    lastMove = now
  }
  updateScores()
  viewHandler.sleep(DELAY.toLong())
```

```
}
private fun updateScores() {
  activity?.tvCurrentScore?.text = "${model?.score}"
  activity?.tvHighScore?.text =
"${activity?.appPreferences?.getHighScore()}"
}
```

setGameCommand()方法设置游戏所执行当前运动命令。如果 DOWN 运动命令处于执行状态，应用程序模型将生成图块下移区域。在 setGameCommand()中调用的 invalidate()方法可视作一个请求，进而绘制屏幕上的变化内容。相应地，invalidate()方法最终将调用 onDraw()方法。

onDraw()方法继承于 View 类，并在视图渲染其中的相关内容时被调用。针对当前视图，应对此提供一个自定义实现，并将其添加至 TetrisView 类中，如下所示：

```
override fun onDraw(canvas: Canvas) {
  super.onDraw(canvas)
  drawFrame(canvas)

  if (model != null) {
    for (i in 0 until FieldConstants.ROW_COUNT.value) {
      for (j in 0 until FieldConstants.COLUMN_COUNT.value) {
        drawCell(canvas, i, j)
      }
    }
  }
}

private fun drawFrame(canvas: Canvas) {
  paint.color = Color.LTGRAY

  canvas.drawRect(frameOffset.width.toFloat(),
          frameOffset.height.toFloat(), width -
frameOffset.width.toFloat(),
          height - frameOffset.height.toFloat(), paint)
}

private fun drawCell(canvas: Canvas, row: Int, col: Int) {
  val cellStatus = model?.getCellStatus(row, col)

  if (CellConstants.EMPTY.value != cellStatus) {
    val color = if (CellConstants.EPHEMERAL.value == cellStatus) {
```

```kotlin
      model?.currentBlock?.color
    } else {
      Block.getColor(cellStatus as Byte)
    }
    drawCell(canvas, col, row, color as Int)
  }
}

private fun drawCell(canvas: Canvas, x: Int, y: Int, rgbColor: Int) {
  paint.color = rgbColor

  val top: Float = (frameOffset.height + y * cellSize.height +
BLOCK_OFFSET).toFloat()
  val left: Float = (frameOffset.width + x * cellSize.width +
BLOCK_OFFSET).toFloat()
  val bottom: Float = (frameOffset.height + (y + 1) * cellSize.height -
BLOCK_OFFSET).toFloat()
  val right: Float = (frameOffset.width + (x + 1) * cellSize.width -
BLOCK_OFFSET).toFloat()
  val rectangle = RectF(left, top, right, bottom)

  canvas.drawRoundRect(rectangle, 4F, 4F, paint)
}

override fun onSizeChanged(width: Int, height: Int, previousWidth:
Int,previousHeight: Int) {
  super.onSizeChanged(width, height, previousWidth, previousHeight)

  val cellWidth = (width - 2 * FRAME_OFFSET_BASE) /
FieldConstants.COLUMN_COUNT.value
  val cellHeight = (height - 2 * FRAME_OFFSET_BASE) /
FieldConstants.ROW_COUNT.value
  val n = Math.min(cellWidth, cellHeight)
  this.cellSize = Dimension(n, n)
  val offsetX = (width - FieldConstants.COLUMN_COUNT.value * n) / 2
  val offsetY = (height - FieldConstants.ROW_COUNT.value * n) / 2
  this.frameOffset = Dimension(offsetX, offsetY)
}
```

TetrisView 中的 onDraw()方法覆写了其超类中的 onDraw()方法。onDraw()方法接收画布（canvas）对象作为其唯一的参数，且需要在其超类中调用 onDraw()函数。这一操作是通过调用 super.onDraw()完成的，并传递画布实例作为参数。

在调用了 super.onDraw()后，TetrisView 中的 onDraw()将调用 drawFrame()，这将绘制 TetrisView 的图框。随后，个体单元格将在画布中进行绘制，即使用所生成的 drawCell() 函数。

setGameCommandWithDelay()的工作方式类似于 setGameCommand()，唯一差别在于，前者更新游戏的积分值，并在执行了游戏命令后，将 viewHandler 置于睡眠状态。updateScore()函数用于更新游戏活动中的当前积分值，以及最高分值的文本视图。

onSizeChanged()函数在视图尺寸发生变化时被调用，该函数可访问视图的当前宽度、高度，以及之前的宽度和高度值。类似于之前所用的其他覆写函数，这里同样可调用超类中的对应函数。相应地，可利用宽度和高度参数计算、设置每个单元格的尺寸维度，即 cellSize。最后，在 onSizeChanged()方法中，还将计算 offsetX 和 offsetY，并用于设置 frameOffset。

4．完成 GameActivity

至此，我们已经成功地实现了视图、句柄、帮助函数、类以及模型，并可将其整合至俄罗斯方块游戏中。具体来讲，我们需要将相关内容整合至 GameActivity 中。首先，需要将新创建的 tetris 视图添加至游戏活动布局中。通过<com.mydomain.tetris.views.TetrisView>布局标签，可方便地将 TetrisView 作为子元素添加至布局文件中，如下所示：

```xml
<?xml version="1.0" encoding="utf-8"?>
<android.support.constraint.ConstraintLayout
xmlns:android="http://schemas.android.com/apk/res/android"
    xmlns:app="http://schemas.android.com/apk/res-auto"
    xmlns:tools="http://schemas.android.com/tools"
    android:layout_width="match_parent"
    android:layout_height="match_parent"
    tools:context="com.mydomain.tetris.GameActivity">
<LinearLayout
        android:layout_width="match_parent"
        android:layout_height="match_parent"
        android:orientation="horizontal"
        android:weightSum="10"
        android:background="#e8e8e8">
    <LinearLayout
        android:layout_width="wrap_content"
        android:layout_height="match_parent"
        android:orientation="vertical"
        android:gravity="center"
        android:paddingTop="32dp"
```

```xml
            android:paddingBottom="32dp"
            android:layout_weight="1">
        <LinearLayout
            android:layout_width="wrap_content"
            android:layout_height="0dp"
            android:layout_weight="1"
            android:orientation="vertical"
            android:gravity="center">
            <TextView
                android:layout_width="wrap_content"
                android:layout_height="wrap_content"
                android:text="@string/current_score"
                android:textAllCaps="true"
                android:textStyle="bold"
                android:textSize="14sp"/>
            <TextView
                android:id="@+id/tv_current_score"
                android:layout_width="wrap_content"
                android:layout_height="wrap_content"
                android:textSize="18sp"/>
            <TextView
                android:layout_width="wrap_content"
                android:layout_height="wrap_content"
                android:layout_marginTop="@dimen/layout_margin_top"
                android:text="@string/high_score"
                android:textAllCaps="true"
                android:textStyle="bold"
                android:textSize="14sp"/>
            <TextView
                android:id="@+id/tv_high_score"
                android:layout_width="wrap_content"
                android:layout_height="wrap_content"
                android:textSize="18sp"/>
        </LinearLayout>
        <Button
            android:id="@+id/btn_restart"
            android:layout_width="wrap_content"
            android:layout_height="wrap_content"
            android:text="@string/btn_restart"/>
    </LinearLayout>
    <View
        android:layout_width="1dp"
```

```xml
            android:layout_height="match_parent"
            android:background="#000"/>
    <LinearLayout
            android:layout_width="0dp"
            android:layout_height="match_parent"
            android:layout_weight="9">
        <!-- Adding TetrisView -->
        <com.mydomain.tetris.views.TetrisView
                android:id="@+id/view_tetris"
                android:layout_width="match_parent"
                android:layout_height="match_parent" />

    </LinearLayout>
  </LinearLayout>
</android.support.constraint.ConstraintLayout>
```

一旦将 tetris 视图添加至 activity_game.xml，即可打开 GameActivity 类，并添加下列代码：

```kotlin
package com.mydomain.tetris

import android.os.Bundle
import android.support.v7.app.AppCompatActivity
import android.view.MotionEvent
import android.view.View
import android.widget.Button
import android.widget.TextView
import com.mydomain.tetris.models.AppModel
import com.mydomain.tetris.storage.AppPreferences
import com.mydomain.tetris.views.TetrisView

class GameActivity: AppCompatActivity() {

  var tvHighScore: TextView? = null
  var tvCurrentScore: TextView? = null
  private lateinit var tetrisView: TetrisView

  var appPreferences: AppPreferences? = null
  private val appModel: AppModel = AppModel()

  public override fun onCreate(savedInstanceState: Bundle?) {
    super.onCreate(savedInstanceState)
    setContentView(R.layout.activity_game)
```

```kotlin
    appPreferences = AppPreferences(this)
    appModel.setPreferences(appPreferences)

    val btnRestart = findViewById<Button>(R.id.btn_restart)
    tvHighScore = findViewById<TextView>(R.id.tv_high_score)
    tvCurrentScore = findViewById<TextView>(R.id.tv_current_score)

    tetrisView = findViewById<TetrisView>(R.id.view_tetris)
    tetrisView.setActivity(this)
    tetrisView.setModel(appModel)

    tetrisView.setOnTouchListener(this::onTetrisViewTouch)
    btnRestart.setOnClickListener(this::btnRestartClick)

    updateHighScore()
    updateCurrentScore()
}

private fun btnRestartClick(view: View) {
    appModel.restartGame()
}

private fun onTetrisViewTouch(view: View, event: MotionEvent):
        Boolean {
    if (appModel.isGameOver() || appModel.isGameAwaitingStart()) {
        appModel.startGame()
        tetrisView.setGameCommandWithDelay(AppModel.Motions.DOWN)

    } else if(appModel.isGameActive()) {
        when (resolveTouchDirection(view, event)) {
            0 -> moveTetromino(AppModel.Motions.LEFT)
            1 -> moveTetromino(AppModel.Motions.ROTATE)
            2 -> moveTetromino(AppModel.Motions.DOWN)
            3 -> moveTetromino(AppModel.Motions.RIGHT)
        }
    }
    return true
}

private fun resolveTouchDirection(view: View, event: MotionEvent):
        Int {
    val x = event.x / view.width
    val y = event.y / view.height
```

```
    val direction: Int

  direction = if (y > x) {
    if (x > 1 - y) 2 else 0
  }
  else {
    if (x > 1 - y) 3 else 1
  }
    return direction
}

private fun moveTetromino(motion: AppModel.Motions) {
  if (appModel.isGameActive()) {
    tetrisView.setGameCommand(motion)
  }
}

private fun updateHighScore() {
  tvHighScore?.text = "${appPreferences?.getHighScore()}"
}

private fun updateCurrentScore() {
tvCurrentScore?.text = "0"
  }
}
```

上述代码以 tetrisView 属性的形式，向 activity_game.xml 中的 tetris 视图布局元素添加了一个对象引用。除此之外，还创建了 AppModel 实例，并通过 GameActivity 加以使用。在 oncreate() 方法中，我们将 tetrisView 使用的活动设置为 GameActivity 的当前实例，并将 tetrisView 使用的模型设置为 appModel——即所创建的 AppModel 实例属性。除此之外，tetrisView 的触摸监听器设置为 onTetrisViewTouch() 函数。

如果单击（触摸）了 tetrisView，且游戏位于 AWAITING_START 或 OVER 状态，则启动一个新游戏。如果单击（触摸）了 tetrisView，且游戏处于 ACTIVE 状态，在 resolveTouchDirection() 的帮助下，可解决 tetrisView 上触摸产生的方向。moveTetromino() 根据传递于其中的相关动作移动俄罗斯方块。如果出现左向触摸，则通过作为参数传递的 AppModel.Motions.LEFT 调用 moveTetromino()，这将在游戏区域内向左移动俄罗斯方块。tetrisView 的右、下、上方向上的触摸则导致右向、下向以及旋转运动。

随后，可构建并运行当前项目。一旦项目在设备上启动，可导航至游戏活动，并触摸屏幕右边的俄罗斯方块视图。游戏的启动状态如图 3.8 所示。

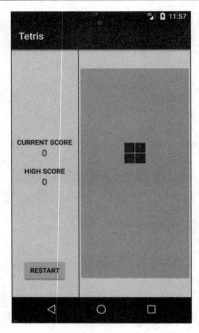

图 3.8　游戏的启动状态

3.2　MVP 模式简介

在 Tetris 应用程序开发过程中,我们曾尝试在代码库中添加结构,根据执行的任务将程序文件分离到不同的包中。此外,还将应用程序逻辑抽象至 AppModel 类中,与游戏相关的用户交互将由 TetrisView 视图类处理。据此,代码库体现了某种秩序,而不是将所有的逻辑放入一个较大的类文件中。

在 Android 应用程序中存在更好的方法来分离所关注的内容,MVP 模式便是其中之一。

3.2.1　MVP 的含义

MVP 是 Android 中的一种常见模式,并源自模型-视图-控制器模式。MVP 试图从应用程序逻辑中查看相关的关注内容。MVP 模式的优点主要体现在以下几方面:

- ❑ 增加代码库的可维护性。
- ❑ 改善应用程序的可读性。

MVP 模式如图 3.9 所示。

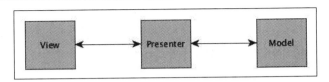

图 3.9 MVP 模式

下面考察 MVP 模式中的各项内容。

1．模型

在 MVP 模式中，模型表示为持有数据管理任务的接口，其责任包括与数据库交互、执行 API 调用、网络通信，以及协同对象和其他编程执行特定任务。

2．视图

视图表示为应用程序实体，用于向用户显示相关内容，并视作用户输入的界面。视图可以是一个活动、片段或者是 Android 微件。视图通过显示机制渲染数据。

3．显示

显示机制饰演了视图和模型的中间组件，主要负责查询模型并更新视图。简单地讲，显示机制中包含了相关显示逻辑，且与视图间包含一一对应的关系。

3.2.2 MVP 实现

在实际操作过程中，MVP 模式实现方式也有所变化。例如，一些 MVP 实现采用了合约描述视图和显示之间的接口。

除此之外，某些 MVP 实现还采用了显示机制中的生命周期回调方法，例如 onCreate() 方法。这试图镜像存在于活动生命周期内的回调。其他一些实现方案则完全启用了这一类回调方法。

实际上，在 Android 应用程序中并没有真正实现 MVP，但是在实现 MVP 的过程中可以遵循一些最佳实践方案。第 5 章将对此加以深入讨论。

3.3 本 章 小 结

本章实现了经典的俄罗斯方块游戏，其中涉及了较多内容。例如，如何利用类对应用程序的逻辑组件建模、访问操作和可见性修饰符，如何创建 Android 应用程序中的视图和句柄、数据类的访问（以方便地创建数据模型），以及 MVP 模式。

第 4 章将考察 Kotlin 语言在 Web 方面的应用，并实现通信应用程序的后端内容。

第 4 章 设计并实现 Messenger 后端应用程序

在前述章节中，通过经典的俄罗斯方块游戏，讨论了 Kotlin 程序设计语言的一些基本知识。第 3 章实现了该游戏的应用程序逻辑，并针对图块、形状、图框以及应用程序构建了相应的模型。除此之外，我们还通过 Tetris 视图创建了自定义视图，即应用程序用户与游戏体验之间的视图。

本章将进一步讨论 Kotlin 应用程序开发技巧，即实现一款基于 Android 平台的通信应用程序。其间，首先将介绍 RESTful API 开发，进而向后台应用程序提供 Web 内容。该应用程序编程接口将采用 Spring Boot 2.0 构建。在应用程序保持接口开发完毕后，会将其部署至远程服务器上。本章主要涉及以下内容：

- 基本的系统设计方案。
- 利用状态示意图对系统行为建模。
- 数据库设计基础知识。
- 利用实体关系（E-R）示意图对数据库建模。
- 利用 Spring Boot 2.0 构建后台微服务。
- 与 PostgreSQL 协同工作。
- 利用 Maven 实现依赖关系管理。
- 亚马逊 Web 服务（AWS）。

下面首先讨论 Messenger 应用程序编程接口。

4.1 设计 Messenger API

对于 MessengerAndroid 应用程序，当设计全功能的 RESTful 编程接口时，需要理解应用程序编程接口、表述性状态转移（REST）以及 RESTful 服务。

4.1.1 应用程序编程接口

应用程序编程接口表示为函数、例程、程序、协议以及资源的集合，并用以构建软件。换言之，应用程序编程接口（简称 API）表示为设计良好的结构化方法，或者是软件组件之间的通信渠道。

应用程序编程接口的开发涉及较多的领域，例如 Web 系统开发、操作系统、计算机硬件，以及嵌入式系统的交互。

4.1.2 REST

Restful 状态传输是一种通过互联网促进两个或多个不同系统（或子系统）之间的功能操作和交互的方法。基于 REST 的 Web 服务使得交互系统可访问 Web 内容；同时，还将针对所访问的 Web 内容执行授权操作。这些系统间通信是使用一组定义良好的无状态操作来完成的。RESTful Web 基于 REST，并通过预定义的无状态操作向通信系统提供 Web 内容。

当今，许多与 Web 服务通信的系统都使用了 REST。基于 REST 的系统均采用了客户机-服务器架构。我们将要开发的 API 是基于 REST 的，因此将使用到表征状态转移。

4.1.3 设计 Messenger API 系统

本节将简要介绍 Messenger API 系统。当前，读者可能尚不了解系统设计的真实含义及其相关职责，稍后将对此予以详细介绍。

系统设计是指定义体系结构、模块、接口以及系统的数据，以满足前期系统分析阶段提出的各项要求。系统设计包含多个处理流程以及不同的设计定位。除此之外，系统的深度设计还将涉及多个话题，例如耦合和聚合，这一类内容则超出了本书的讨论范围。针对于此，下面将对系统的交互行为和应用数据给出基本的定义，并采用渐进式系统设计方案。

1. 增量式开发

增量式开发常出现于系统开发过程中，并采用了递增的构建模块。这里，递增构建模块是一种软件开发方法，其中，产品采用渐进方式设计、实现和测试。本章中的 Messenger API 即采用了增量式开发方案。下面将逐步开发 Messenger API。在开始编码之前，我们并不会尝试指定 Messenger API 所需的全部内容。相应地，将确定一组规范，并持续进行开发，然后再添加某些功能，随后将重复这一过程。

为了轻松地使用增量开发方法，须通过相关软件以避免开发过程中内容更改所导致的负面影响，例如需要更改系统所提供的数据类型。Spring Boot 是增量开发系统的完美候选者，因为它支持对系统进行快速和简单的更改。

到目前为止，我们已经多次提到 Spring Boot，但并未言及其内容和用途，下面将对此予以简要介绍。

2．Spring Boot

Spring Boot 是一个 Web 应用程序框架，主要是针对 Spring 应用程序的引导指令和开发而设计的。Spring 是一个 Web 应用程序框架，并为 Java 平台开发 Web 应用程序提供了便利。另外，Spring Boot 大大降低了工业强度产品级 Spring 应用的开发难度。

本章将探讨如何使用 Spring Boot 创建 Web 应用程序。在开始开发应用程序之前，需要首先指定应用程序的实际功能（毕竟，在不了解其工作方式之前，尚无法构建具体内容）。

3．Messenger 系统的任务

下面将确定 Messenger 系统的初始需求条件，以及系统中可能出现的活动。此外，还将考察 Messenger 应用程序中的高级用例。

（1）用例

用例描述了实体与系统之间的应用方式。这里，实体表示为与系统交互的用户或组件的类型。在用例定义中，实体也称为参与者（actor）。

下面尝试定义 Messenger 系统中的参与者。显然，应用程序用户可视为参与者（使用应用程序满足其消息机制的用户）。另一个参与者则是管理员。不过，针对当前简单的 Messenger 应用程序，此处仅考察单一用户参与者。用户的用例包括以下内容：

- ❏ 用户使用 Messenger 平台发送、接收消息。
- ❏ 用户使用 Messenger 平台查看 Messenger 应用程序中的其他用户。
- ❏ 用户使用 Messenger 平台设置、更新其状态。
- ❏ 用户可注册 Messenger 平台。
- ❏ 用户可注销 Messenger 平台。

如果在系统开发过程中遇到了一个新的用例，那么，应可很方便地将其添加到系统中。在确定了系统的用例后，还需要恰当地描述系统在满足这些用例时的行为。

（2）系统行为

系统行为的定义旨在更加准确地描述系统执行的任务，以及清晰地展示系统组件之间的交互行为。考虑到当前应用程序较为简单，因而可借助示意图清楚地描述应用程序的行为。对此，可利用状态图对此予以描述。

状态图用于描述系统的行为，也就是说，描述基于不同状态的系统。在状态图中，系统中包含了有限数量的状态。

图 4.1 显示了当前系统中的状态图，其中包含了所定义的多个用例。

示意图中的每个圆圈代表某一时刻系统的执行状态。另外，每个箭头表示为用户请求的动作，并由系统予以执行。在初始状态，API 等待来自客户端应用程序的请求，该

行为在图中显示为"等待动作"状态。当动作请求被源自客户端应用程序的 API 接收后，系统退出"等待动作"状态和服务，请求通过适当的进程发送。

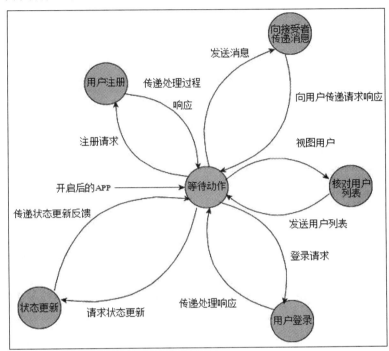

图 4.1　当前系统中的状态图

例如，当用户从 Android 应用程序中请求状态更新时，服务器将退出"等待动作"状态并执行"状态更新"过程，然后返回"等待动作"状态。

（3）识别数据

需要注意的是，在实现系统之前，应了解所需的数据类型。对此，可以很容易地从前面给出的用例定义中识别这些数据。基于当前用例规范，可以确定需要两种基本类型的数据，即用户数据和消息数据。顾名思义，用户数据是每个用户所需的数据；消息数据则是与发送的消息相关的数据。当前，可暂不考虑模式、实体或实体关系图等内容，只需对系统所需的数据有一个大致的概念。

对于 Messenger 应用，用户需要使用到用户名、电话号码、密码和状态信息。除此之外，还有必要跟踪账户的状态，以了解某个特定用户的账户是否被激活，或由于某种原因被停用。当前，暂不需要对所发送的消息予以太多关注，我们需要跟踪消息的发送方和消息的预期接收者。

关于所需数据，相关内容大致如此。随着开发过程的不断深入，还将识别更多的所需数据。下面讨论具体的编码过程。

4.2 实现 Messenger 后端

前述内容整体探讨了 Messenger 系统的用例、系统所需的数据和系统的行为，本节将着手开发系统的后端。如前所述，作为增量式开发的最佳候选者，我们将使用 Spring Boot 开发 messenger API。除此之外，Kotlin 和 Spring Boot 功能之间也实现了较好的整合。

考虑到将在 Messenger API 中处理数据，因而需要一种较为适宜的数据库，以存储 Messenge 系统所需的数据。对此，我们将使用 PostgreSQL 数据库。下面对 PostgreSQL 做一个简要的介绍。

4.2.1 PostgreSQL

PostgreSQL 是一个对象关系数据库管理系统，特别强调可扩展性和标准的遵从性。PostgreSQL 也称为 Postgres，通常被用作数据库服务器。当以这种方式使用时，其主要功能是安全地存储数据并返回软件应用程序请求存储的数据。

作为一种数据库，PostgreSQL 拥有诸多优点，其中包括：

❑ 可扩展性。PostgreSQL 特性可方便、可靠地由其用户进行扩展——对应的源代码免费向用户提供。
❑ 可移植性。PostgreSQL 适用于全部主流平台。几乎每个 UNIX 版本都可以使用 PostgreSQL。另外，Windows 兼容性也可以通过 Cygwin 框架实现。
❑ 完整性。包含了许多 GUI 工具，并可方便地与 PostgreSQL 进行交互。

PostgreSQL 在各种平台上的安装过程也较为简单，本节主要讨论 Windows、macOS 和 Linux 环境下的安装过程。

1. Windows 环境下安装

当在 Windows 环境下安装 PostgreSQL 时，须执行以下步骤：

（1）读者可访问 https://www.enterprisedb.com/downloads/postgres-postgresql-downloads#windows，下载并运行适当版本的 Windows PostgreSQL 安装程序。

（2）将 PostgreSQL 安装为 Windows 服务，并确保持有 PostgreSQL Windows 服务账户和密码。在后续安装过程中，将会使用到这些细节信息。

（3）当安装程序显示提示信息时，选择 PL/pgsql 过程式语言。

（4）在 Installation options 窗口中，可选择安装 pgAdmin。如果安装了 pgAdmin，当在安装程序显示提示信息时，应启用 Adminpack 构建软件捐赠（contrib）模块。

在执行了上述各步骤后，PostgreSQL 将被安装至用户的系统中。

2. macOS 安装

利用 Homebrew，PostgreSQL 可方便地安装在 macOS 上。如果系统上尚未安装 Homebrew，读者可参考第 1 章中的相关内容。若系统中已经安装了 Homebrew，则可打开终端，并运行下列命令：

```
brew search postgres
```

当终端内显示提示信息时，只需遵循相关安装指令即可。在安装过程中，可能会被询问管理员密码，读者可输入对应的密码，并等待安装结束。

3. Linux 安装

通过 PostgreSQL Linux 安装程序，PostgreSQL 可方便地安装于 Linux 环境下，相关步骤如下：

（1）访问 PostgreSQL 安装程序下载页面，对应网址为 https://www.enterprisedb.com/downloads/postgres-postgresqldownloads。

（2）选择希望安装的 PostgreSQL 版本。

（3）针对 PostgreSQL 选择相应的 Linux 安装程序。

（4）单击下载按钮并下载安装程序。

（5）下载完毕后运行安装程序，并遵循其间的安装指令。

（6）在提供了安装程序所需的信息后，PostgreSQL 即在系统中安装完毕。

当 PostgreSQL 在系统中设置完毕后，即可开始着手创建 Messenger API。

4.2.2 创建新的 Spring Boot 应用程序

通过 IntelliJ IDE 和 Spring 初始化器，很容易创建 Spring Boot 应用程序。打开 IntelliJ IDE，并利用 Spring Initializer 创建新的项目。对此，可单击 Create New Project，并选择 New Project 屏幕左侧操作栏的 Spring Initializer，如图 4.2 所示。

在选择了 Spring Initializer 之后，单击 Next 按钮。随后，在显示下一个窗口之前，IDE 将检索 Spring 初始化器，这需要占用几分钟的时间。

> **注意：**
> Spring 插件只能在 IntelliJ IDEA 的最终版本中使用，并附带付费订阅。

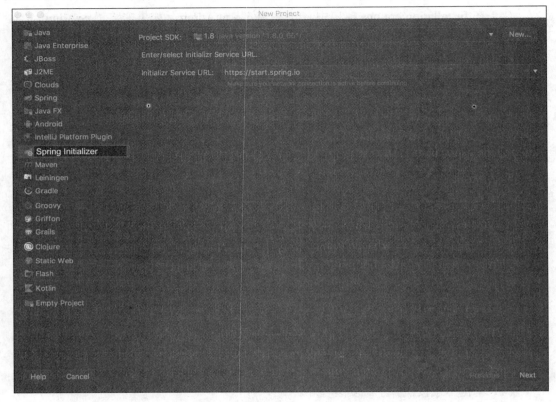

图 4.2　创建 Spring Boot 应用程序

一旦检索到 Spring Initializer，用户将被询问提供某些项目细节信息。在填写完毕后，读者即可开发本书中的应用程序；或者，读者也可提供自己的相关信息。在此之前，读者还需执行以下各项步骤：

（1）输入 com.example 作为组 ID。

（2）输入 messenger-api 作为工作 ID。

（3）选择 Maven Project 作为项目类型。

（4）保留包机制选项以及 Java 版本。

（5）选择 Kotlin 作为当前语言，这一项十分重要，后续程序将在 Kotlin 语言的基础上进行开发。

（6）保持 SNAPSHOT 值不变。

（7）输入选择描述。

（8）输入 com.example.messenger.api 作为包名。

在填写完所需的项目信息后，单击 Next 按钮，随后将显示如图 4.3 所示的窗口。

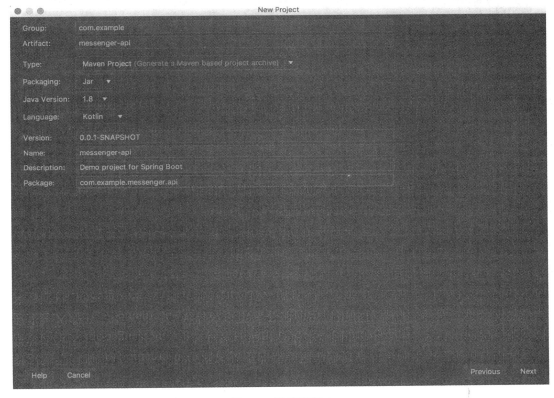

图 4.3　新项目窗口

在图 4.4 所示窗口中，用户将被询问选择项目依赖关系。对于初学者来说，可选择 Security、Web、JPA 和 PostgreSQL 依赖关系。其中，Security 位于 Core 分类下；Web 位于 Web 分类下；JPA 和 PostgreSQL 则位于 SQL 分类下。另外，在窗口上方的 Spring Boot Version 下拉菜单中，可选取 2.0.0 M5 作为当前版本。

在选取了必要的依赖关系后，当前内容如图 4.4 所示。

在选取了依赖关系后，单击 Next 按钮。在接下来的窗口中，用户将会被询问提供项目名称以及项目位置。这里，可填写 messenger-api 作为项目名称，并选择项目在计算机上的保存位置。最后，单击 Finish 按钮结束项目的设置过程。此时，会显示一个包含初始项目文件的 IDE 窗口。

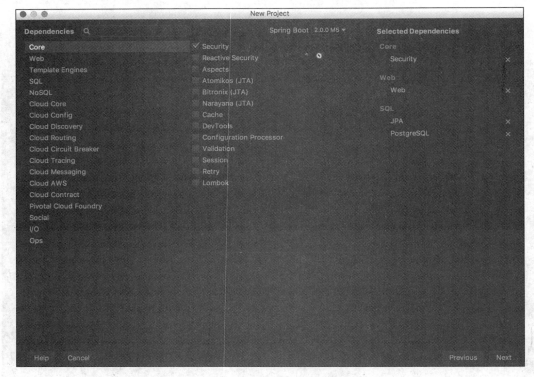

图 4.4 选取依赖关系

4.2.3 Spring Boot 概述

对于当前 Spring Boot 应用程序，下面考察初始程序文件的结构。图 4.5 显示了 Spring Boot 应用程序文件的结构。

全部源文件均位于 scr 目录中，该目录包含了核心应用程序文件，以及针对给程序编写的测试程序。核心应用程序文件置于 src/main 目录中；而测试程序则位于 src/test 中。另外，主目录中还设置了两个子目录，即 kotlin 目录和 resources 目录，全部数据包和主源文件均置于该目录中。特别地，当前程序文件和数据包将置于 com.example.messenger.api 数据包中。下面将快速查看一下 MessengerApiAplication.kt 文件，如图 4.6 所示。

MessengerApiAplication.kt 文件包含了主函数，这也是 Spring Boot 应用程序的入口点。当应用程序启动时，将调用该函数。一旦被调用，该函数将调用 SpringApplication.run() 函数。SpringApplication.run()函数接收两个参数。其中，第一个参数表示为类引用；第二个参数在启动时被传递至应用程序中。

第 4 章　设计并实现 Messenger 后端应用程序

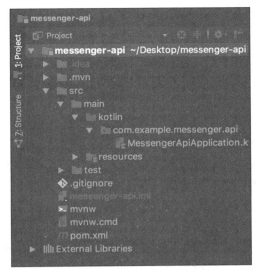

图 4.5　Spring Boot 应用程序文件的结构

图 4.6　MessengerApiAplication.kt 文件

在同一文件中，还定义了一个 MessengerApiApplication 类，并利用@SpringBootApplication 进行注解。这里，该注解等同于@Configuration、@EnableAutoConfiguration 和@ComponentScan 的组合效果。采用@Configuration 注解的类表示为 bean 定义的来源。

🛈 注意：

bean 表示为一个对象，并通过 Spring IoC 容器进行初始化和装配。

@EnableAutoConfiguration 属性通知 Spring Boot，Spring 应用程序将根据所提供的 jar 依赖关系自动配置。@ComponentScan 注解将配置（与@Configuration 类协同使用的）组

件扫描目录。

在 Spring Boot 应用程序开发过程中，注解的使用存在多种原因。最初使用这些注释可能会让人不知所措，但随着时间的推移，注解将变成一种自然的行为方式。

除了 MessengerApplication.kt 文件之外，另一个较为重要的文件是 application.properties，该文件位于 src/main/resources 中。当打开该文件时，将会发现其中未包含任何内容，其原因在于：当前尚未定义任何应用程序配置或属性。下面尝试添加一组配置信息，并向 application.properties 文件中输入下列内容：

```
spring.jpa.generate-ddl=true
spring.jpa.hibernate.ddl-auto=create-drop
```

spring.jpa.generate-ddl 属性用于指定在应用程序启动时是否生成数据库模式。当该属性设置为 true 时，将在应用程序启动时生成数据库模式。spring.jpa.hibernate.ddl-auto 属性则用于指定 DDL 模式，考虑到此处需要在应用程序启动时创建模式，并在程序结束时对其进行销毁，因而可使用 create-drop。

上述内容通过属性定义了数据库模式，但尚未针对 messenger-api 创建最终的数据库。如果安装了 pgAdmin 和 PostgreSQL，那么，可方便地利用这些软件创建数据库。如果读者还未安装 pgAdmin，则可通过 PostgreSQL 的 createdb 命令针对当前应用程序创建数据库。对此，读者可在终端中输入下列命令：

```
createdb -h localhost --username=<username> --password messenger-api
```

其中，-h 标记用于确定运行数据库服务器的主机名称。--username 标记则用于指定连接到服务器的用户名。--password 则要求执行相应的密码规范。messenger-api 表示为数据库设置的名称。对此，读者可利用自己的服务器用户名替换<username>。在输入了上述命令行后，按 Enter 键并运行该命令，同时输入对应的密码。随后，PostgreSQL 将创建名为 messenger-api 的数据库。

在数据库配置完毕后，需要将 Spring Boot 应用程序连接至该数据库中。针对于此，可使用 spring.datasource.url、spring.datasource.username 和 spring.datasource.password 属性。随后，将下列配置添加至 application.properties 文件中。

```
spring.jpa.generate-ddl=true
spring.jpa.hibernate.ddl-auto=create-drop
spring.datasource.url=jdbc:postgresql://localhost:5432/messenger-api
spring.datasource.username=<username>
spring.datasource.password=<password>
```

spring.datasource.url 属性通过 Spring Boot 连接的数据库确定 JDBC URL。

spring.datasource.username 和 spring.datasource.password 属性分别用于指定服务器用户名以及与其关联的密码。相应地，读者可利用自己的用户名和密码替换<username>和<password>。

在上述属性配置完毕后，即可准备启动 Spring Boot 应用程序。

通过单击 MessengerApiApplication.kt 中主函数一侧的 Kotlin 图标，并选择 Run 选项，即可运行 messenger-api 应用程序，如图 4.7 所示。

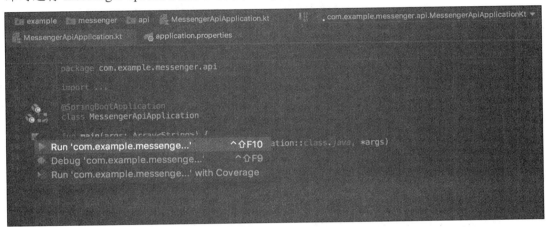

图 4.7　运行 messenger-api 应用程序

当项目构建处理过程完毕后（可能会等待少许时间），当前应用程序将在 Tomcat 服务器上启动。

下面继续考察项目中的 pom.xml 文件，该文件位于项目的根目录中。这里，POM 是项目对象模型的简写。在 Apache Maven 网站中，POM 解释如下：项目对象模型或 POM 表示为 Maven 中的基本工作单元，并表示为一个 XML 文件，其中包含了 Maven 所用的项目和配置细节信息，并以此构建项目。出于完整性考虑，下面简要地介绍一下 Maven。

1. Maven

Apache Maven 是一种基于 POM 概念的软件项目管理和理解工具。Maven 可以用于多种用途，例如项目构建管理和文档。

在理解了项目文件之后，我们将通过一些模型以满足之前确定的数据。

2. 构建模型

下面将数据建模至相应的实体类中，经 Spring Boot 处理后构建适宜的数据库模式。这里第一个考察的模型是用户模型。对此，在当前数据包中创建 User.kt 文件，并输入下列代码：

```kotlin
package com.example.messenger.api.models

import org.hibernate.validator.constraints.Length
import org.springframework.format.annotation.DateTimeFormat
import java.time.Instant
import java.util.*
import javax.persistence.*
import javax.validation.constraints.Pattern
import javax.validation.constraints.Size

@Entity
@Table(name = "`user`")
@EntityListeners(UserListener::class)
class User(
  @Column(unique = true)
  @Size(min = 2)
  var username: String = "",
  @Size(min = 11)
  @Pattern(regexp="^\\(?(\\d{3})\\)?[- ]?(\\d{3})[- ]?(\\d{4})$")
  var phoneNumber: String = "",
  @Size(min = 60, max = 60)
  var password: String = "",
  var status: String = "",
  @Pattern(regexp = "\\A(activated|deactivated)\\z")
  var accountStatus: String = "activated"
)
```

  上述代码块中使用了大量的注解。首先，@Entity 表明当前类为 Java 持久化 API（JPA）。针对类所表达的实体，@Table 注解指定了表名，这在方案生成期间十分有用。如果未使用@Table 注解，所生成的表名则表示为类名。在 PostgreSQL 中，数据库表将采用名称 user 加以创建。顾名思义，@EntityListener 针对实体类定义了一个实体监听器，稍后将对 UserListener 类加以讨论。

  下面考察 User 类属性，其中包含了共计 7 个类属性。其中，前 5 个属性分别是 username、password、phoneNumber、accountStatus 和 status，代表了用户所需的数据类型。之前曾定义了 4 个用户实体，但当前仍存在一个问题——我们需要一种方法可唯一辨识所创建的每个用户。除此之外，还应记录新用户何时添加至 Messenger 平台中，这对于后续参考十分重要。经过仔细考虑后，当前实体中应定义 id 和 createdAt 属性——读者可能对此感到疑问，为何要向用户实体中加入 id 和 createdAt 属性？毕竟，我们之前并没有明确指出需要使用到这类属性。但这一决定是正确的。但是，随着开发过程的逐步进行，可以在必要时进行更改和添加。下面尝试添加这两个属性，如下所示：

```
@Entity
@Table(name = "`user`")
@EntityListeners(UserListener::class)
class User(
  @Column(unique = true)
  @Size(min = 2)
  var username: String = "",
  @Size(min = 8, max = 15)
  @Column(unique = true)
  @Pattern(regexp = "^\\(?(\\d{3})\\)?[- ]?(\\d{3})[- ]?(\\d{4})$")
  var phoneNumber: String = "",
  @Size(min = 60, max = 60)
  var password: String = "",
  var status: String = "available",
  @Pattern(regexp = "\\A(activated|deactivated)\\z")
  var accountStatus: String = "activated",
  @Id
  @GeneratedValue(strategy = GenerationType.AUTO)
  var id: Long = 0,
  @DateTimeFormat
  var createdAt: Date = Date.from(Instant.now())
)
```

读者需要理解每项注解的含义。其中，@Column 属性用于指定表中的各列。在实际操作过程中，所有的实体属性均体现了表中的一列。相应地，可在代码中使用@Column (unique = true)，并对属性设置唯一性约束。若不希望多条记录共用某一特定属性值时，这将十分有用。相信读者已经猜到@Size，用于指定表中属性的尺寸。@Pattern 用于确定表属性匹配的模式。

@Id 属性用于唯一识别实体（此处为 id 属性）。@GeneratedValue(strategy = GenerationType.AUTO)负责确定自动生成的 id 值。@DateTimeFormat 将对数值设置时间戳，该值存储于用户表的 created_at 列中。

下面将定义 UserListener 类。对此，须创建一个名为 listeners 的数据包，并向该包中添加 UserListener 类，如下所示：

```
package com.example.messenger.api.listeners

import com.example.messenger.api.models.User
```

```
import org.springframework.security.crypto.bcrypt.BCryptPasswordEncoder
import javax.persistence.PrePersist
import javax.persistence.PreUpdate

class UserListener {

  @PrePersist
  @PreUpdate
  fun hashPassword(user: User) {
    user.password = BCryptPasswordEncoder().encode(user.password)
  }
}
```

注意,在数据库中,密码不可存储为明文。出于安全原因,密码在存储前须执行哈希计算。相应地,hashPassword()函数负责执行这项任务,也就是说,利用哈希对应值(使用 BCrypt)替换用户对象密码属性所持有的字符串。@PrePersist 和@PreUpdate 用于指定该函数应在数据库用户记录持久化或更新之前被调用。

下面创建一个消息实体,在 models 数据包中定义 Message 类,并向该类中添加下列代码:

```
package com.example.messenger.api.models

import org.springframework.format.annotation.DateTimeFormat
import java.time.Instant
import java.util.*
import javax.persistence.*

@Entity
class Message(
  @ManyToOne(optional = false)
  @JoinColumn(name = "user_id", referencedColumnName = "id")
  var sender: User? = null,
  @ManyToOne(optional = false)
  @JoinColumn(name = "recipient_id", referencedColumnName = "id")
  var recipient: User? = null,
  var body: String? = "",
  @ManyToOne(optional = false)
  @JoinColumn(name="conversation_id", referencedColumnName = "id")
  var conversation: Conversation? = null,
  @Id @GeneratedValue(strategy = GenerationType.AUTO) var id: Long = 0,
  @DateTimeFormat
```

```
    var createdAt: Date = Date.from(Instant.now())
)
```

除了之前所讨论的注解之外，代码中还新加入了两个注解。如前所述，每条消息均包含了一个发送者和一个接收者，且均表示为平台中的用户。因此，消息实体中包含了 User 类型的发送者和接收者属性。这里，用户可表示为多条消息的发送者，以及多条消息的接收者，这一关系需要在实际操作过程中予以实现。此外，还可使用@ManyToOne 注解实现这一功能。对于多对一关系，可使用@ManyToOne(optional = false)。@JoinColumn 则指定了一个列，用于连接实体关联或元素集合，如下所示：

```
@JoinColumn(name = "user_id", referencedColumnName = "id")
var sender: User? = null
```

上述代码片段加入了一个 user_id 属性，该属性将用户的 id 引用到消息表。

读者可能已经注意到，Message 类中使用了会话属性。这是因为用户之间发送的消息发生在会话线程中。简单地说，每个消息都属于一个线程。我们需要在 models 包中添加一个 Conversation 类，代表对话实体，如下所示：

```
package com.example.messenger.api.models

import org.springframework.format.annotation.DateTimeFormat
import java.time.Instant
import java.util.*
import javax.persistence.*

@Entity
class Conversation(
  @ManyToOne(optional = false)
  @JoinColumn(name = "sender_id", referencedColumnName = "id")
  var sender: User? = null,
  @ManyToOne(optional = false)
  @JoinColumn(name = "recipient_id", referencedColumnName = "id")
  var recipient: User? = null,
  @Id
  @GeneratedValue(strategy = GenerationType.AUTO)
  var id: Long = 0,
  @DateTimeFormat
  val createdAt: Date = Date.from(Instant.now())
) {

  @OneToMany(mappedBy = "conversation", targetEntity =
```

```
    Message::class)
    private var messages: Collection<Message>? = null
}
```

其中，多个消息均属于一个会话，因此我们在 Conversation 类体中设置一个 messages 集合。

实体模型的创建过程已基本完成，剩下的工作是为用户发送和接收的消息添加适当的集合，如下所示：

```
package com.example.messenger.api.models

import com.example.messenger.api.listeners.UserListener
import org.springframework.format.annotation.DateTimeFormat
import java.time.Instant
import java.util.*
import javax.persistence.*
import javax.validation.constraints.Pattern
import javax.validation.constraints.Size

@Entity
@Table(name = "`user`")
@EntityListeners(UserListener::class)
class User(
  @Column(unique = true)
  @Size(min = 2)
  var username: String = "",
  @Size(min = 8, max = 15)
  @Column(unique = true)
  @Pattern(regexp = "^\\(?(\\d{3})\\)?[- ]?(\\d{3})[- ]?(\\d{4})$")
  var phoneNumber: String = "",
  @Size(min = 60, max = 60)
  var password: String = "",
  var status: String = "available",
  @Pattern(regexp = "\\A(activated|deactivated)\\z")
  var accountStatus: String = "activated",
  @Id
  @GeneratedValue(strategy = GenerationType.AUTO)
  var id: Long = 0,
  @DateTimeFormat
  var createdAt: Date = Date.from(Instant.now())
) {
  //collection of sent messages
```

```
@OneToMany(mappedBy = "sender", targetEntity = Message::class)
private var sentMessages: Collection<Message>? = null

//collection of received messages
@OneToMany(mappedBy = "recipient", targetEntity = Message::class)
private var receivedMessages: Collection<Message>? = null
}
```

至此，我们完成了实体创建过程。为了帮助读者理解所创建的实体及其关系，图 4.8 显示了一幅实体关系图（E-R 图），并体现了已经建立的实体以及实体间的关系。

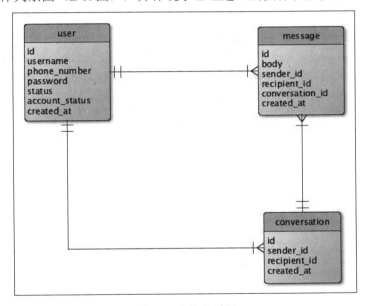

图 4.8　实体关系图

根据 E-R 图，不难发现，用户可包含多个消息；而一条消息可隶属于某个用户；消息可属于某个会话；某个会话可包含多个消息。除此之外，用户也可包含多个会话。

在模型创建完毕后，目前仅剩下一个问题——尚无法访问存储于此类实体中的数据。下面将创建存储库以解决这一问题。

3．创建存储库

Spring Data JPA 可在运行时从存储库接口中自动生成存储库实现。下面将创建一个存储库，并访问 User 实体，进而查看其工作方式。首先，创建 repositories 数据包，并将其纳入 UserRepository.kt 文件中，如下所示：

```kotlin
package com.example.messenger.api.repositories

import com.example.messenger.api.models.User
import org.springframework.data.repository.CrudRepository

interface UserRepository : CrudRepository<User, Long> {

  fun findByUsername(username: String): User?

  fun findByPhoneNumber(phoneNumber: String): User?
}
```

其中，UserRepository 扩展了 CrudRepository 接口，其所协同工作的 entity 类型和 id 类型通过 CrudRepository 的泛型参数指定。通过扩展 CrudRepository，UserRepository 继承了 User 持久化方法，例如保存方法、搜索方法以及删除 User 实体的方法。

Spring JPA 允许使用方法签名声明其他查询函数。我们可利用这一功能创建 findByUsername()和 findByPhoneNumber()函数。

当前定义了 3 个实体，因而需要设置 3 个存储库对其进行查询。对此，可在 repositories 中定义 MessageRepository 接口，如下所示：

```kotlin
package com.example.messenger.api.repositories

import com.example.messenger.api.models.Message
import org.springframework.data.repository.CrudRepository

interface MessageRepository : CrudRepository<Message, Long> {
  fun findByConversationId(conversationId: Long): List<Message>
}
```

需要注意的是，上述方法签名将 List<Message>指定为返回类型。Spring JPA 可对此自动识别，并在调用 findByConversationId()时返回 Message 元素列表。

最后，还需实现 ConversationRepository 接口，如下所示：

```kotlin
package com.example.messenger.api.repositories

import com.example.messenger.api.models.Conversation
import org.springframework.data.repository.CrudRepository

interface ConversationRepository : CrudRepository<Conversation, Long> {
    fun findBySenderId(id: Long): List<Conversation>
```

```
    fun findByRecipientId(id: Long): List<Conversation>

    fun findBySenderIdAndRecipientId(senderId: Long,
        recipientId: Long): Conversation?
}
```

由于已经定义了实体集和必要的存储库来查询这些实体，所以可以开始执行 Messenger 后端的业务逻辑。对此，需要了解服务和服务实现。

4. 服务和服务实现

服务实现定义为一个 Spring bean，并通过@Service 予以标记。Spring 应用程序的业务逻辑一般置于服务实现中。另外一方面，服务也表示为一个接口，并包含了须进一步实现的应用程序函数签名。简而言之，服务定义为一个接口，而服务实现则表示为实现了该服务的类。

下面尝试创建服务以及服务实现。对此，可构建 service 数据包，并将服务和服务实现添加于其中。在该数据包中，可定义 UserService 接口，对应代码如下所示：

```
package com.example.messenger.api.services

import com.example.messenger.api.models.User

interface UserService {
    fun attemptRegistration(userDetails: User): User

    fun listUsers(currentUser): List<User>

    fun retrieveUserData(username: String): User?

    fun retrieveUserData(id: Long): User?

    fun usernameExists(username: String): Boolean
}
```

在上述 UserService 接口中，所定义的函数须在实现了 UserService 的类中加以声明。下面着手创建服务实现。相应地，将 UserServiceImpl 类添加至 services 包中，当实现 UserService 时，需要覆写 attemptRegistration()、listUsers()、retrieveUserData()以及 usernameExists()函数，如下所示：

```
package com.example.messenger.api.services

import com.example.messenger.api.exceptions.InvalidUserIdException
```

```kotlin
import com.example.messenger.api.exceptions.UserStatusEmptyException
import com.example.messenger.api.exceptions.UsernameUnavailableException
import com.example.messenger.api.models.User
import com.example.messenger.api.repositories.UserRepository
import org.springframework.stereotype.Service

@Service
class UserServiceImpl(val repository: UserRepository) : UserService {
  @Throws(UsernameUnavailableException::class)
  override fun attemptRegistration(userDetails: User): User {
    if (!usernameExists(userDetails.username)) {
      val user = User()
      user.username = userDetails.username
      user.phoneNumber = userDetails.phoneNumber
      user.password = userDetails.password
      repository.save(user)
      obscurePassword(user)
      return user
    }
    throw UsernameUnavailableException("The username
            ${userDetails.username} is unavailable.")
  }

  @Throws(UserStatusEmptyException::class)
  fun updateUserStatus(currentUser: User, updateDetails: User): User {
    if (!updateDetails.status.isEmpty()) {
      currentUser.status = updateDetails.status
      repository.save(currentUser)
      return currentUser
    }
    throw UserStatusEmptyException()
  }

  override fun listUsers(currentUser: User): List<User> {
    return repository.findAll().mapTo(ArrayList(), { it })
                .filter{ it != currentUser }
  }

  override fun retrieveUserData(username: String): User? {
    val user = repository.findByUsername(username)
    obscurePassword(user)
    return user
  }
```

```
@Throws(InvalidUserIdException::class)
override fun retrieveUserData(id: Long): User {
  val userOptional = repository.findById(id)
  if (userOptional.isPresent) {
    val user = userOptional.get()
    obscurePassword(user)
    return user
  }
  throw InvalidUserIdException("A user with an id of '$id'
                                does not exist.")
}

override fun usernameExists(username: String): Boolean {
  return repository.findByUsername(username) != null
}

private fun obscurePassword(user: User?) {
  user?.password = "XXX XXXX XXX"
  }
}
```

在 UserServiceImpl 的主构造函数中，UserRepository 的实例定义为所需的参数。读者无须对传递此类参数而感到担心。Spring 意识到 UserServiceImpl 需要一个 UserRepository 实例，并通过依赖注入为该类提供一个实例。除了所实现的函数之外，此处还声明了一个 obscurePassword() 函数，该函数使用 XXX XXXX XXX 对 User 实体中的密码进行哈希处理。

依据之前所讨论的服务和服务实现构建原则，下面继续为消息和对话增加一些内容，并向 service 中添加 MessageService 接口，如下所示：

```
package com.example.messenger.api.services

import com.example.messenger.api.models.Message
import com.example.messenger.api.models.User

interface MessageService {

  fun sendMessage(sender: User, recipientId: Long,
                  messageText: String): Message
}
```

此处针对 sendMessage() 添加了一个方法签名，该签名必须由 MessageServiceImpl 所覆写。以下是消息服务的实现过程：

```kotlin
package com.example.messenger.api.services

import com.example.messenger.api.exceptions.MessageEmptyException
import com.example.messenger.api.exceptions.MessageRecipientInvalidException
import com.example.messenger.api.models.Conversation
import com.example.messenger.api.models.Message
import com.example.messenger.api.models.User
import com.example.messenger.api.repositories.ConversationRepository
import com.example.messenger.api.repositories.MessageRepository
import com.example.messenger.api.repositories.UserRepository
import org.springframework.stereotype.Service

@Service
class MessageServiceImpl(val repository: MessageRepository,
            val conversationRepository: ConversationRepository,
            val conversationService: ConversationService,
            val userRepository: UserRepository) : MessageService {

  @Throws(MessageEmptyException::class,
      MessageRecipientInvalidException::class)
  override fun sendMessage(sender: User, recipientId: Long,
                messageText: String): Message {
    val optional = userRepository.findById(recipientId)

    if (optional.isPresent) {
      val recipient = optional.get()

      if (!messageText.isEmpty()) {
        val conversation: Conversation = if (conversationService
            .conversationExists(sender, recipient)) {
          conversationService.getConversation(sender, recipient)
                      as Conversation
        } else {
          conversationService.createConversation(sender, recipient)
        }
        conversationRepository.save(conversation)

        val message = Message(sender, recipient, messageText,
                    conversation)
        repository.save(message)
        return message
```

```
      }
    } else {
      throw MessageRecipientInvalidException("The recipient id
                         '$recipientId' is invalid.")
    }
    throw MessageEmptyException()
  }
}
```

上述 sendMessage()实现首先判断消息内容是否为空。若否，该函数将检测是否存在发送者和接收者之间的一个活动会话。若是，则该会话存储于 conversation 中；否则，则创建两个用户间的一个会话，并将其存储于 conversation 中。随后，会话将被保存，并创建、保存当前消息。

下面实现 ConversationService 和 ConversationServiceImpl。对此，可在 services 中定义 ConversationService 接口，并添加下列代码：

```
package com.example.messenger.api.services

import com.example.messenger.api.models.Conversation
import com.example.messenger.api.models.User

interface ConversationService {

  fun createConversation(userA: User, userB: User): Conversation
  fun conversationExists(userA: User, userB: User): Boolean
  fun getConversation(userA: User, userB: User): Conversation?
  fun retrieveThread(conversationId: Long): Conversation
  fun listUserConversations(userId: Long): List<Conversation>
  fun nameSecondParty(conversation: Conversation, userId: Long): String
}
```

其中包含 6 个函数签名，即 createConversation()、conversationExists()、getConversation()、retrieveThread()、listUserConversations()和 nameSecondParty()。当前，还应向 services 中添加 ConversationServiceImpl，并实现前 3 个方法，即 createConversation()、conversationExists()和 getConversation()方法。对应实现过程如下所示：

```
package com.example.messenger.api.services

import com.example.messenger.api.exceptions.ConversationIdInvalidException
import com.example.messenger.api.models.Conversation
import com.example.messenger.api.models.User
```

```kotlin
import com.example.messenger.api.repositories.ConversationRepository
import org.springframework.stereotype.Service

@Service
class ConversationServiceImpl(val repository:
       ConversationRepository) :ConversationService {

  override fun createConversation(userA: User, userB: User):
           Conversation {
    val conversation = Conversation(userA, userB)
    repository.save(conversation)
    return conversation
  }

  override fun conversationExists(userA: User, userB: User): Boolean {
    return if (repository.findBySenderIdAndRecipientId
           (userA.id, userB.id) != null)
    true
    else repository.findBySenderIdAndRecipientId
           (userB.id, userA.id) != null
  }

  override fun getConversation(userA: User, userB: User): Conversation? {
    return when {
      repository.findBySenderIdAndRecipientId(userA.id,
                     userB.id) != null ->
      repository.findBySenderIdAndRecipientId(userA.id, userB.id)
      repository.findBySenderIdAndRecipientId(userB.id,
                     userA.id) != null ->
      repository.findBySenderIdAndRecipientId(userB.id, userA.id)
      else -> null
    }

  }
}
```

在添加了前 3 个方法后，下面继续向 ConversationServiceImpl 中加入 retrieveThread()、listUserConversations()和 nameSecondParty()方法，如下所示：

```kotlin
override fun retrieveThread(conversationId: Long): Conversation {
  val conversation = repository.findById(conversationId)

  if (conversation.isPresent) {
```

```kotlin
        return conversation.get()
    }
    throw ConversationIdInvalidException("Invalid conversation id
                                    '$conversationId'")
}

override fun listUserConversations(userId: Long):
            ArrayList<Conversation> {
    val conversationList: ArrayList<Conversation> = ArrayList()
    conversationList.addAll(repository.findBySenderId(userId))
    conversationList.addAll(repository.findByRecipientId(userId))

    return conversationList
}

override fun nameSecondParty(conversation: Conversation,
                             userId: Long): String {
    return if (conversation.sender?.id == userId) {
        conversation.recipient?.username as String
    } else {
        conversation.sender?.username as String
    }
}
```

读者可能已经注意到，在服务实现类中，多次抛出了不同类型的异常。对此，应对此类异常加以定义。另外，还需针对每个异常定义一个 ExceptionHandler。这一类异常处理程序将在抛出异常时向客户端发送适当的错误响应。

接下来，可创建 exceptions 数据包，向其中添加 AppExceptions.kt 文件，并在该文件中加入下列代码：

```kotlin
package com.example.messenger.api.exceptions

class UsernameUnavailableException(override val message: String) :
RuntimeException()

class InvalidUserIdException(override val message: String) :
RuntimeException()

class MessageEmptyException(override val message: String = "A message
cannot be empty.") : RuntimeException()

class MessageRecipientInvalidException(override val message: String) :
```

```
RuntimeException()

class ConversationIdInvalidException(override val message: String) :
RuntimeException()

class UserDeactivatedException(override val message: String) :
RuntimeException()

class UserStatusEmptyException(override val message: String = "A user's
status cannot be empty") : RuntimeException()
```

对于服务器运行期内出现的异常，每个异常将扩展 RuntimeException。同时，所有异常都具有 message 属性。顾名思义，这一类消息称作异常消息。在异常添加完毕后，还需要创建控制器设备类。ControllerAdvice 类用于处理出现于 Spring 应用程序中的错误，并通过 @ControllerAdvice 注解创建。另外，控制器设备表示为 Spring 组件类型。下面将构建控制器设备类，并对上述异常进行处理。

下面考察 UsernameUnavailableException、InvalidUserIdException 和 UserStatusEmptyException，需要注意的是，这 3 个异常均与用户相关。因此，可将当前控制器设备命名为 UserControllerAdvice。下面创建 components 数据包，并将 UserControllerAdvice 类添加于其中，如下所示：

```
package com.example.messenger.api.components

import com.example.messenger.api.constants.ErrorResponse
import com.example.messenger.api.constants.ResponseConstants
import com.example.messenger.api.exceptions.InvalidUserIdException
import com.example.messenger.api.exceptions.UserStatusEmptyException
import com.example.messenger.api.exceptions.UsernameUnavailableException
import org.springframework.http.ResponseEntity
import org.springframework.web.bind.annotation.ControllerAdvice
import org.springframework.web.bind.annotation.ExceptionHandler

@ControllerAdvice
class UserControllerAdvice {

  @ExceptionHandler(UsernameUnavailableException::class)
  fun usernameUnavailable(usernameUnavailableException:
                     UsernameUnavailableException):
  ResponseEntity<ErrorResponse> {
    val res = ErrorResponse(ResponseConstants.USERNAME_UNAVAILABLE
```

```kotlin
                          .value, usernameUnavailableException.message)
    return ResponseEntity.unprocessableEntity().body(res)
}

@ExceptionHandler(InvalidUserIdException::class)
fun invalidId(invalidUserIdException: InvalidUserIdException):
    ResponseEntity<ErrorResponse> {
  val res = ErrorResponse(ResponseConstants.INVALID_USER_ID.value,
                          invalidUserIdException.message)
    return ResponseEntity.badRequest().body(res)
}

@ExceptionHandler(UserStatusEmptyException::class)
fun statusEmpty(userStatusEmptyException: UserStatusEmptyException):
    ResponseEntity<ErrorResponse> {
  val res = ErrorResponse(ResponseConstants.EMPTY_STATUS.value,
                          userStatusEmptyException.message)
    return ResponseEntity.unprocessableEntity().body(res)
}
}
```

我们刚刚定义了相关函数，以满足可能出现的 3 个异常，并使用@ExceptionHanlder()注释每个函数。@ExceptionHanlder()接收一个指向异常的类引用。每个函数都接收一个参数，该参数是所抛出的异常实例。除此之外，全部定义后的函数均返回 ResponseEntity<ErrorResponse>实例。响应实体表示发送到客户端的整个 HTTP 响应。

当前并未创建 ErrorResponse。下面生成 constants 数据包，并将 ErrorResponse 添加至其中，如下所示：

```kotlin
package com.example.messenger.api.constants

class ErrorResponse(val errorCode: String, val errorMessage: String)
```

ErrorResponse 类包含两个属性，即 errorCode 和 errorMessage。在进行讨论之前，首先需要将 ResponseConstants 枚举类添加至 constants 数据包中，如下所示：

```kotlin
package com.example.messenger.api.constants

enum class ResponseConstants(val value: String) {
  SUCCESS("success"), ERROR("error"),
  USERNAME_UNAVAILABLE("USR_0001"),
  INVALID_USER_ID("USR_002"),
```

```
    EMPTY_STATUS("USR_003"),
    MESSAGE_EMPTY("MES_001"),
    MESSAGE_RECIPIENT_INVALID("MES_002"),
    ACCOUNT_DEACTIVATED("GLO_001")
}
```

下面创建 3 个控制器设备类，即 MessageControllerAdvice、ConversationControllerAdvice 和 RestControllerAdvice。其中，RestControllerAdvice 针对服务器运行过程中的错误定义了异常处理机制。

MessageControllerAdvice 类定义如下所示：

```
package com.example.messenger.api.components

import com.example.messenger.api.constants.ErrorResponse
import com.example.messenger.api.constants.ResponseConstants
import com.example.messenger.api.exceptions.MessageEmptyException
import com.example.messenger.api.exceptions.MessageRecipientInvalidException
import org.springframework.http.ResponseEntity
import org.springframework.web.bind.annotation.ControllerAdvice
import org.springframework.web.bind.annotation.ExceptionHandler

@ControllerAdvice
class MessageControllerAdvice {
  @ExceptionHandler(MessageEmptyException::class)
  fun messageEmpty(messageEmptyException: MessageEmptyException):
      ResponseEntity<ErrorResponse> {
    //ErrorResponse object creation
    val res = ErrorResponse(ResponseConstants.MESSAGE_EMPTY.value,
                    messageEmptyException.message)

    // Returning ResponseEntity containing appropriate ErrorResponse
    return ResponseEntity.unprocessableEntity().body(res)
  }

   @ExceptionHandler(MessageRecipientInvalidException::class)
    fun messageRecipientInvalid(messageRecipientInvalidException:
                        MessageRecipientInvalidException):
        ResponseEntity<ErrorResponse> {
   val res = ErrorResponse(ResponseConstants.MESSAGE_RECIPIENT_INVALID
               .value, messageRecipientInvalidException.message)
```

```
    return ResponseEntity.unprocessableEntity().body(res)
  }
}
```

ConversationControllerAdvice 类定义如下所示:

```
package com.example.messenger.api.components

import com.example.messenger.api.constants.ErrorResponse
import com.example.messenger.api.exceptions.ConversationIdInvalidException
import org.springframework.http.ResponseEntity
import org.springframework.web.bind.annotation.ControllerAdvice
import org.springframework.web.bind.annotation.ExceptionHandler

@ControllerAdvice
class ConversationControllerAdvice {
  @ExceptionHandler
  fun conversationIdInvalidException(conversationIdInvalidException:
          ConversationIdInvalidException): ResponseEntity<ErrorResponse> {
    val res = ErrorResponse("", conversationIdInvalidException.message)
    return ResponseEntity.unprocessableEntity().body(res)
  }
}
```

最后, RestControllerAdvice 类定义如下所示:

```
package com.example.messenger.api.components

import com.example.messenger.api.constants.ErrorResponse
import com.example.messenger.api.constants.ResponseConstants
import com.example.messenger.api.exceptions.UserDeactivatedException
import org.springframework.http.HttpStatus
import org.springframework.http.ResponseEntity
import org.springframework.web.bind.annotation.ControllerAdvice
import org.springframework.web.bind.annotation.ExceptionHandler

@ControllerAdvice
class RestControllerAdvice {

  @ExceptionHandler(UserDeactivatedException::class)
  fun userDeactivated(userDeactivatedException:
```

```
                    UserDeactivatedException):
    ResponseEntity<ErrorResponse> {
  val res = ErrorResponse(ResponseConstants.ACCOUNT_DEACTIVATED
                    .value, userDeactivatedException.message)

  // Return an HTTP 403 unauthorized error response
  return ResponseEntity(res, HttpStatus.UNAUTHORIZED)
  }
}
```

上述内容实现了业务逻辑，在通过 REST 端点将 HTTP 请求置入 API 之前，须确保 API 的安全性。

4.2.4　限制 API 访问

允许任何人访问 RESTful API 资源可视为一种安全大忌。针对于此，须设计一种方法，将服务器的访问限定为注册用户和登录用户。我们将使用 Spring Security 和 JSON Web Tokens（JWTs）来实现这一功能。

1. Spring Security

Spring Security 是一个针对 Spring 应用程序的、高度可定制的访问控制框架，同时也是一种被公认的标准，用于保护 Spring 构建的应用程序。通常在项目的开始阶段被选择用来添加安全依赖。当前项目并不需要向 pom 添加 Spring 安全依赖（已添加完毕）。

2. JSON Web Tokens

JSON Web Tokens 是一种开放的行业标准方法，用于在双方之间安全地表示声明。JWTs 允许解码、验证和生成 JWT 等操作。JWTs 可以很容易地使用 Spring Boot 来实现应用程序中的身份验证。下面将演示如何使用 JWTs 和 Spring Security 组合来保护 Messenger 后端程序。

当在 Spring 应用程序中使用 JWTs 时，首先需要将 JWTs 的依赖关系添加到项目的 pom.xml 文件中，如下所示：

```
<dependencies>
 ...
 <dependency>
   <groupId>io.jsonwebtoken</groupId>
   <artifactId>jjwt</artifactId>
   <version>0.7.0</version>
```

```
        </dependency>
    </dependencies>
```

当在 pom.xml 中设置了新的 Maven 依赖关系时，IntelliJ 将提示导入新的依赖关系，如图 4.9 所示。

图 4.9　IntelliJ 提示导入新的依赖关系

在显示提示信息后，单击 Import Changes 按钮，JWT 依赖关系将导入当前项目中。

3．配置 Web 安全

首先需要构建自定义 Web 安全配置。对此，可在 com.example.messenger.api 中创建 config 数据包，将 WebSecurityConfig 类添加至该数据包中，并输入下列代码：

```
package com.example.messenger.api.config

import com.example.messenger.api.filters.JWTAuthenticationFilter
import com.example.messenger.api.filters.JWTLoginFilter
import com.example.messenger.api.services.AppUserDetailsService
import org.springframework.context.annotation.Configuration
import org.springframework.http.HttpMethod
import org.springframework.security.config.annotation.authentication.builders.AuthenticationManagerBuilder
import org.springframework.security.config.annotation.web.builders.HttpSecurity
import org.springframework.security.config.annotation.web.configuration.EnableWebSecurity
import org.springframework.security.config.annotation.web.configuration.WebSecurityConfigurerAdapter
```

```kotlin
import org.springframework.security.core.userdetails.UserDetailsService
import org.springframework.security.crypto.bcrypt.BCryptPasswordEncoder
import org.springframework.security.web.authentication.UsernamePasswordAuthenticationFilter

@Configuration
@EnableWebSecurity
class WebSecurityConfig(val userDetailsService: AppUserDetailsService)
    : WebSecurityConfigurerAdapter() {

  @Throws(Exception::class)
  override fun configure(http: HttpSecurity) {
    http.csrf().disable().authorizeRequests()
        .antMatchers(HttpMethod.POST, "/users/registrations")
        .permitAll()
        .antMatchers(HttpMethod.POST, "/login").permitAll()
        .anyRequest().authenticated()
        .and()
```

过滤/login 请求,如下所示:

```kotlin
.addFilterBefore(JWTLoginFilter("/login",
        authenticationManager()),
        UsernamePasswordAuthenticationFilter::class.java)
```

过滤其他请求,并检测数据头中是否存在 JWT,如下所示:

```kotlin
.addFilterBefore(JWTAuthenticationFilter(),
        UsernamePasswordAuthenticationFilter::class.java)
}

@Throws(Exception::class)
override fun configure(auth: AuthenticationManagerBuilder) {
  auth.userDetailsService<UserDetailsService>(userDetailsService)
      .passwordEncoder(BCryptPasswordEncoder())
  }
}
```

其中,WebSecurityConfig 利用@EnableWebSecurity 进行标注,这将开启 Spring Security 的 Web 安全支持。除此之外,WebSecurityConfig 扩展了 WebSecurityConfigurerAdapter,并覆写了其中的 configure()方法,并向 Web 安全配置中加入了某些自定义内容。

configure(HttpSecurity)方法负责配置安全的 URL 路径。在 WebSecurityConfig 中，允许所有对/users/registration 和/login 路径的 POST 请求。这两个端点不需要受到保护，因为用户在登录或在平台上注册之前不能进行身份验证。此外，我们还为请求添加了过滤器。对/loign 的请求将被 JWTLoginFilter 过滤（当前尚未实现这一功能）；所有未经身份验证和所禁止的请求都将被 JWTAuthenticationFilter 过滤（当前尚未实现这一功能）。

configure(AuthenticationManagerBuilder)负责设置 UserDetailsService，并指定所采用的密码编码器。

如前所述，目前，尚有多个类还未予实现。下面首先定义 JWTLoginFilter。对此，可创建名为 filters 的新数据包，并添加 JWTLoginFilter 类，如下所示：

```
package com.example.messenger.api.filters

import com.example.messenger.api.security.AccountCredentials
import com.example.messenger.api.services.TokenAuthenticationService
import com.fasterxml.jackson.databind.ObjectMapper

import org.springframework.security.authentication.AuthenticationManager
import org.springframework.security.authentication.UsernamePasswordAuthenticationToken
import org.springframework.security.core.Authentication
import org.springframework.security.core.AuthenticationException
import org.springframework.security.web.authentication.AbstractAuthenticationProcessingFilter
import org.springframework.security.web.util.matcher.AntPathRequestMatcher

import javax.servlet.FilterChain
import javax.servlet.ServletException
import javax.servlet.http.HttpServletRequest
import javax.servlet.http.HttpServletResponse
import java.io.IOException

class JWTLoginFilter(url: String, authManager:
    AuthenticationManager)  :AbstractAuthenticationProcessingFilter
(AntPathRequestMatcher(url)){

  init {
    authenticationManager = authManager
```

```kotlin
    }

    @Throws(AuthenticationException::class, IOException::class,
            ServletException::class)
    override fun attemptAuthentication( req: HttpServletRequest,
                        res: HttpServletResponse): Authentication{
      val credentials = ObjectMapper()
          .readValue(req.inputStream, AccountCredentials::class.java)
      return authenticationManager.authenticate(
        UsernamePasswordAuthenticationToken(
          credentials.username,
          credentials.password,
          emptyList()
        )
      )
    }

    @Throws(IOException::class, ServletException::class)
    override fun successfulAuthentication(
            req: HttpServletRequest,
            res: HttpServletResponse, chain: FilterChain,
            auth: Authentication) {
      TokenAuthenticationService.addAuthentication(res, auth.name)
    }
}
```

JWTLoginFilter 接收字符串 URL，以及一个 AuthenticationManager 实例作为主构造函数的参数，同时扩展了 AbstractAuthenticationProcessingFilter。该过滤器拦截发向服务器的 HTTP 请求，并对其进行验证。相应地，attemptAuthentication()负责执行实际的验证过程，并利用 ObjectMapper()实例读取 via HTTP 请求中的证书。随后，authenticationManager 用于验证当前请求。AccountCredentials 则是尚未实现的一个类，对此，创建名为 security 的新数据包，并将 AccountCredentials.kt 文件添加至其中，如下所示：

```kotlin
package com.example.messenger.api.security

class AccountCredentials {
  lateinit var username: String
  lateinit var password: String
}
```

由于需要对用户进行验证，因而此处定义了 username 和 password 变量。

当用户验证成功后，即调用 SuccessfulAuthentication()方法，对应任务是向 HTTP 响应头 Authorization 中添加验证标记，实际的添加过程则由 TokenAuthenticationService.addAuthentication()完成。下面向 services 数据包中加入该服务，如下所示：

```
package com.example.messenger.api.services
import io.jsonwebtoken.Jwts
import io.jsonwebtoken.SignatureAlgorithm
import org.springframework.security.authentication.UsernamePasswordAuthenticationToken
import org.springframework.security.core.Authentication
import org.springframework.security.core.GrantedAuthority

import javax.servlet.http.HttpServletRequest
import javax.servlet.http.HttpServletResponse
import java.util.Date

import java.util.Collections.emptyList

internal object TokenAuthenticationService {
  private val TOKEN_EXPIRY: Long = 864000000
  private val SECRET = "$78gr43g7g8feb8we"
  private val TOKEN_PREFIX = "Bearer"
  private val AUTHORIZATION_HEADER_KEY = "Authorization"

  fun addAuthentication(res: HttpServletResponse, username: String) {
    val JWT = Jwts.builder()
              .setSubject(username)
              .setExpiration(Date(System.currentTimeMillis() +
                             TOKEN_EXPIRY))
              .signWith(SignatureAlgorithm.HS512, SECRET)
              .compact()
    res.addHeader(AUTHORIZATION_HEADER_KEY, "$TOKEN_PREFIX $JWT")
  }

  fun getAuthentication(request: HttpServletRequest): Authentication? {
    val token = request.getHeader(AUTHORIZATION_HEADER_KEY)
    if (token != null) {
```

标记的解析过程如下所示：

```
    val user = Jwts.parser().setSigningKey(SECRET)
            .parseClaimsJws(token.replace(TOKEN_PREFIX, ""))
            .body.subject

    if (user != null)
      return UsernamePasswordAuthenticationToken(user, null,
                  emptyList<GrantedAuthority>())
  }
  return null
}
```

顾名思义，addAuthentication()向 HTTP 响应头 Authorization 中加入验证标记；getAuthentication()负责对用户进行验证。

下面向 filters 数据包中添加 JWTAuthenticationFilter。对此，可向 filters 数据包中加入 JWTAuthenticationFilter 类，如下所示：

```
package com.example.messenger.api.filters

import com.example.messenger.api.services.TokenAuthenticationService
import org.springframework.security.core.context.SecurityContextHolder
import org.springframework.web.filter.GenericFilterBean
import javax.servlet.FilterChain
import javax.servlet.ServletException
import javax.servlet.ServletRequest
import javax.servlet.ServletResponse
import javax.servlet.http.HttpServletRequest
import java.io.IOException

class JWTAuthenticationFilter : GenericFilterBean() {

  @Throws(IOException::class, ServletException::class)
  override fun doFilter(request: ServletRequest,
                        response: ServletResponse,
                        filterChain: FilterChain) {
    val authentication = TokenAuthenticationService
            .getAuthentication(request as HttpServletRequest)
    SecurityContextHolder.getContext().authentication = authentication
    filterChain.doFilter(request, response)
  }
}
```

每次请求/响应作为资源请求传递至过滤器链时，JWTAuthenticationFilter 的 doFilter() 函数将被容器调用。传递至 doFilter()中的 FilterChain 实例使得过滤器将请求和响应传递至过滤器链中的下一个实体中。

最后，像以往一样，还需要实现 AppUserDetailsService 类，并将该类置于项目的 services 包中，如下所示：

```
package com.example.messenger.api.services

import com.example.messenger.api.repositories.UserRepository
import org.springframework.security.core.GrantedAuthority
import org.springframework.security.core.authority.SimpleGrantedAuthority
import org.springframework.security.core.userdetails.User
import org.springframework.security.core.userdetails.UserDetails
import org.springframework.security.core.userdetails.UserDetailsService
import org.springframework.security.core.userdetails.UsernameNotFoundException
import org.springframework.stereotype.Component
import java.util.ArrayList

@Component
class AppUserDetailsService(val userRepository: UserRepository) : UserDetailsService {

@Throws(UsernameNotFoundException::class)
override fun loadUserByUsername(username: String): UserDetails {
  val user = userRepository.findByUsername(username) ?:
             throw UsernameNotFoundException("A user with the
                        username $username doesn't exist")
    return User(user.username, user.password,
                ArrayList<GrantedAuthority>())
  }
}
```

对于传递至 loadUsername(String)函数中的用户名，该函数尝试加载与其匹配的 UserDetails。如果不存在此类用户，则抛出 UsernameNotFoundException。

至此，我们已经成功配置了 Spring Security。下面准备利用控制器并通过 RESTful 端点展示一些 API 功能。

4．通过 RESTful 端点访问服务器资源

截至目前，我们已经创建了模型、组件、服务以及服务实现，并将 Spring Security

整合至消息应用程序中,但尚未实现外部客户端与消息 API 之间的通信方式。下面将构建控制器类,进而处理来自不同 HTTP 请求路径的请求。与往常一样,首先需要创建一个数据包,以包含即将创建的控制器。

这里,第一个需要实现的控制器是 UserController,该控制器将与用户资源相关的 HTTP 请求映射为类中的动作,进而处理、响应 HTTP 请求。首先,需要一个端点以简化新用户的注册。随后,可调用这一动作处理此类 create 注册请求。下列代码表示为包含 create 动作的 UserController 代码。

```kotlin
package com.example.messenger.api.controllers

import com.example.messenger.api.models.User
import com.example.messenger.api.repositories.UserRepository
import com.example.messenger.api.services.UserServiceImpl
import org.springframework.http.ResponseEntity
import org.springframework.validation.annotation.Validated
import org.springframework.web.bind.annotation.*
import javax.servlet.http.HttpServletRequest

@RestController
@RequestMapping("/users")
class UserController(val userService: UserServiceImpl,
                    val userRepository: UserRepository) {

  @PostMapping
  @RequestMapping("/registrations")

  fun create(@Validated @RequestBody userDetails: User):
          ResponseEntity<User> {
    val user = userService.attemptRegistration(userDetails)
    return ResponseEntity.ok(user)
  }
}
```

控制器类利用 @RestController 和 @RequestMapping 进行标注。其中,注解 @RestController 指定某个类为 REST 控制器;而与 UserController 一同使用的 @RequestMapping,则将全部请求(以/users 路径开始)映射至 UserController。

create 函数利用@PostMapping 和@RequestMapping("/registrations")加以标注。这两个注解的组合将全部 POST 请求(包含/users/registrations 路径)映射至 create 函数。通过 @Validated 和@RequestBody 标注的 User 实例将被传递至 create 中。@RequestBody 将

POST 请求体中发送的 JSON 值绑定到 userDetails 中。@Validated 确保 JSON 参数将被验证。在建立了端点并成功运行后,下面将对其进行测试。启动应用程序并导航至终端窗口,使用 CURL 向 Messenger API 发送请求,如下所示:

```
curl -H "Content-Type: application/json" -X POST -d
'{"username":"kevin.stacey",
  "phoneNumber":"5472457893",
  "password":"Hello123"}'
http://localhost:8080/users/registrations
```

服务器将创建一个用户并发送一个响应,如下所示:

```
{
  "username":"kevin.stacey",
  "phoneNumber":"5472457893",
  "password":"XXX XXXX XXX",
  "status":"available",
  "accountStatus":"activated",
  "id":6,"createdAt":1508579448634
}
```

一切均工作良好,但是我们可以看到,HTTP 响应中包含了许多不需要的值,例如 password 和 accountStatus 响应参数。除此之外,我们还希望 createdAt 中包含一个人类可读的日期。下面将使用装配器和值对象来完成所有这些工作。

首先将定义一个值对象,其中包含了用户数据,并以相应的格式传递至客户端。相应地,可创建 helpers.objects 数据包,其中设置了 ValueObjects.kt 文件,如下所示:

```
package com.example.messenger.api.helpers.objects

data class UserVO(
  val id: Long,
  val username: String,
  val phoneNumber: String,
  val status: String,
  val createdAt: String
)
```

不难发现,UserVO 表示为一个数据类,并对传递至用户的信息进行建模。相应地,可针对后续操作中的其他响应添加值对象,以避免再次返回至当前文件中,如下所示:

```
package com.example.messenger.api.helpers.objects
```

```kotlin
data class UserVO(
  val id: Long,
  val username: String,
  val phoneNumber: String,
  val status: String,
  val createdAt: String
)

data class UserListVO(
  val users: List<UserVO>
)

data class MessageVO(
  val id: Long,
  val senderId: Long?,
  val recipientId: Long?,
  val conversationId: Long?,
  val body: String?,
  val createdAt: String
)

data class ConversationVO(
  val conversationId: Long,
  val secondPartyUsername: String,
  val messages: ArrayList<MessageVO>
)

data class ConversationListVO(
  val conversations: List<ConversationVO>
)
```

在所请求的值对象设置完毕后，下面针对 UserVO 创建一个装配器。这里，装配器可简单地表示为一个组件，并装配所需的对象值。当创建 UserAssembler 时，该装配器将被调用。考虑到装配器表示为一个组件，因而其隶属于 components 数据包中，如下所示：

```kotlin
package com.example.messenger.api.components

import com.example.messenger.api.helpers.objects.UserListVO
import com.example.messenger.api.helpers.objects.UserVO
import com.example.messenger.api.models.User
import org.springframework.stereotype.Component
```

```kotlin
@Component
class UserAssembler {

  fun toUserVO(user: User): UserVO {
    return UserVO(user.id, user.username, user.phoneNumber,
                  user.status, user.createdAt.toString())
  }

  fun toUserListVO(users: List<User>): UserListVO {
    val userVOList = users.map { toUserVO(it) }
    return UserListVO(userVOList)
  }
}
```

装配器定义了单一函数 toUserVO()，并接收 User 作为参数，同时返回对应的 UserVO。类似地，toUserListVO() 接收一个 User 实例列表，并返回对应的 UserListVO。

下面编辑 create 端点，并使用 UserAssembler 和 UserVO，如下所示：

```kotlin
package com.example.messenger.api.controllers

import com.example.messenger.api.components.UserAssembler
import com.example.messenger.api.helpers.objects.UserVO
import com.example.messenger.api.models.User
import com.example.messenger.api.repositories.UserRepository
import com.example.messenger.api.services.UserServiceImpl
import org.springframework.http.ResponseEntity
import org.springframework.validation.annotation.Validated
import org.springframework.web.bind.annotation.*
import javax.servlet.http.HttpServletRequest

@RestController
@RequestMapping("/users")
class UserController(val userService: UserServiceImpl,
                     val userAssembler: UserAssembler,
                     val userRepository: UserRepository) {

  @PostMapping
  @RequestMapping("/registrations")
  fun create(@Validated @RequestBody userDetails: User):
          ResponseEntity<UserVO> {
    val user = userService.attemptRegistration(userDetails)
```

```
        return ResponseEntity.ok(userAssembler.toUserVO(user))
    }
}
```

重启服务器,并发送一个新的请求以注册一个 User。这里,我们将从 API 中得到更为适宜的响应结果,如下所示:

```
{
  "id":6,
  "username":"kevin.stacey",
  "phoneNumber":"5472457893",
  "status":"available",
  "createdAt":"Sat Oct 21 11:11:36 WAT 2017"
}
```

下面为 Messenger Android 应用程序创建所有必要的端点,以结束端点创建过程。首先,可添加端点显示用户的详细信息、列出所有用户、获取当前用户的详细信息,并将 User 的状态更新为 UserController,如下所示:

```
package com.example.messenger.api.controllers

import com.example.messenger.api.components.UserAssembler
import com.example.messenger.api.helpers.objects.UserListVO
import com.example.messenger.api.helpers.objects.UserVO
import com.example.messenger.api.models.User
import com.example.messenger.api.repositories.UserRepository
import com.example.messenger.api.services.UserServiceImpl
import org.springframework.http.ResponseEntity
import org.springframework.validation.annotation.Validated
import org.springframework.web.bind.annotation.*
import javax.servlet.http.HttpServletRequest

@RestController
@RequestMapping("/users")
class UserController(val userService: UserServiceImpl,
                     val userAssembler: UserAssembler,
                     val userRepository: UserRepository) {

    @PostMapping
    @RequestMapping("/registrations")
    fun create(@Validated @RequestBody userDetails: User):
            ResponseEntity<UserVO> {
```

```kotlin
  val user = userService.attemptRegistration(userDetails)
  return ResponseEntity.ok(userAssembler.toUserVO(user))
}

@GetMapping
@RequestMapping("/{user_id}")
fun show(@PathVariable("user_id") userId: Long):
         ResponseEntity<UserVO> {
  val user = userService.retrieveUserData(userId)
  return ResponseEntity.ok(userAssembler.toUserVO(user))
}

@GetMapping
@RequestMapping("/details")
fun echoDetails(request: HttpServletRequest):
                ResponseEntity<UserVO>{
  val user = userRepository.findByUsername
             (request.userPrincipal.name) as User
  return ResponseEntity.ok(userAssembler.toUserVO(user))
}

@GetMapping
fun index(request: HttpServletRequest): ResponseEntity<UserListVO> {
  val user = userRepository.findByUsername
             (request.userPrincipal.name) as User
  val users = userService.listUsers(user)

  return ResponseEntity.ok(userAssembler.toUserListVO(users))
}

@PutMapping
fun update(@RequestBody updateDetails: User,
    request: HttpServletRequest): ResponseEntity<UserVO> {
  val currentUser = userRepository.findByUsername
                    (request.userPrincipal.name)
  userService.updateUserStatus(currentUser as User, updateDetails)
  return ResponseEntity.ok(userAssembler.toUserVO(currentUser))
}
}
```

下面创建控制器并处理消息资源和会话资源，即MessageController和ConversationController。

在构建控制器之前,将会使用到装配器,进而装配源自 JPA 实体中的值对象。相应地,MessageAssembler 的定义如下所示:

```kotlin
package com.example.messenger.api.components

import com.example.messenger.api.helpers.objects.MessageVO
import com.example.messenger.api.models.Message
import org.springframework.stereotype.Component

@Component
class MessageAssembler {
  fun toMessageVO(message: Message): MessageVO {
    return MessageVO(message.id, message.sender?.id,
            message.recipient?.id, message.conversation?.id,
            message.body, message.createdAt.toString())
  }
}
```

接下来,将创建 ConversationAssembler,如下所示:

```kotlin
package com.example.messenger.api.components

import com.example.messenger.api.helpers.objects.ConversationListVO
import com.example.messenger.api.helpers.objects.ConversationVO
import com.example.messenger.api.helpers.objects.MessageVO
import com.example.messenger.api.models.Conversation
import com.example.messenger.api.services.ConversationServiceImpl
import org.springframework.stereotype.Component

@Component
class ConversationAssembler(val conversationService:
                    ConversationServiceImpl,
                    val messageAssembler: MessageAssembler) {

  fun toConversationVO(conversation: Conversation, userId: Long):
ConversationVO {
    val conversationMessages: ArrayList<MessageVO> = ArrayList()
    conversation.messages.mapTo(conversationMessages) {
      messageAssembler.toMessageVO(it)
    }
```

```kotlin
        return ConversationVO(conversation.id, conversationService
                    .nameSecondParty(conversation, userId),
                    conversationMessages)
    }

    fun toConversationListVO(conversations: ArrayList<Conversation>,
                    userId: Long): ConversationListVO {
        val conversationVOList = conversations.map { toConversationVO(it,
                                    userId) }
        return ConversationListVO(conversationVOList)
    }
}
```

对于 MessageController 和 ConversationController 来说，一切均已就绪。对于当前相对简单的 Messenger 应用程序，只需为 MessageController 提供一个消息创建动作。下列代码展示了基于消息创建动作的 MessageController，即 create。

```kotlin
package com.example.messenger.api.controllers

import com.example.messenger.api.components.MessageAssembler
import com.example.messenger.api.helpers.objects.MessageVO
import com.example.messenger.api.models.User
import com.example.messenger.api.repositories.UserRepository
import com.example.messenger.api.services.MessageServiceImpl
import org.springframework.http.ResponseEntity
import org.springframework.web.bind.annotation.*
import javax.servlet.http.HttpServletRequest

@RestController
@RequestMapping("/messages")
class MessageController(val messageService: MessageServiceImpl,
                        val userRepository: UserRepository,
                        val messageAssembler: MessageAssembler) {

    @PostMapping
    fun create(@RequestBody messageDetails: MessageRequest,
            request: HttpServletRequest): ResponseEntity<MessageVO> {
        val principal = request.userPrincipal
        val sender = userRepository.findByUsername(principal.name) as User
        val message = messageService.sendMessage(sender,
```

```
                    messageDetails.recipientId, messageDetails.message)
    return ResponseEntity.ok(messageAssembler.toMessageVO(message))
}

data class MessageRequest(val recipientId: Long, val message: String)
}
```

最后，还需要生成ConversationController，且仅需要使用到两个端点。其中，一个端点列出用户的所有活动会话；另一个端点用于获取会话线程中的消息。这一类端点分别通过list()和show()动作被创建。ConversationController类定义如下所示：

```
package com.example.messenger.api.controllers

import com.example.messenger.api.components.ConversationAssembler
import com.example.messenger.api.helpers.objects.ConversationListVO
import com.example.messenger.api.helpers.objects.ConversationVO
import com.example.messenger.api.models.User
import com.example.messenger.api.repositories.UserRepository
import com.example.messenger.api.services.ConversationServiceImpl
import org.springframework.http.ResponseEntity
import org.springframework.web.bind.annotation.*
import javax.servlet.http.HttpServletRequest

@RestController
@RequestMapping("/conversations")
class ConversationController(
  val conversationService: ConversationServiceImpl,
  val conversationAssembler: ConversationAssembler,
  val userRepository: UserRepository
) {

  @GetMapping
  fun list(request: HttpServletRequest): ResponseEntity<ConversationListVO>
{
    val user = userRepository.findByUsername(request
                .userPrincipal.name) as User
    val conversations = conversationService.listUserConversations
                (user.id)
    return ResponseEntity.ok(conversationAssembler
                .toConversationListVO(conversations, user.id))
```

```kotlin
}

@GetMapping
@RequestMapping("/{conversation_id}")
fun show(@PathVariable(name = "conversation_id") conversationId: Long,
        request: HttpServletRequest): ResponseEntity<ConversationVO> {
    val user = userRepository.findByUsername(request
                    .userPrincipal.name) as User
    val conversationThread = conversationService.retrieveThread
                    (conversationId)
    return ResponseEntity.ok(conversationAssembler
                .toConversationVO(conversationThread, user.id))
}
}
```

回忆一下，用户包含了相应的账户状态，并有可能注销该账户。此时，作为 API 的定义者，我们并不希望注销用户使用当前平台。因此，应避免此类用户与 API 进行交互。对此，存在多种方式可实现这一功能。当前示例将使用拦截器截取 HTTP 请求，并在其继续沿着请求链下行之前，执行一项或多项操作。类似于装配器，拦截器也表示为一类组件。相应地，可调用拦截器检测账户 AccountValidityInterceptor 的有效性。下列代码定义了拦截器类（位于 components 包中）。

```kotlin
package com.example.messenger.api.components

import com.example.messenger.api.exceptions.UserDeactivatedException
import com.example.messenger.api.models.User
import com.example.messenger.api.repositories.UserRepository
import org.springframework.stereotype.Component
import org.springframework.web.servlet.handler.HandlerInterceptorAdapter
import java.security.Principal
import javax.servlet.http.HttpServletRequest
import javax.servlet.http.HttpServletResponse

@Component
class AccountValidityInterceptor(val userRepository:
        UserRepository) : HandlerInterceptorAdapter() {

    @Throws(UserDeactivatedException::class)
    override fun preHandle(request: HttpServletRequest,
```

```
            response: HttpServletResponse, handler: Any?): Boolean {
  val principal: Principal? = request.userPrincipal

  if (principal != null) {
    val user = userRepository.findByUsername(principal.name)
            as User

    if (user.accountStatus == "deactivated") {
      throw UserDeactivatedException("The account of this user has
                                     been deactivated.")
    }
  }
  return super.preHandle(request, response, handler)
 }
}
```

AccountValidityInterceptor 类覆写了其超类的 preHandle()函数,在将请求路由至控制器动作之前,该函数将被调用并执行某些操作。在生成拦截器后,该拦截器需要通过 Spring 应用程序注册,这一配置操作可通过 WebMvcConfigurer 完成。接下来,将 AppConfig 文件添加至项目的 config 数据包中,并在该文件中输入下列代码:

```
package com.example.messenger.api.config

import com.example.messenger.api.components.AccountValidityInterceptor
import org.springframework.beans.factory.annotation.Autowired
import org.springframework.context.annotation.Configuration
import org.springframework.web.servlet.config.annotation.InterceptorRegistry
import org.springframework.web.servlet.config.annotation.WebMvcConfigurer

@Configuration
class AppConfig : WebMvcConfigurer {

  @Autowired
  lateinit var accountValidityInterceptor: AccountValidityInterceptor

  override fun addInterceptors(registry: InterceptorRegistry) {
    registry.addInterceptor(accountValidityInterceptor)
    super.addInterceptors(registry)
  }
}
```

AppConfig 表示为 WebMvcConfigurer 的子类，并覆写了其超类中的 addInterceptor(InterceptorRegistry)函数。通过 registry.addInterceptor()，可将 accountValidityInterceptor 添加至注册表中。

至此，我们完成了向 Messenger Android 应用程序提供 Web 资源所需的全部代码，下面将代码部署至远程服务器上。

4.3 将 Messenger API 部署至 AWS 上

将 Spring Boot 应用程序部署至 AWS 的过程十分简单，甚至可以在 10 分钟之内完成。本节介绍如何将基于 Spring 的应用程序部署至 AWS 上。在部署应用程序之前，首先配置应用程序需要连接的、AWS 上的 PostgreSQL 数据库。

4.3.1 配置 AWS 上的 PostgreSQL

首先需要创建 AWS 账户，对此，读者可访问 https://portal.aws.amazon.com/billing/signup#/start。待注册完毕后，可登录 AWS 并访问 Amazon Relational Database Service(RDS)（或者在导航栏中单击 Services | Database | RDS），如图 4.10 所示。在 RDS 窗口中，单击 Get Started Now 按钮。

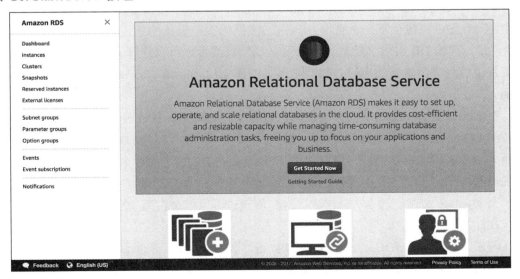

图 4.10　Amazon Relational Database Service 窗口

浏览 Launch DB instance Web 页面，读者需要选取与 DB 配置相关的选项。这里，选择 PostgreSQL 作为可用的 DB 引擎，如图 4.11 所示。

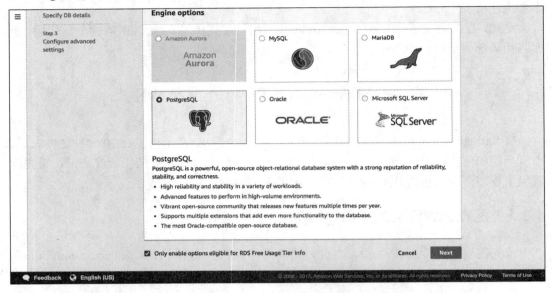

图 4.11　选择 PostgreSQL 作为可用的 DB 引擎

这里，应确保选中 Only enable options eligible for RDS Free Usage Tier info 复选框。单击 Next 按钮将转至下一个页面。其中，应保留实例规范的默认设置，并输入必要的 DB 设置，例如 DB 实例名称以及主密码。当然，读者也可以选择其他名称。无论如何，应确保谨慎处理所输入的内容。随后可进入如图 4.12 所示的 Configure advanced settings 页面。

在 Network & Security 部分中，确保启用了公共访问，并选取 VPC Security Groups 下方的 Create new VPC security group 选项。随后，将页面滚动至 Database options 部分，并输入 DB 名称。再次强调，这里使用了 MessengerDB 作为 DB 名称。其他选项则保留默认设置，单击页面下方的 Launch DB instance 按钮。

至此，DB 实例通过 AWS 创建。这一创建过程可能会占用 10 分钟左右的时间，读者可在此期间稍做歇息。

随后，单击 View DB instance details 按钮，接下来所显示的页面包含了 DB 部署实例的详细信息。将页面滚动至 Connect 部分，并查看 DB 实例的连接信息，如图 4.13 所示。

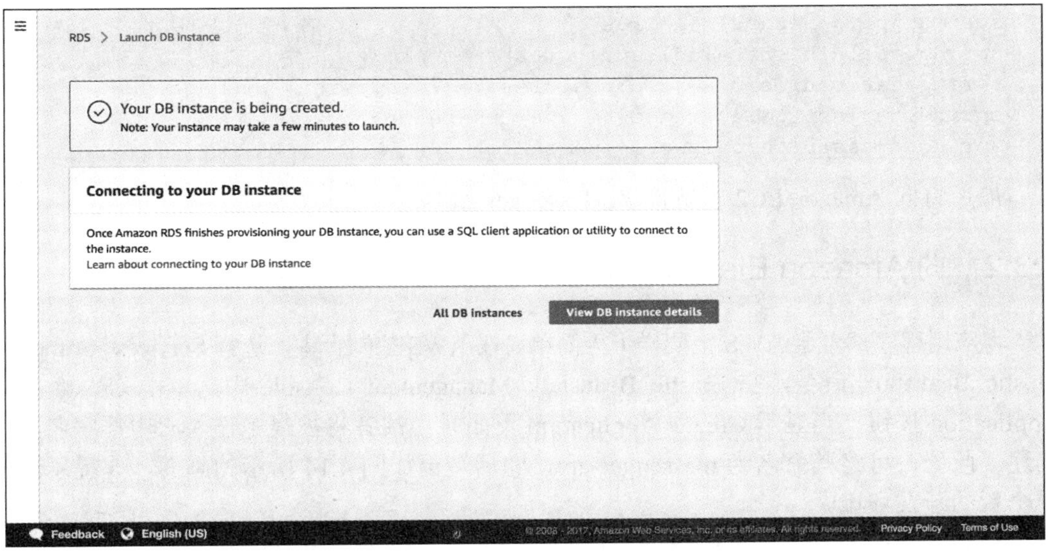

图 4.12　Configure advanced settings 页面

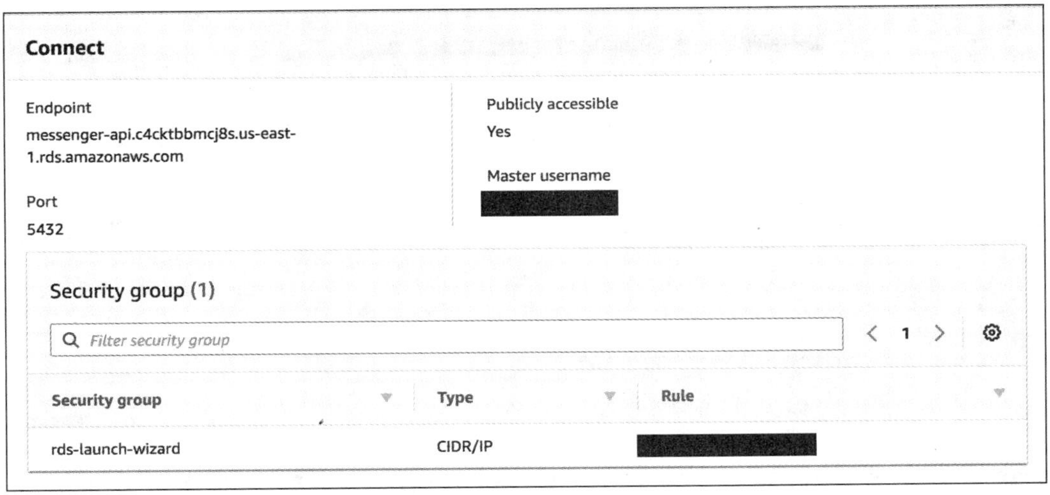

图 4.13　查看 DB 实例的连接信息

通过上述信息，可成功地连接至 PostgreSQL DB 实例上的 MessengerDB。当启用 messenger-api 并连接至 MessengerDB 时，需要编辑 application.properties 文件中的 spring.datasource.url、spring.datasource.username 以及 spring.datasource.password 属性。随后，application.properties 中的对应内容如下所示：

```
spring.jpa.generate-ddl=true
spring.jpa.hibernate.ddl-auto=create-drop
spring.datasource.url=jdbc:postgresql://<endpoint>/MessengerDB
spring.datasource.username=<master_username>
spring.datasource.password=<password>
```

最后将向 Amazon EC2 实例部署 Messenger API。

4.3.2 向 Amazon Elastic Beanstalk 部署 Messenger API

将应用程序部署至 AWS 上同样十分简单。在 AWS 控制台中，选择 Services | Compute | Elastic Beanstalk 命令。在 Elastic Beanstalk Management Console 中，单击 Create New Application 按钮。当显示 Create Application 页面时，读者将被提示输入应用程序名称和描述。此处，可将其命名为 messenger-api，随后转至如图 4.14 所示的画面，并提示创建一个新的开发环境。

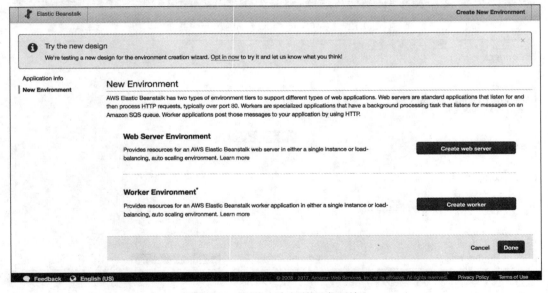

图 4.14　创建一个新的开发环境

其中，读者可构建一个 Web 服务器环境，并于随后配置环境类型。这里，可选取 Tomcat 预定义配置环境，并将当前环境类型调整为 Single instance，如图 4.15 所示。

待设置完毕后，将转至如图 4.16 所示画面，此时需要针对当前应用程序选取相关来源。此处可选中 Upload your own。

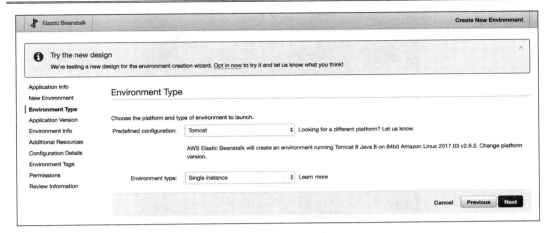

图 4.15　构建一个 Web 服务器环境

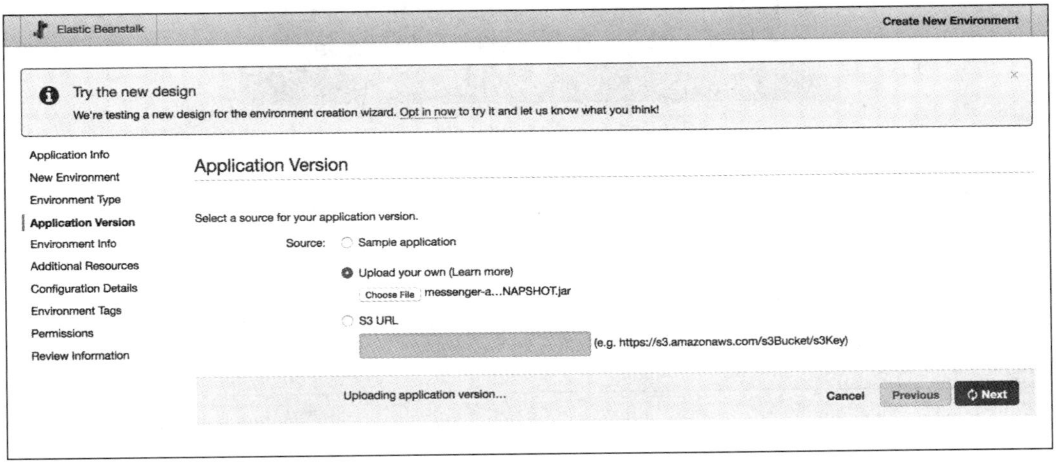

图 4.16　针对当前应用程序选取相关来源

当前需要创建相应的项目压缩包并予以上传。通过 Maven，可将 Messenger API 打包，如图 4.17 所示。

单击项目 IDE 窗口右侧的 Maven Projects 按钮，并选择 messenger-api | Lifecycle | package。相应地，项目压缩包（jar）将被打包，并存储于项目的目标目录中。

随后返回至 AWS，并选取该 jar 文件作为上传源文件。其他属性则保持默认选项，并单击 Next 按钮。在打包 jar 文件上传过程中，读者可能需要等待少许时间。待上传完毕后，新窗口中将显示当前环境信息。在随后的几个窗口中，读者可简单地单击 Next 按钮，直至到达 Configuration Details 窗口。其中，需要将实例类型修改为 t2.micro。

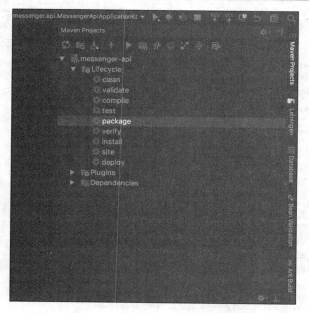

图4.17 通过 Maven，可将 Messenger API 打包

在 Review Information 窗口中，将页面滚动至 Environment Info，如图 4.18 所示。

图4.18 Environment Info 窗口

注意，当前环境 URL 将发生变化，读者应对此留意，稍后将会使用到此类信息。随后，将 Web 页面滚动至下方，并单击 Launch 按钮。此时，Elastic Beanstalk 将启用当前新环境。

至此，messenger-api 已经成功地部署至 AWS 中。

4.4 本章小结

本章介绍了如何利用 Kotlin 构建 Spring Boot REST 应用程序编程接口。其间，我们学习了系统设计的基本知识，通过状态图显示了 Messenger API 系统的行为，同时还了解到状态图中新型的拦截方式。随后，我们还进一步构建了 E-R 图，并以图形方式详细描述了系统实体及其关系。

此外，本章还讨论了如何在本地机器上配置 PostgreSQL，并创建新的 PostgreSQL 数据库、通过 Spring Boot 2.0 构建微服务、将微服务连接至数据库，并利用 Spring Data 与数据库中的数据进行交互。

除此之外，本章还利用 Spring Security 和 JSON Web Tokens (JWTs)实现了 RESTful Spring Boot Web 应用程序的安全性。其中生成了自定义 Spring Security 配置，并借助于 JWTs 实现了用户验证，同时针对验证处理创建了自定义过滤器。最后，本章还探讨了如何向 AWS 部署 Spring Boot 应用程序。

第 5 章将进一步完善 Android Messenger 应用程序，并深入讨论 Kotlin 方面的知识。

第 5 章　构建 Messenger Android App（第 1 部分）

第 4 章通过设计和实现 REST 应用程序编程接口的方式（客户端 Messenger 应用程序将与其进行通信）开始构建 Messenger 应用程序。在实现后端 API 的过程中，涉及了大量内容，包括与 Spring Boot 的协同工作方式、RESTful 应用程序编程接口及其工作方式、基于 PostgreSQL 的数据库搭建，以及向 AWS 部署 Spring Boot Web 应用程序等。

本章将进一步揭秘应用程序开发之旅，实现 Android Messenger 应用程序，并将其与 RESTful API 进行整合。在 Messenger Android App 开发过程中，将会涉及大量的新鲜话题，其中包括：

- 设计 MVP Android 应用程序。
- 基于 HTTP 的服务器通信。
- 响应式编程。
- 在 Android App 中使用基于令牌的身份验证。

本章将会展示 Kotlin 语言在 Android 应用程序开发领域中强大的功能，并深入分析 Messenger App 的开发细节内容。

5.1　开发 Messenger App

首先需要针对应用程序构建新的 Android Studio 项目，并将其命名为 Messenger，对应的包名为 com.example.messenger。在项目构建过程中，当提示创建新的启动程序活动时，可将对应活动命名为 LoginActivity，并生成一个空活动。

5.1.1　纳入项目依赖关系

本章将使用大量的外部应用程序依赖关系。因此，需要将其纳入当前项目中。打开 build.gradle 模块文件，并向其添加下列依赖关系：

```
dependencies {
  implementation fileTree(dir: 'libs', include: ['*.jar'])
  implementation "org.jetbrains.kotlin:kotlin-stdlib-jre7
                 :$kotlin_version"
```

```
    implementation 'com.android.support:appcompat-v7:26.1.0'
    implementation 'com.android.support.constraint:constraint-layout:1.0.2'
    implementation 'com.android.support:recyclerview-v7:26.1.0'
    implementation 'com.android.support:design:26.1.0'

    implementation "android.arch.persistence.room:runtime:1.0.0-alpha9-1"
    implementation "android.arch.persistence.room:rxjava2:1.0.0-alpha9-1"
    implementation 'com.android.support:support-v4:26.1.0'
    implementation 'com.android.support:support-vector-drawable:26.1.0'
    annotationProcessor "android.arch.persistence.room:compiler
                        :1.0.0-alpha9-1"

    implementation "com.squareup.retrofit2:retrofit:2.3.0"
    implementation "com.squareup.retrofit2:adapter-rxjava2:2.3.0"
    implementation "com.squareup.retrofit2:converter-gson:2.3.0"
    implementation "io.reactivex.rxjava2:rxandroid:2.0.1"

    implementation 'com.github.stfalcon:chatkit:0.2.2'

    testImplementation 'junit:junit:4.12'
    androidTestImplementation 'com.android.support.test:runner:1.0.1'
    androidTestImplementation 'com.android.support.test.espresso
                              :espresso-core:3.0.1'
}
```

此处，应确保 Android 支持库版本间不存在任何冲突。下面调整 build.gradle 项目文件，并包含 jcenter、Google 存储库以及 Android 构建工具依赖关系，如下所示：

```
buildscript {
  ext.kotlin_version = '1.1.4-3'
  repositories {
    google()
    jcenter()
  }
  dependencies {
    classpath 'com.android.tools.build:gradle:3.0.0-alpha9'
    classpath "org.jetbrains.kotlin:kotlin-gradle-plugin:$kotlin_version"
  }
}

allprojects {
  repositories {
    google()
```

```
    jcenter()
  }
}

task clean(type: Delete) {
  delete rootProject.buildDir
}
```

本章稍后还将对依赖关系所添加的内容再次予以分析。

5.1.2 开发登录 UI

当项目创建完毕后,即可构建新的数据包,并在 com.example.messenger 应用程序源数据包中将其命名为 ui。该包将加载所有与用户界面相关的类和 Android 应用程序逻辑。下面生成一个包含 ui 的 login 包。相信读者已经猜到,该包将加载与用户登录处理相关的类和逻辑。接下来,可将 LoginActivity 移至 login 包中,并针对登录活动创建相应的布局。

在 activity_login.xml 布局源文件中,输入下列内容:

```
<?xml version="1.0" encoding="utf-8"?>
<LinearLayout xmlns:android="http://schemas.android.com/apk/res/android"
    xmlns:tools="http://schemas.android.com/tools"
    android:layout_width="match_parent"
    android:layout_height="match_parent"
    tools:context=".ui.login.LoginActivity"
    android:orientation="vertical"
    android:paddingTop="32dp"
    android:paddingBottom="@dimen/default_margin"
    android:paddingStart="@dimen/default_padding"
    android:paddingEnd="@dimen/default_padding"
    android:gravity="center_horizontal">
    <EditText
        android:id="@+id/et_username"
        android:layout_width="match_parent"
        android:layout_height="wrap_content"
        android:inputType="text"
        android:hint="@string/username"/>
    <EditText
        android:id="@+id/et_password"
        android:layout_width="match_parent"
        android:layout_height="wrap_content"
        android:layout_marginTop="@dimen/default_margin"
```

```xml
        android:inputType="textPassword"
        android:hint="@string/password"/>
    <Button
        android:id="@+id/btn_login"
        android:layout_width="wrap_content"
        android:layout_height="wrap_content"
        android:layout_marginTop="@dimen/default_margin"
        android:text="@string/login"/>
    <Button
        android:id="@+id/btn_sign_up"
        android:layout_width="wrap_content"
        android:layout_height="wrap_content"
        android:layout_marginTop="@dimen/default_margin"
        android:background="@android:color/transparent"
        android:text="@string/sign_up_solicitation"/>
    <ProgressBar
        android:id="@+id/progress_bar"
        android:layout_width="wrap_content"
        android:layout_height="wrap_content"
        android:visibility="gone"/>
</LinearLayout>
```

此处使用了字符串和尺寸资源，这一类资源尚未在.xml 文件中创建，因而须对其予以添加。除此之外，还应该加入后续应用程序开发阶段所需的各种资源，从而避免在程序和资源文件之间往复跳转。打开项目的字符串资源文件（strings.xml 文件），并确保加入了以下资源：

```xml
<resources>
  <string name="app_name">Messenger</string>
  <string name="username">Username</string>
  <string name="password">Password</string>
  <string name="login">Login</string>
  <string name="sign_up_solicitation">
    Don\'t have an account? Sign up!
  </string>
  <string name="sign_up">Sign up</string>
  <string name="phone_number">Phone number</string>
  <string name="action_settings">settings</string>
  <string name="hint_enter_a_message">Type a message…</string>

  <!-- Account settings -->
  <string name="title_activity_settings">Settings</string>
```

```xml
  <string name="pref_header_account">Account</string>
  <string name="action_logout">logout</string>
</resources>
```

下面创建尺寸维度资源文件（dimens.xml 文件），并添加下列尺寸维度资源：

```xml
<?xml version="1.0" encoding="utf-8"?>
<resources>
  <dimen name="default_margin">16dp</dimen>
  <dimen name="default_padding">16dp</dimen>
</resources>
```

在添加了所需的项目资源后，返回至 activity_login.xml 文件中，并切换至设计预览窗口，以查看所创建的布局效果，如图 5.1 所示。

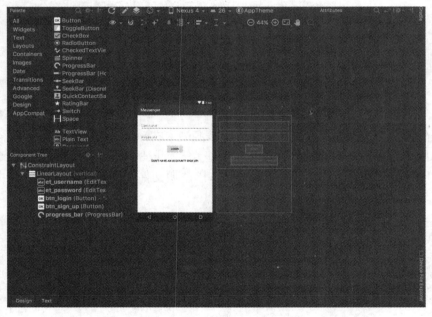

图 5.1　查看所创建的布局效果

该布局相对简单，但是却很实用，对于当前的 Messenger 应用程序来说已然足够。

1．创建登录视图

下面讨论 LoginActivity 中的工作流程。考虑到采用 MVP 模式构建该应用程序，LoginActivity 此处将视为视图。显然，LoginActivity 与其他通用视图有所不同。该视图更多地关注于逻辑过程。对此，应确定视图向用户提供登录界面时须具备的一组必要行为，其中包括：
- 登录时需要向用户显示进度栏。
- 必要时应可隐藏进度栏。
- 向用户显示字段错误。
- 用户应可返回值首页。
- 未注册用户须移至注册页面。

在确认了上述行为后，还应确保登录视图 LoginActivity 具备此类行为。对此，一种较好的方式是使用接口。下面在 login 包中定义 LoginView 接口，并包含下列内容：

```
package com.example.messenger.ui.login

interface LoginView {
  fun showProgress()
  fun hideProgress()
  fun setUsernameError()
  fun setPasswordError()
  fun navigateToSignUp()
  fun navigateToHome()
}
```

当前，LoginView 工作良好，但接口仍然存在些许问题。LoginView 需要将其布局视图绑定至相应的对象表达上。除此之外，如果出现验证错误，LoginView 应可向用户提供反馈信息。读者可能会认为这两种行为对于 LoginView 来说并没有什么不同，事实也的确如此。所有的视图都可将布局元素绑定至程序对象上。另外，如果验证过程中出现问题，注册视图还应向用户提供某种反馈信息。

接下来将构建两个独立的接口，以强调上述行为。其中，第一个接口命名为 BaseView。相应地，在 com.example.messenger.ui 中构建 base 包，将 BaseView 接口添加至该包中，并输入以下内容：

```
package com.example.messenger.ui.base

import android.content.Context

interface BaseView {
  fun bindViews()
  fun getContext(): Context
}
```

BaseView 接口要求，针对视图绑定和上下文检索，所实现的类须声明 bindViews()

和 getContext()函数。

下面在 com.example.messenger.ui 中定义 auth 包,并将 AuthView 接口添加至该包中,如下所示:

```
package com.example.messenger.ui.auth

interface AuthView {
  fun showAuthError()
}
```

下面返回 LoginView 接口,并确保扩展了 BaseView 和 AuthView,如下所示:

```
package com.example.messenger.ui.login

import com.example.messenger.ui.auth.AuthView
import com.example.messenger.ui.base.BaseView

interface LoginView : BaseView, AuthView {
  fun showProgress()
  fun hideProgress()
  fun setUsernameError()
  fun setPasswordError()
  fun navigateToSignUp()
  fun navigateToHome()
}
```

通过将 LoginView 接口声明为 BaseView 和 AuthView 的扩展,可确保实现了 LoginView 的每个类须声明 bindViews()、getContext()和 showAuthError()函数(除了声明于 LoginView 中的函数之外)。需要注意的是,实现了 LoginView 的任意类也可视作 LoginView、BaseView 和 AuthView 类型。当某个类涉及多种类型时,这一特性称作多态。

当 LoginView 设置完毕后,即可与 LoginActivity 协同工作。首先,可创建 LoginActivity,以实现声明于 BaseView 和 AuthView 中的方法;随后,可将此类特定方法添加至 LoginView 中。LoginActivity 代码如下所示:

```
package com.example.messenger.ui.login

import android.content.Context
import android.content.Intent
import android.support.v7.app.AppCompatActivity
import android.os.Bundle
import android.view.View
import android.widget.Button
```

```kotlin
import android.widget.EditText
import android.widget.ProgressBar
import android.widget.Toast
import com.example.messenger.R

class LoginActivity : AppCompatActivity(), LoginView, View.OnClickListener
{

  private lateinit var etUsername: EditText
  private lateinit var etPassword: EditText
  private lateinit var btnLogin: Button
  private lateinit var btnSignUp: Button
  private lateinit var progressBar: ProgressBar

  override fun onCreate(savedInstanceState: Bundle?) {
    super.onCreate(savedInstanceState)
    setContentView(R.layout.activity_login)

    bindViews()
  }
}
```

当调用时，可将布局视图对象引用绑定至视图元素上，如下所示：

```kotlin
override fun bindViews() {
  etUsername = findViewById(R.id.et_username)
  etPassword = findViewById(R.id.et_password)
  btnLogin = findViewById(R.id.btn_login)
  btnSignUp = findViewById(R.id.btn_sign_up)
  progressBar = findViewById(R.id.progress_bar)
  btnLogin.setOnClickListener(this)
  btnSignUp.setOnClickListener(this)
}

/**
 * Shows an appropriate Authentication error message when invoked.
 */
override fun showAuthError() {
   Toast.makeText(this, "Invalid username and password combination.",
Toast.LENGTH_LONG).show()
  }

  override fun onClick(view: View) {
  }
```

```kotlin
override fun getContext(): Context {
  return this
}
```

目前为止，一切均较为顺利。我们成功地实现了 LoginActivity 中的 BaseView 和 AuthView。但仍需要处理 LoginView 的特定方法，包括 showProgress()、hideProgress()、setUsernameError()、setPasswordError()、navigateToSignUp()以及 navigateToHome()方法。下列代码显示了这些方法的实现过程，随后可将其添加至 LoginActivity 中。

```kotlin
override fun hideProgress() {
  progressBar.visibility = View.GONE
}

override fun showProgress() {
  progressBar.visibility = View.VISIBLE
}

override fun setUsernameError() {
  etUsername.error = "Username field cannot be empty"
}

override fun setPasswordError() {
  etPassword.error = "Password field cannot be empty"
}

override fun navigateToSignUp() {
}

override fun navigateToHome() {
}
```

在添加了上述所有定义的方法后，我们即完成了 LoginActivity 类，同时实现了 LoginView 以及 View.OnClickListener 接口。因此，LoginActivity 提供了声明于此类接口中所声明的方法实现。注意，当前 LoginActivity 示例作为参数并通过 this 传递至 btnLogin.setOnClickListener()中——此处声明了 LoginActivity 并实现了 View.OnClickListener。因此，LoginActivity 表示为一类有效的 View.OnClickListener 实例（这也是多态工作方式的完美展示）。

前述内容讨论了登录视图的工作流程，下面将构建相应的模型以处理登录逻辑。除此之外，还需创建必要的服务以及数据存储库，以供该模型通信使用。针对于此，下面首先介绍所需的相关服务，随后将开发必要的数据存储库。最后，还将设置相应的截取器。

2. 生成 Messenger API 服务以及数据存储库

在深入讨论应用程序开发之前，首先需要考察数据存储问题，其中涉及两个较为重要的话题，即数据的存储位置，以及存储数据的访问方式。

关于数据的存储位置，可采用本地（在 Android 设备上）或远程（Messenger API）方式实现。对于后者，当访问数据时，需要创建合适的模型、服务以及存储库，以便于数据检索。

（1）通过 SharedPreferences 实现本地数据存储

当前示例相对简单，且无须在本地存储大量数据。设备上须存储的数据主要涉及标记的访问以及用户信息。对此，可通过 SharedPreferences 予以实现。

相应地，可在应用程序的 source 包中创建 data 包。这里，我们计划与本地存储和远程存储的数据协同工作。因此，可在 data 中分别创建两个额外的数据包，即 local 和 remote。类似于之前讨论的俄罗斯方块游戏程序，此处将通过 AppPreferences 类实现本地持久化存储。因此，可在 local 包中定义 AppPreferences 类，如下所示：

```
package com.example.messenger.data.local

import android.content.Context
import android.content.SharedPreferences
import com.example.messenger.data.vo.UserVO

class AppPreferences private constructor() {

  private lateinit var preferences: SharedPreferences

  companion object {
    private val PREFERENCE_FILE_NAME = "APP_PREFERENCES"

    fun create(context: Context): AppPreferences {
      val appPreferences = AppPreferences()
      appPreferences.preferences = context
        .getSharedPreferences(PREFERENCE_FILE_NAME, 0)
      return appPreferences
    }
  }
}
```

```kotlin
val accessToken: String?
  get() = preferences.getString("ACCESS_TOKEN", null)

fun storeAccessToken(accessToken: String) {
  preferences.edit().putString("ACCESS_TOKEN", accessToken).apply()
}

val userDetails: UserVO
get(): UserVO {
```

该方法返回包含用户信息的 UserVO 实例，如下所示：

```kotlin
return UserVO(
  preferences.getLong("ID", 0),
  preferences.getString("USERNAME", null),
  preferences.getString("PHONE_NUMBER", null),
  preferences.getString("STATUS", null),
)
}
```

相应地，可将传递至 UserVO 的用户信息存储至应用程序的 haredPreferences 文件中，如下所示：

```kotlin
fun storeUserDetails(user: UserVO) {
  val editor: SharedPreferences.Editor = preferences.edit()

  editor.putLong("ID", user.id).apply()
  editor.putString("USERNAME", user.username).apply()
  editor.putString("PHONE_NUMBER", user.phoneNumber).apply()
  editor.putString("STATUS", user.status).apply()
  editor.putString("CREATED_AT", user.createdAt).apply()
}

fun clear() {
  val editor: SharedPreferences.Editor = preferences.edit()
  editor.clear()
  editor.apply()
}
}
```

在 AppPreferences 中，分别定义了 storeAccessToken(String)、storeUserDetails(UserVO) 以及 clear() 函数。其中，storeAccessToken(String) 用于存储远程服务器和本地预置文件之

间的访问令牌。storeUserDetails(UserVO)接收用户值对象（包含用户信息的数据对象）作为其唯一参数，并将值对象包含的信息存储于预置文件中。顾名思义，clear()方法负责清空存储于预置文件中的全部数值。此外，AppPreferences 还包含了 accessToken 和 userDetails 属性，且分别包含了特定的 getter 函数，以检索对应值。除了该函数以及定义于 AppPreferences 中的属性之外，此处还定义了伴随对象，并设置了 create(Context)函数。create()方法则用于创建和返回新的 AppPreferences 实例。由于使用 AppPreferences 的任何类仅针对 AppPreferences 的实例化方调用 create()方法，因而 AppPreferences 的主构造函数定义为私有类型。

（2）生成值对象

类似于构建 Messenger 后端程序，此处也需要创建值对象并对较为常见的数据类型建模。对此，可在 data 包中设置一个 vo 包。实际上，读者已对值对象有所了解，在前述 API 开发时已对此有所讨论。下面将分别创建 ConversationListVO、ConversationVO、UserListVO、UserVO 以及 MessageVO，生成 Kotlin 文件并在 vo 包中加载这一类值对象。在定义值对象数据模型之前，首先需要设定基本模型，包括 UserVO、MessageVO 以及 ConversationVO。

UserVO 数据类的构建过程如下所示：

```kotlin
package com.example.messenger.data.vo

data class UserVO(
  val id: Long,
  val username: String,
  val phoneNumber: String,
  val status: String,
  val createdAt: String
)
```

考虑到之前曾创建过值对象，因而相关代码无须赘述。下面将 MessageVO 加入 MessageVO.kt 文件中，如下所示：

```kotlin
package com.example.messenger.data.vo

data class MessageVO(
  val id: Long,
  val senderId: Long,
  val recipientId: Long,
  val conversationId: Long,
  val body: String,
```

```
    val createdAt: String
)
```

接下来,在 ConversationVO.kt 文件中定义 ConversationVO 数据类,如下所示:

```
package com.example.messenger.data.vo

data class ConversationVO(
  val conversationId: Long,
  val secondPartyUsername: String,
  val messages: ArrayList<MessageVO>
)
```

在定义了基本的值对象后,即可创建 ConversationListVO 和 UserListVO。ConversationListVO 定义如下:

```
package com.example.messenger.data.vo

data class ConversationListVO(
  val conversations: List<ConversationVO>
)
```

ConversationListVO 数据类定义了单一的 List 类型的 conversations 属性,其中仅包含 ConversationVO 类型元素。UserListVO 类与 ConversationVO 基本相同,唯一差别在于前者未包含用户属性,且仅涵盖了 UserVO 属性元素(而非 conversations 属性)。UserListVO 数据类如下所示:

```
package com.example.messenger.data.vo

data class UserListVO(
  val users: List<UserVO>
)
```

3. 检索远程数据

前述内容讨论了函数所需的重要数据,Messenger Android App 则采用远程方式存储于 Messenger 后端。对此,Android 应用程序应可通过某种方式访问后端加载的数据,Messenger 应用程序需要通过 HTTP 与 API 进行通信。

(1)与远程服务器通信

在 Android 中,存在多种方式可与远程服务器进行通信。Android 社区中较为常用的网络库是 Retrofit、OkHttp 和 Volley。每种库均包含各自的优缺点。当前项目将采用 Retrofit,但出于完整性考虑,本章也将介绍基于 OkHttp 的远程服务器通信方式。

(2)利用 OkHttp 与服务器通信

OkHttp 是一种高效、便捷的 HTTP 客户端，支持同步和异步网络调用。基于 Android 的 OkHttp 应用十分简单。对此，可将其依赖关系添加至项目的 build.gradle 模块文件中，如下所示：

```
implementation 'com.squareup.okhttp3:okhttp:3.9.0'
```

(3)利用 OkHttp 向服务器发送请求

如前所述，OkHttp API 的设计宗旨是易于使用，据此，基于 OkHttp 的发送请求十分快捷、可靠。

下列 post(String,String)方法接收 URL 和 JSON 请求体作为参数，并利用 JSON 主体向特定的 URL 发送 POST 请求。

```
fun post(url: String, json: String): String {
  val mediaType: MediaType = MediaType.parse("application/json; 
                                              charset=utf-8")
  val client:OkHttpClient = OkHttpClient()
  val body: RequestBody = RequestBody.create(mediaType, json)

  val request: Request = Request.Builder()
                                .url(url)
                                .post(body)
                                .build()

  val response: Response = client.newCall(request).execute()
  return response.body().string()
}
```

上述函数的使用方式十分简单，读者可通过适当值对其进行调用，就像调用任何其他函数一样，如下所示：

```
val fullName: String = "John Wayne"
val response = post("http://example.com", "{ \"full_name\": $fullName")
println(response)
```

下面尝试通过 Retrofit 实现与远程服务器之间的通信。在讨论 Retrofit 之前，需要对 HTTP 中所发送的数据建模。

(4)对请求数据建模

这里将使用数据类，并对发送至 API 的 HTTP 请求数据建模。下面在 remote 包中创建一个 request 包，其中存在 4 种请求且包含了数据负载，并被发送至 API 中，即登录请求、消息请求、状态更新请求，以及包含用户数据的请求。这 4 种请求分别通过 LoginRequestObject、MessageRequestObject、StatusUpdateRequestObject 以及 UserRequest 对象进行建模。

下列代码片段显示了 LoginRequestObject 数据类，可将其添加至 request 包中，以对其他请求执行相同的操作，如下所示：

```
package com.example.messenger.data.remote.request

data class LoginRequestObject(
  val username: String,
  val password: String
)
```

LoginRequestObject 数据类包含了 username 和 password 属性，此类属性表示为须提供至 API 登录端点的证书。MessageRequestObject 数据类定义如下所示：

```
package com.example.messenger.data.remote.request

data class MessageRequestObject(val recipientId: Long, val message: String)
```

MessageRequestObject 也包含了两个属性，其中，recipientId 表示接收消息的用户 ID；message 则表示被发送的消息体，如下所示：

```
package com.example.messenger.data.remote.request

data class StatusUpdateRequestObject(val status: String)
```

StatusUpdateRequestObject 数据类包含了单一 status 属性。顾名思义，该属性表示当前消息所更新的状态，如下所示：

```
package com.example.messenger.data.remote.request

data class UserRequestObject(
  val username: String,
  val password: String,
  val phoneNumber: String = ""
)
```

UserRequestObject 类似于 LoginRequestObject，但包含了额外的 phoneNumber 属性。该请求对象包含多种用例，例如发送至 API 的用户注册数据。

在生成了所需的请求对象后，下面定义实际的 MessengerApiService。

(5)创建 Messenger API 服务

这里将创建一项服务,并与第 4 章构建的 Messenger API 进行通信,并使用到 Retrofit 以及 Retrofit 的 RxJava 适配器。对于 Android 和 Java 来说,Retrofit 是 Square Inc.发布的类型安全的 HTTP 客户端;RxJava 则是面向 ReactiveX 的开源实现。

在本章开始处,曾通过下列代码将 Retrofit 添加至 Android 项目中:

```
implementation "com.squareup.retrofit2:retrofit:2.3.0"
```

除此之外,还向 build.gradle 模块脚本文件中添加了 Retrofit 的 RxJava 适配器,如下所示:

```
implementation "com.squareup.retrofit2:adapter-rxjava2:2.3.0"
```

当利用 Retrofit 构建服务时,首先需要定义一个接口,以描述 HTTP API。对此,可在应用程序 source 包中设置一个 service 包,并添加 MessengerApiService 接口,如下所示:

```
package com.example.messenger.service

import com.example.messenger.data.remote.request.LoginRequestObject
import com.example.messenger.data.remote.request.MessageRequestObject
import com.example.messenger.data.remote.request.
StatusUpdateRequestObject
import com.example.messenger.data.remote.request.UserRequestObject
import com.example.messenger.data.vo.*
import io.reactivex.Observable
import okhttp3.ResponseBody
import retrofit2.Retrofit
import retrofit2.adapter.rxjava2.RxJava2CallAdapterFactory
import retrofit2.converter.gson.GsonConverterFactory
import retrofit2.http.*

interface MessengerApiService {

  @POST("login")
  @Headers("Content-Type: application/json")
  fun login(@Body user: LoginRequestObject):
        Observable<retrofit2.Response<ResponseBody>>

  @POST("users/registrations")
  fun createUser(@Body user: UserRequestObject): Observable<UserVO>

  @GET("users")
```

```kotlin
  fun listUsers(@Header("Authorization") authorization: String):
          Observable<UserListVO>

@PUT("users")
fun updateUserStatus(
  @Body request: StatusUpdateRequestObject,
  @Header("Authorization") authorization: String): Observable<UserVO>

@GET("users/{userId}")
fun showUser(
  @Path("userId") userId: Long,
  @Header("Authorization") authorization: String): Observable<UserVO>

@GET("users/details")
fun echoDetails(@Header("Authorization") authorization: String):
Observable<UserVO>

@POST("messages")
fun createMessage(
  @Body messageRequestObject: MessageRequestObject,
  @Header("Authorization") authorization: String): Observable<MessageVO>

@GET("conversations")
fun listConversations(@Header("Authorization") authorization: String):
              Observable<ConversationListVO>

@GET("conversations/{conversationId}")
fun showConversation(
  @Path("conversationId") conversationId: Long,
  @Header("Authorization") authorization:
String):Observable<ConversationVO>
}
```

从上述代码中可以看到,Retrofit 依赖于注解的使用,进而描述被发送的 HTTP 请求。例如,考察下列代码片段:

```kotlin
@POST("login")
@Headers("Content-Type: application/json")
fun login(@Body user: LoginRequestObject):
Observable<retrofit2.Response<ResponseBody>>
```

@POST 注解通知 Retrofit,对应函数描述了映射至/login 路径的 HTTP POST 请求。@Headers 注解用于指定 HTTP 请求头。在前述代码片段的 HTTP 请求中,Content-Type

头被设置为application/json。因此，该请求发送的内容表示为JSON。

@Body注解表明，传递至login()的user包含了JSON请求体中的数据，并发送至API。这里，user定义为LoginRequestObject类型（之前曾创建了该请求对象）。最后，该函数被声明并返回Observable对象，其中包含了retrofit2.Response对象。

除了@POST、@Headers和@Body注解之外，我们还使用了@GET、@PUT、@Path和@Header。其中，@GET和@PUT分别用于指定GET和PUT请求。@Path注解则用于声明一个值，并作为所发送的HTTP请求的路径参数。例如，考察下列showUser()函数：

```
@GET("users/{userId}")
fun showUser(
  @Path("userId") userId: Long,
  @Header("Authorization") authorization: String): Observable<UserVO>
```

函数showUser()描述了包含users/{userId}路径的GET请求。这里，{userId}并非是HTTP请求路径中的实际内容。Retrofit将利用传递至showUser()中的userId值。此处应注意userId如何采用@Path("userId")进行注解。这将通知Retrofit，userId加载了一个数值，且应置于HTTP请求URL路径中{userId}所处位置处。

@Header与@Headers类似，唯一差别在于，前者用于指定所发送的HTTP请求中的键-值对。基于@Header("Authorization")的注解验证将设置HTTP请求的Authorization头（发送至验证中的数值）。

前述内容创建了MessengerApiService接口，并对应用程序通信的HTTP API建模。接下来，我们需要检索对应服务的实例。对此，可定义伴生对象，负责创建MessengerApiService实例，如下所示：

```
package com.example.messenger.service

import com.example.messenger.data.remote.request.LoginRequestObject
import com.example.messenger.data.remote.request.MessageRequestObject
import com.example.messenger.data.remote.request.
StatusUpdateRequestObject
import com.example.messenger.data.remote.request.UserRequestObject
import com.example.messenger.data.vo.*
import io.reactivex.Observable
import okhttp3.ResponseBody
import retrofit2.Retrofit
import retrofit2.adapter.rxjava2.RxJava2CallAdapterFactory
import retrofit2.converter.gson.GsonConverterFactory
import retrofit2.http.*
```

```kotlin
interface MessengerApiService {
    …
    …
    companion object Factory {
        private var service: MessengerApiService? = null
```

当调用时,该对象返回 MessengerApiService 实例。若之前未曾创建,此时将生成一个新的 MessengerApiService 实例,如下所示:

```kotlin
        fun getInstance(): MessengerApiService {
            if (service == null) {

                val retrofit = Retrofit.Builder()
                        .addCallAdapterFactory(RxJava2CallAdapterFactory.create())
                        .addConverterFactory(GsonConverterFactory.create())
                        .baseUrl("{AWS_URL}")
                        // replace AWS_URL with URL of AWS EC2
                        // instance deployed in the previous chapter
                        .build()

                service = retrofit.create(MessengerApiService::class.java)
            }

            return service as MessengerApiService
        }
    }
}
```

Factory 定义了单一的 getInstance()函数,并在调用时构建和返回 MessengerApiService 实例。Retrofit.Builder 实例用于创建该接口。其中,可将 CallAdapterFactory 设置为 RxJava2CallAdapterFactory,将 ConverterFactory 设置为 GsonConverterFactory(这将处理 JSON 序列化和反序列化操作)。注意,期间不要忘记将"{AWS_URL}"替换为 Messenger API AWS EC2 实例的 URL(参见第 4 章)。

在成功地构建了 Retrofit.Builder()实例后,即可以此生成 MessengerApiService 实例,如下所示:

```kotlin
service = retrofit.create(MessengerApiService::class.java)
```

尽管已经创建了一个相关服务与 Messenger API 通信,但如果未在 AndroidManifest 中指定必要的权限,仍无法使用该服务与网络通信。打开项目的 AndroidManifest 文件,在<manifest></manifest>标签之间添加下列代码:

```xml
<uses-permission android:name="android.permission.INTERNET" />
<uses-permission android:name="android.permission.ACCESS_NETWORK_STATE" />
```

当前,Messenger 服务已可使用,下面将创建相应的存储库并使用该服务。

(6)实现数据存储库

相信读者已对存储库有所了解,此处不再赘述。此处将要生成的存储库与第 4 章类似,唯一的差别在于,这里所实现的存储库数据源为原创服务器,而非驻留于主机上的数据库。

下面在 remote 包中创建 repository 包,并实现用户存储库,以检索与应用程序用户相关的数据。下列代码将 UserRepository 接口添加至当前存储库中。

```kotlin
package com.example.messenger.data.remote.repository

import com.example.messenger.data.vo.UserListVO
import com.example.messenger.data.vo.UserVO
import io.reactivex.Observable

interface UserRepository {

  fun findById(id: Long): Observable<UserVO>
  fun all(): Observable<UserListVO>
  fun echoDetails(): Observable<UserVO>
}
```

考虑到接口的特性,我们需要定义一个 UserRepositoryImpl 类,并实现 UserRepository 中函数。下面在 repository 包中定义 UserRepositoryImpl 类,如下所示:

```kotlin
package com.example.messenger.data.remote.repository

import android.content.Context
import com.example.messenger.service.MessengerApiService
import com.example.messenger.data.local.AppPreferences
import com.example.messenger.data.vo.UserListVO
import com.example.messenger.data.vo.UserVO
import io.reactivex.Observable

class UserRepositoryImpl(ctx: Context) : UserRepository {

  private val preferences: AppPreferences = AppPreferences.create(ctx)
  private val service: MessengerApiService =
MessengerApiService.getInstance()
```

```kotlin
override fun findById(id: Long): Observable<UserVO> {
  return service.showUser(id, preferences.accessToken as String)
}

override fun all(): Observable<UserListVO> {
  return service.listUsers(preferences.accessToken as String)
}

override fun echoDetails(): Observable<UserVO> {
  return service.echoDetails(preferences.accessToken as String)
}
}
```

UserRepositoryImpl 类定义了两个实例变量，即 preferences 和 service。其中，preferences 变量表示为之前生成的 AppPreferences 类实例；service 则表示为 MessengerApiService 实例，该实例通过定义于 MessengerApiService 接口的、Factory 伴随对象中的 getInstance() 函数得到。

UserRepositoryImpl 类提供了定义于 UserRepository 中的 findById()、all() 和 echoDetails()函数实现。这 3 个函数使用了 service 并通过 HTTP 请求检索服务器上的所需数据。findById()调用了服务中的 showUser()函数，向 Messenger API 发送请求，并利用特定的用户 ID 检索用户的详细信息。showUser()函数需要使用到当前登录用户的授权令牌作为第二个参数。该令牌可通过 AppPreferences 实例获得，即作为第二个参数向函数传递 preferences.accessToken。

all()函数利用 MessengerApiService#listUsers()检索消息服务器上的所有用户。echoDetails()则通过 MessengerApiService#echoDetails()获取当前登录用户的详细信息。

下面创建会话存储库，以简化与会话相关的数据访问操作。下列代码将 ConversationRepository 接口添加至 com.example.messenger.data.remote.repository 中。

```kotlin
package com.example.messenger.data.remote.repository

import com.example.messenger.data.vo.ConversationListVO
import com.example.messenger.data.vo.ConversationVO
import io.reactivex.Observable

interface ConversationRepository {
  fun findConversationById(id: Long): Observable<ConversationVO>

  fun all(): Observable<ConversationListVO>
}
```

接下来在当前包中构建对应的 ConversationRepositoryImpl，如下所示：

```
package com.example.messenger.data.remote.repository

import android.content.Context
import com.example.messenger.service.MessengerApiService
import com.example.messenger.data.local.AppPreferences
import com.example.messenger.data.vo.ConversationListVO
import com.example.messenger.data.vo.ConversationVO
import io.reactivex.Observable

class ConversationRepositoryImpl(ctx: Context) : ConversationRepository {

  private val preferences: AppPreferences = AppPreferences.create(ctx)
  private val service: MessengerApiService = MessengerApiService
                                                .getInstance()
```

这将利用 Messenger API 中的请求会话 ID 检索与会话相关的信息，如下所示：

```
override fun findConversationById(id: Long): Observable<ConversationVO> {
  return service.showConversation(id, preferences.accessToken as String)
}
```

当被调用时，该函数将从 API 中检索当前用户的所有活动会话，如下所示：

```
override fun all(): Observable<ConversationListVO> {
    return service.listConversations(preferences.accessToken as String)
  }
}
```

findConversationById(Long)函数利用传递至该函数中的对应 ID 检索会话线程。all()函数简单地检索当前用户的全部活动会话。

4．构建登录交互器

下面创建登录交互器，即与登录显示相交互的模型。在 login 包中定义 LoginInteractor 接口，如下所示：

```
package com.example.messenger.ui.login

import com.example.messenger.data.local.AppPreferences
import com.example.messenger.ui.auth.AuthInteractor

interface LoginInteractor : AuthInteractor {
```

```
interface OnDetailsRetrievalFinishedListener {
  fun onDetailsRetrievalSuccess()
  fun onDetailsRetrievalError()
}

fun login(username: String, password: String,
    listener: AuthInteractor.onAuthFinishedListener)

fun retrieveDetails(preferences: AppPreferences,
    listener: OnDetailsRetrievalFinishedListener)
}
```

相信读者已经注意到，LoginInteractor 扩展了 AuthInteractor，这与 LoginView 和 AuthView 之间的扩展方式类似。AuthInteractor 中声明的行为和特性需要通过处理验证逻辑的交互器予以实现，下面对此予以实现。

相应地，将 AuthInteractor 接口添加至 com.exampla.messenger.auth 包中，如下所示：

```
package com.example.messenger.ui.auth

import com.example.messenger.data.local.AppPreferences
import com.example.messenger.data.remote.vo.UserVO

interface AuthInteractor {

  var userDetails: UserVO
  var accessToken: String
  var submittedUsername: String
  var submittedPassword: String

  interface onAuthFinishedListener {
    fun onAuthSuccess()
    fun onAuthError()
    fun onUsernameError()
    fun onPasswordError()
  }

  fun persistAccessToken(preferences: AppPreferences)

  fun persistUserDetails(preferences: AppPreferences)

}
```

其中，表示为 AuthInteractor 的某一个交互器须包含下列字段：userDetails、

accessToken、submittedUsername 以及 submittedPassword。除此之外,实现了 AuthInteractor 的交互器还需定义 persistAccessToken(AppPreferences)和 persistUserDetails(AppPreferences) 方法,这些方法负责访问令牌和用户信息与应用程序 SharedPreferences 文件之间的持久化操作。可能读者已经猜测到,这里需要针对 LoginInteractor 定义一个实现类,即 LoginInteractorImpl 类。

在下列代码中,LoginInteractorImpl 类定义了实现后的 login()方法,并将该类添加至 ui 包的 login 包中。

```
package com.example.messenger.ui.login

import com.example.messenger.data.local.AppPreferences
import com.example.messenger.data.remote.request.LoginRequestObject
import com.example.messenger.data.vo.UserVO
import com.example.messenger.service.MessengerApiService
import com.example.messenger.ui.auth.AuthInteractor
import io.reactivex.android.schedulers.AndroidSchedulers
import io.reactivex.schedulers.Schedulers

class LoginInteractorImpl : LoginInteractor {

  override lateinit var userDetails: UserVO
  override lateinit var accessToken: String
  override lateinit var submittedUsername: String
  override lateinit var submittedPassword: String

  private val service: MessengerApiService = MessengerApiService
                                             .getInstance()

  override fun login(username: String, password: String,
                     listener: AuthInteractor.onAuthFinishedListener) {
    when {
```

如果登录框输入了空 username,则该 username 视为无效。当出现这一情况时,将调用监听器的 onUsernameError()函数,如下所示:

```
username.isBlank() -> listener.onUsernameError()
```

同样,若输入空 password,则调用监听器的 onPasswordError()函数,如下所示:

```
  password.isBlank() -> listener.onPasswordError()
else -> {
```

下列代码初始化模型的 submittedUsername 和 submittedPassword 字段，并生成相应的 LoginRequestObject。

```
submittedUsername = username
submittedPassword = password
val requestObject = LoginRequestObject(username, password)
```

通过 MessengerApiService，可向 Messenger API 发送登录请求，如下所示：

```
service.login(requestObject)
    .subscribeOn(Schedulers.io())
      // subscribing Observable to Scheduler thread
    .observeOn(AndroidSchedulers.mainThread())
      // setting observation to be done on the main thread
    .subscribe({ res ->
      if (res.code() != 403) {
        accessToken = res.headers()["Authorization"] as String
        listener.onAuthSuccess()
      } else {
```

当服务器返回一个 HTTP 403（禁用）状态码时，则表明登录无效，且用户未被授权访问服务器，如下所示：

```
        listener.onAuthError()
      }
    }, { error ->
      listener.onAuthError()
      error.printStackTrace()
    })
   }
  }
 }
}
```

login()首先检测 username 和 password 参数是否为空。当出现空用户名时，将调用 onAuthFinishedListener 的 onUsernameError() 函数；若出现空密码，则调用 onPasswordError()。否则，将利用 MessengerApiService 向 Messenger API 发送登录请求。若成功，则将 ccessToken 属性设置为 API Authorization 响应头中的访问令牌，并于随后调用监听器的 onAuthSuccess()函数。当登录无效时，将调用 onAuthError()监听器函数。

在理解了登录流程后，即可将 retrieveDetails()、persistAccessToken() 和 persistUserDetails() 方法添加至 LoginInteractorImpl 中，如下所示：

```
override fun retrieveDetails(preferences: AppPreferences,
         listener: LoginInteractor.OnDetailsRetrievalFinishedListener)
{
```

这将在首次登录时检索用户信息，如下所示：

```
    service.echoDetails(preferences.accessToken as String)
        .subscribeOn(Schedulers.io())
        .observeOn(AndroidSchedulers.mainThread())
        .subscribe({ res ->
          userDetails = res
          listener.onDetailsRetrievalSuccess()},
        { error ->
          listener.onDetailsRetrievalError()
          error.printStackTrace()})
}

override fun persistAccessToken(preferences: AppPreferences) {
  preferences.storeAccessToken(accessToken)
}

override fun persistUserDetails(preferences: AppPreferences) {
  preferences.storeUserDetails(userDetails)
}
```

读者应理解上述代码片段中的注释内容，其中解释了 LoginInteractor 的工作机制。下面处理 LoginPresenter 的实现过程。

5. 构建登录 presenter

presente 表示为视图和模型之间的中间人，且有必要针对视图构建相应的 presenter，以实现更为清晰的视图-模型交互。该过程较为简单，首先需要定义一个接口，声明 presenter 所展示的各种行为。对此，可在 login 包中定义一个 LoginPresenter 接口，如下所示：

```
package com.example.messenger.ui.login

interface LoginPresenter {
  fun executeLogin(username: String, password: String)
}
```

不难发现，在上述代码片段中，需要定义一个类，该类针对 LoginView 视作 LoginPresenter，并传递一个 executeLogin(String, String)函数。该函数被视图所调用，并于随后与处理应用程序登录逻辑的模型交互。此处,需要定义一个实现了 LoginPresenter 的 LoginPresenterImpl 类,如下所示:

```kotlin
package com.example.messenger.ui.login

import com.example.messenger.data.local.AppPreferences
import com.example.messenger.ui.auth.AuthInteractor

class LoginPresenterImpl(private val view: LoginView) :
    LoginPresenter, AuthInteractor.onAuthFinishedListener,
    LoginInteractor.OnDetailsRetrievalFinishedListener {

 private val interactor: LoginInteractor = LoginInteractorImpl()
 private val preferences: AppPreferences =
AppPreferences.create(view.getContext())

 override fun onPasswordError() {
   view.hideProgress()
   view.setPasswordError()
 }

 override fun onUsernameError() {
   view.hideProgress()
   view.setUsernameError()
 }

 override fun onAuthSuccess() {
   interactor.persistAccessToken(preferences)
   interactor.retrieveDetails(preferences, this)
 }

 override fun onAuthError() {
   view.showAuthError()
   view.hideProgress()
 }

 override fun onDetailsRetrievalSuccess() {
   interactor.persistUserDetails(preferences)
   view.hideProgress()
   view.navigateToHome()
 }
```

```
override fun onDetailsRetrievalError() {
  interactor.retrieveDetails(preferences, this)
}

override fun executeLogin(username: String, password: String) {
  view.showProgress()
  interactor.login(username, password, this)
}
}
```

LoginPresenterImpl 类实现了 LoginPresenter、AuthInteractor.onAuthFinishedListener 以及 LoginInteractor.OnDetailsRetrievalFinishedListener，因而也实现了接口所需的全部行为。LoginPresenterImpl 共计覆写了 7 个函数，分别是 onPasswordError()、onUsernameError()、onAuthSuccess()、onAuthError()、onDetailsRetrievalSuccess()、onDetailsRetrievalError()和 executeLogin(String, String)。LoginPresenter 和 LoginInteractor 之间的交互行为可视为位于 onAuthSuccess()和 executeLogin(String, String)函数中。当用户提交其登录信息时，LoginView 将调用 LoginPresenter 中的 executeLogin(String, String)函数。相应地，LoginPresenter 利用 LoginInteractor 处理处理实际的登录过程，即调用 LoginInteractor 中的 login(String, String)函数。

如果用户登录成功，LoginPresenter 的 onAuthSuccess()回调函数将被 LoginInteractor 所调用。随后，将存储服务器返回的访问令牌，以及登录用户的信息检索。若登录请求被服务器所拒绝，onAuthError()将被调用，并向用户显示错误消息。

当用户账户信息成功地被交互器所检索，将调用 LoginPresenter 的 onDetailsRetrievalSuccess()回调函数，从而引发账户信息的存储操作。随后，登录过程中向用户显示的进程栏将通过 view.hideProgress()被隐藏；接下来，用户将通过 view.navigateToHome()导航至主屏幕中。如果用户信息检索失败，onDetailsRetrievalError()将被 LoginInteractor 调用。presenter 将再次请求并检索用户账户信息，即再次调用 interactor.retrieveDetails(preferences, this)。

6. 完成 LoginView

回忆一下，前述内容尚未完成 LoginView；navigateToSignUp()、navigateToHome() 和 onClick(view: View)函数中的相关内容也未予定义。除此之外，LoginView 也未采用任何方式与 LoginPresenter 进行交互，下面对此加以讨论。

首先，当用户进入注册页面或主屏幕时，需要针对此类画面提供相应的视图。当前，我们暂不需要考虑此类视图的实现过程（后续内容将对此加以分析）。对此，可在 com.example.messenger.ui 下设置 signup 和 main 包，并在 signup 包中生成名为 SignUpActivity

的空活动，以及在 main 包中生成名为 MainActivity 的空活动。

下面打开 LoginActivity.kt 文件，调整之前所定义的函数以执行相应的任务。另外，还需针对 LoginPresenter 实例和 AppPreferences 实例添加私有属性，相关变化内容将在后续代码中体现。

首先，在 LoginActivity 类开始处添加下列属性：

```
private lateinit var progressBar: ProgressBar
private lateinit var presenter: LoginPresenter
private lateinit var preferences: AppPreferences
```

随后，调整 navigateToSignUp()、navigateToHome()和 onClick(view: View)方法，如下所示：

```
override fun navigateToSignUp() {
  startActivity(Intent(this, SignUpActivity::class.java))
}

override fun navigateToHome() {
  finish()
  startActivity(Intent(this, MainActivity::class.java))
}

override fun onClick(view: View) {
  if (view.id == R.id.btn_login) {
    presenter.executeLogin(etUsername.text.toString(),
                           etPassword.text.toString())
  } else if (view.id == R.id.btn_sign_up) {
    navigateToSignUp()
  }
}
```

当调用 navigateToSignUp()方法时，该方法使用一个显式意图，并启动 SignUpActivity。navigateToHome() 的操作方式类似于 navigateToSignUp()，即启动 MainActivity。navigateToHome()和 navigateToSignUp()之间的主要差别在于，navigateToHome()在启用 MainActivity 之前通过调用 finish()销毁当前 LoginActivity 实例。

单击登录按钮后，onClick()方法使用 LoginPresenter 开始执行登录处理操作。否则，当单击注册按钮时，SignUpActivity 则通过 navigateToSignUp()启动。

至此，与登录应用程序相关的视图、presenter 已经构建完毕。需要注意的是，在用户登录前，需要注册为该平台的用户。因此，还需实现注册逻辑，下面将对此加以讨论。

5.1.3 设计注册 UI

本节讨论用户注册界面。首先，应在 SignUpActivity 布局基础上实现相关的视图。SignUpActivity 布局并未涉及太多内容，其中包括 3 个输入框，用以输入用户名、密码以及注册用户的电话号码。除此之外，还包含了一个按钮，以提交注册表单，以及注册处理过程中的进度栏。

activity_sign_up.xml 布局文件如下所示：

```xml
<?xml version="1.0" encoding="utf-8"?>
<android.support.constraint.ConstraintLayout
    xmlns:android="http://schemas.android.com/apk/res/android"
    xmlns:tools="http://schemas.android.com/tools"
    android:layout_width="match_parent"
    android:layout_height="match_parent"
    tools:context=".ui.signup.SignUpActivity"
    android:paddingTop="@dimen/default_padding"
    android:paddingBottom="@dimen/default_padding"
    android:paddingStart="@dimen/default_padding"
    android:paddingEnd="@dimen/default_padding"
    android:orientation="vertical"
    android:gravity="center_horizontal">
    <EditText
        android:id="@+id/et_username"
        android:layout_width="match_parent"
        android:layout_height="wrap_content"
        android:hint="@string/username"
        android:inputType="text"/>
    <EditText
        android:id="@+id/et_phone"
        android:layout_width="match_parent"
        android:layout_height="wrap_content"
        android:layout_marginTop="@dimen/default_margin"
        android:hint="@string/phone_number"
        android:inputType="phone"/>
    <EditText
        android:id="@+id/et_password"
        android:layout_width="match_parent"
        android:layout_height="wrap_content"
        android:layout_marginTop="@dimen/default_margin"
        android:hint="@string/password"
```

```xml
        android:inputType="textPassword"/>
    <Button
        android:id="@+id/btn_sign_up"
        android:layout_width="wrap_content"
        android:layout_height="wrap_content"
        android:layout_marginTop="@dimen/default_margin"
        android:text="@string/sign_up"/>
    <ProgressBar
        android:id="@+id/progress_bar"
        android:layout_width="wrap_content"
        android:layout_height="wrap_content"
        android:layout_marginTop="@dimen/default_margin"
        android:visibility="gone"/>
</android.support.constraint.ConstraintLayout>
```

对应的 XML 布局效果如图 5.2 所示。

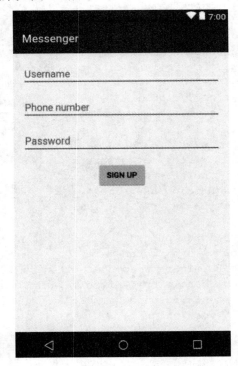

图 5.2　XML 布局效果

不难发现，当前所设计的布局包含了之前所提到的全部元素。

1．构建注册交互器

下面实现注册交互器，并继续讨论之前未完成的模型。对此，在 signup 包中定义 SignUpInteractor 接口，如下所示：

```
package com.example.messenger.ui.signup

import com.example.messenger.ui.auth.AuthInteractor

interface SignUpInteractor : AuthInteractor {

  interface OnSignUpFinishedListener {
    fun onSuccess()
    fun onUsernameError()
    fun onPasswordError()
    fun onPhoneNumberError()
    fun onError()
  }

  fun signUp(username: String, phoneNumber: String, password: String,
             listener: OnSignUpFinishedListener)

  fun getAuthorization(listener: AuthInteractor.onAuthFinishedListener)
}
```

读者可能已经注意到，SignUpInteractor 扩展了 AuthInteractor。类似于 LoginInteractor，SignUpInteractor 需要使用到 userDetails、accessToken、submittedUsername 和 submittedPassword 属性。除此之外，通过 persistAccessToken(AppPreferences)和 persistUserDetails(AppPreferences) 函数，SignUpInteractor 应可实现用户访问令牌和信息的持久化存储，这两个函数声明于 AuthInteractor 接口中。

SignUpInteractor 中定义了接口 OnSignUpFinishedListener，并声明了回调函数，并通过 OnSignUpFinishedListener 予以实现。当对此加以实现时，该监听器记为 SignUpPresenter。

在构建 SignUpInteractorImpl 时，首先应考察其属性声明以及 login()方法实现，并确保将 SignUpInteractorImpl 添加至与 SignUpInteractor 相同的包中，如下所示：

```
package com.example.messenger.ui.signup

import android.text.TextUtils
import android.util.Log
import com.example.messenger.data.local.AppPreferences
import com.example.messenger.data.remote.request.LoginRequestObject
```

```kotlin
import com.example.messenger.data.remote.request.UserRequestObject
import com.example.messenger.data.vo.UserVO
import com.example.messenger.service.MessengerApiService
import com.example.messenger.ui.auth.AuthInteractor
import io.reactivex.android.schedulers.AndroidSchedulers
import io.reactivex.schedulers.Schedulers

class SignUpInteractorImpl : SignUpInteractor {

  override lateinit var userDetails: UserVO
  override lateinit var accessToken: String

  override lateinit var submittedUsername: String
  override lateinit var submittedPassword: String

  private val service: MessengerApiService = MessengerApiService
                                                    .getInstance()

  override fun signUp(username: String,
                phoneNumber: String, password: String,
                listener: SignUpInteractor.OnSignUpFinishedListener){
    submittedUsername = username
    submittedPassword = password
    val userRequestObject = UserRequestObject(username, password,
                                                    phoneNumber)

    when {
      TextUtils.isEmpty(username) -> listener.onUsernameError()
      TextUtils.isEmpty(phoneNumber) -> listener.onPhoneNumberError()
      TextUtils.isEmpty(password) -> listener.onPasswordError()
      else -> {
```

接下来利用 MessengerApiService 在 Messenger 平台中注册新用户，如下所示：

```kotlin
        service.createUser(userRequestObject)
            .subscribeOn(Schedulers.io())
            .observeOn(AndroidSchedulers.mainThread())
            .subscribe({ res ->
      userDetails = res
      listener.onSuccess()
    }, { error ->
      listener.onError()
      error.printStackTrace()
```

```
            })
          }
        }
      }
    }
```

下面将 getAuthorization()、persistAccessToken() 和 persistUserDetails() 添加至 SignUpInteractorImpl 中，如下所示：

```
override fun getAuthorization(listener:
            AuthInteractor.onAuthFinishedListener) {
 val userRequestObject = LoginRequestObject(submittedUsername,
                                            submittedPassword)
```

下面利用 MessengerApiService 将注册用户登录至当前平台中，如下所示：

```
service.login(userRequestObject)
      .subscribeOn(Schedulers.io())
      .observeOn(AndroidSchedulers.mainThread())
      .subscribe( { res ->
accessToken = res.headers()["Authorization"] as String
```

在用户成功登录后，可调用监听器的 onAuthSuccess()回调函数，如下所示：

```
    listener.onAuthSuccess()

  }, { error ->
    listener.onAuthError()
    error.printStackTrace()
  })
}

override fun persistAccessToken(preferences: AppPreferences) {
  preferences.storeAccessToken(accessToken)
}

override fun persistUserDetails(preferences: AppPreferences) {
  preferences.storeUserDetails(userDetails)
}
```

SignUpInteractorImpl 类则是 SignUpInteractor 的直接实现。对于 AuthInteractor 中包含的 userDetails、accessToken、submittedUsername 以及 submittedPassword，第 19~22 行代码包含了属性声明。signUp(String, String, String,SignUpInteractor.OnSignUpFinishedListener)

包含了当前应用程序的注册逻辑。如果用户提交的全部数值均为有效,那么,用户利用 MessengerApiService(通过 Retrofit 创建)的 createUser(UserRequestObject)函数在当前平台上完成注册。

调用 getAuthorization(AuthInteractor.onAuthFinishedListener)函数将对 Messenger 平台上的新用户注册用户授权。此处,读者应留意 SignUpInteractorImpl 中的注释,以了解更多内容。

下面将创建 SignUpPresenter。

2. 创建注册 presenter

正如构建 LoginPresenter 所做的那样,当前也需要定义 SignUpPresenter 接口以及 SignUpPresenterImpl 类。这里,SignUpPresenterImpl 的构建过程并不复杂。对于当前应用程序,注册 presenter 应包含 AppPreferences 类型的属性,以及执行注册处理的相关函数。SignUpPresenter 接口定义如下:

```kotlin
package com.example.messenger.ui.signup

import com.example.messenger.data.local.AppPreferences

interface SignUpPresenter {
    var preferences: AppPreferences

    fun executeSignUp(username: String, phoneNumber: String, password: String)
}
```

SignUpPresenter 的实现代码如下所示:

```kotlin
package com.example.messenger.ui.signup

import com.example.messenger.data.local.AppPreferences
import com.example.messenger.ui.auth.AuthInteractor

class SignUpPresenterImpl(private val view: SignUpView) : SignUpPresenter,
                    SignUpInteractor.OnSignUpFinishedListener,
                    AuthInteractor.onAuthFinishedListener {

    private val interactor: SignUpInteractor = SignUpInteractorImpl()
    override var preferences: AppPreferences = AppPreferences
                                    .create(view.getContext())
```

当用户成功注册后，onSuccess()函数将被调用，如下所示：

```
override fun onSuccess() {
   interactor.getAuthorization(this)
 }
```

若用户在注册过程中出现错误，则调用下列回调函数：

```
override fun onError() {
  view.hideProgress()
  view.showSignUpError()
}

override fun onUsernameError() {
  view.hideProgress()
  view.setUsernameError()
}

override fun onPasswordError() {
  view.hideProgress()
  view.setPasswordError()
}

override fun onPhoneNumberError() {
  view.hideProgress()
  view.setPhoneNumberError()
}

override fun executeSignUp(username: String, phoneNumber: String,
                    password: String) {
  view.showProgress()
   interactor.signUp(username, phoneNumber, password, this)
}

interactor.persistAccessToken(preferences)
   interactor.persistUserDetails(preferences)
   view.hideProgress()
   view.navigateToHome()
 }

override fun onAuthError() {
  view.hideProgress()
```

```
    view.showAuthError()
  }
}
```

上述 SignUpPresenterImpl 类实现了 SignUpPresenter、SignUpInteractor.OnSignUpFinishedListener 以及 AuthInteractor.onAuthFinishedListener 接口，因而针对多个所需的函数提供了相关实现。此类函数包括 onSuccess()、onError()、onUsernameError()、onPasswordError()、onPhoneNumberError()、executeSignUp(String, String, String)、onAuthSuccess()和 onAuthError()。SignUpPresenterImpl 接收单一参数作为其主构造函数，对应参数应为 SignUpView 类型。

在 SignUpView 开始处理用户的注册操作时，executeSignUp(String, String, String)函数将被调用。当用户注册请求成功后，onSuccess()将被调用，该函数即刻调用交互器的 getAuthorization()函数，并获取新注册用户的访问令牌。若注册请求失败，则调用 onError()回调函数，这将隐藏进度条，并显示相应的错误消息。

在提交用户名、密码或电话号码过程中出现错误，onUsernameError()、onPasswordError()以及 onPhoneNumberError()方法将被调用；相应地，若验证成功，onAuthSuccess()将被调用。另外一方面，若验证失败，onAuthError()将被调用。

3. 创建注册视图

下面讨论 SignUpView 的构建过程。首先需要定义 SignUpView 接口，随后令 SignUpActivity 实现该接口。需要注意的是，在当前应用程序中，SignUpView 定义为 BaseView 和 AuthView 的扩展。SignUpView 接口定义如下所示：

```
package com.example.messenger.ui.signup

import com.example.messenger.ui.auth.AuthView
import com.example.messenger.ui.base.BaseView

interface SignUpView : BaseView, AuthView {

  fun showProgress()
  fun showSignUpError()
  fun hideProgress()
  fun setUsernameError()
  fun setPhoneNumberError()
  fun setPasswordError()
  fun navigateToHome()
}
```

下面调整当前项目中的 SignUpActivity 类,进而实现 SignUpView,并使用到 SignUpPresenter。将下列代码片段添加至 SignUpActivity 中,如下所示:

```kotlin
package com.example.messenger.ui.signup

import android.content.Context
import android.content.Intent
import android.support.v7.app.AppCompatActivity
import android.os.Bundle
import android.view.View
import android.widget.Button
import android.widget.EditText
import android.widget.ProgressBar
import android.widget.Toast
import com.example.messenger.R
import com.example.messenger.data.local.AppPreferences
import com.example.messenger.ui.main.MainActivity

class SignUpActivity : AppCompatActivity(), SignUpView,
View.OnClickListener {

  private lateinit var etUsername: EditText
  private lateinit var etPhoneNumber: EditText
  private lateinit var etPassword: EditText
  private lateinit var btnSignUp: Button
  private lateinit var progressBar: ProgressBar
  private lateinit var presenter: SignUpPresenter

  override fun onCreate(savedInstanceState: Bundle?) {
    super.onCreate(savedInstanceState)
    setContentView(R.layout.activity_sign_up)
    presenter = SignUpPresenterImpl(this)
    presenter.preferences = AppPreferences.create(this)
    bindViews()
  }

  override fun bindViews() {
    etUsername = findViewById(R.id.et_username)
    etPhoneNumber = findViewById(R.id.et_phone)
    etPassword = findViewById(R.id.et_password)
    btnSignUp = findViewById(R.id.btn_sign_up)
    progressBar = findViewById(R.id.progress_bar)
```

```kotlin
    btnSignUp.setOnClickListener(this)
}

override fun showProgress() {
  progressBar.visibility = View.VISIBLE
}

override fun hideProgress() {
  progressBar.visibility = View.GONE
}

override fun navigateToHome() {
  finish()
  startActivity(Intent(this, MainActivity::class.java))
}

override fun onClick(view: View) {
  if (view.id == R.id.btn_sign_up) {
    presenter.executeSignUp(etUsername.text.toString(),
                            etPhoneNumber.text.toString(),
                            etPassword.text.toString())
  }
 }
}
```

随后将 setUsernameError()、setPasswordError()、showAuthError()、showSignUpError() 和 getContext()函数添加至 SignUpActivity 中，如下所示：

```kotlin
override fun setUsernameError() {
  etUsername.error = "Username field cannot be empty"
}

override fun setPhoneNumberError() {
  etPhoneNumber.error = "Phone number field cannot be empty"
}

override fun setPasswordError() {
  etPassword.error = "Password field cannot be empty"
}

override fun showAuthError() {
  Toast.makeText(this, "An authorization error occurred. Please try again later.",
```

```
                    Toast.LENGTH_LONG).show()
  }

  override fun showSignUpError() {
    Toast.makeText(this, "An unexpected error occurred.
                         Please try again later.",
                    Toast.LENGTH_LONG).show()
  }

  override fun getContext(): Context {
    return this
  }
```

至此，我们已经完成了 Messenger 应用程序的过半内容，剩余工作主要涉及主 UI 方面的内容，第 6 章将对此展开讨论。

5.2 本章小结

本章介绍了 Messenger Android 应用程序的开发过程，期间涉及了大量的话题，包括 MVP 模式，并根据该模式深入讨论了应用程序的构建过程。

本章深入分析了响应式程序设计，并使用到了 RxJava 和 RxAndroid。其中包括如何利用 OkHttp 和 Retrofit 与远程服务器进行通信，随后实现了 Retrofit 服务，并与第 4 章开发的 Messenger API 进行通信。

第 6 章将结束 Messenger 应用程序的开发之旅。

第 6 章 构建 Messenger Android App（第 2 部分）

第 5 章讨论了 Messenger 应用程序开发，深度讲解了 Kotlin 和 Android 相关内容，其中涉及 MVP 模式，以及如何利用该模式创建功能强大的 Android 应用程序。除此之外，第 5 章还介绍了响应式程序设计，以及如何在应用程序中使用 RxJava 和 RxAndroid。关于与远程服务器之间的通信方式，第 5 章还讲解了 OkHttp 和 Retrofit，并实现了全功能的 Retrofit 服务，以及 Messenger API 之间的通信。在将 Android 和 Kotlin 进行有效的整合后，我们针对 Messenger 应用程序创建了登录和注册用户界面。

本章将结束 Messenger 应用程序的开发之旅，期间主要涉及以下内容：
- 与应用程序配置协同工作。
- 与 ChatKit 协同工作。
- Android 应用程序测试机制。
- 执行后台任务。

下面首先考察主 UI。

6.1 创建主 UI

与登录 UI 和注册 UI 类似，我们也将对主 UI 构建模型、视图以及 presenter。另外，对于前两个 UI 视图，本章解释其中的一些新概念，具体实现过程将不再赘述。

6.1.1 创建 MainView

在讨论主视图之前，下面首先介绍一下将要构建的用户界面。一种较好的方法是利用字面方式描述 MainView 的各项功能，如下所示：
- 主视图可使登录用户创建新的会话。
- 主视图可显示当前登录用户的联系方式（在当前应用程序中，显示 Messenger 平台上所有注册用户列表）。
- 用户可直接从 MainView 中访问设置页面。
- 用户可直接从 MainView 中注销应用程序。

下面简要地介绍一下 MainView 的功能。为了更加清晰地描述 MainView，图 6.1 显示了其示意图。

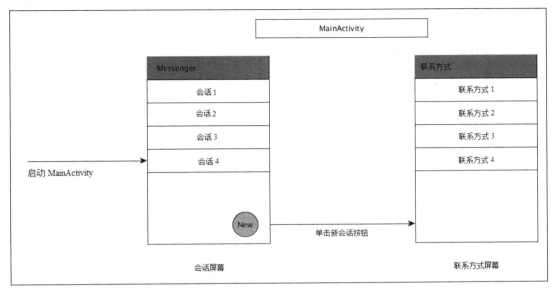

图 6.1 MainView 示意图

从图 6.1 中可以看到，MainActivity 向用户渲染了两个完全独立的视图，第一个视图是绘画视图；第二个视图是联系方式视图。对此，一种较好的实现是在 MainActivity 中使用应用片段（fragment），即会话片段和联系方式片段。

在明晰了 MainView 所包含的内容后，则需要定义需要的接口，并声明 MainView 的行为。MainView 接口的定义如下所示：

```kotlin
package com.example.messenger.ui.main
import com.example.messenger.ui.base.BaseView

interface MainView : BaseView {
    fun showConversationsLoadError()
    fun showContactsLoadError()
    fun showConversationsScreen()
    fun showContactsScreen()
    fun getContactsFragment(): MainActivity.ContactsFragment
    fun getConversationsFragment(): MainActivity.ConversationsFragment
    fun showNoConversations()
    fun navigateToLogin()
    fun navigateToSettings()
}
```

MainView 的实现将通过 MainActivity 予以保存，以供后续使用。下面将处理 MainInteractor。

6.1.2 创建 MainInteractor

用户应可在 Messenger 平台上查看其他用户（的联系方式），及其在主屏幕中的活动会话。除此之外，用户还应可直接从主屏幕中执行注销操作。对此，MainInteractor 需要加载联系方式、加载会话并执行注销操作。下列代码表示为 MainInteractor 接口定义，同时应确保将其与 com.example.messenger.ui.main 包中的其他 Main_文件进行整合。

```kotlin
package com.example.messenger.ui.main

import com.example.messenger.data.vo.ConversationListVO
import com.example.messenger.data.vo.UserListVO

interface MainInteractor {

  interface OnConversationsLoadFinishedListener {
    fun onConversationsLoadSuccess(
    conversationsListVo: ConversationListVO)

      fun onConversationsLoadError()
  }

  interface OnContactsLoadFinishedListener {
    fun onContactsLoadSuccess(userListVO: UserListVO)
    fun onContactsLoadError()
  }

  interface OnLogoutFinishedListener {
    fun onLogoutSuccess()
  }

  fun loadContacts(
  listener: MainInteractor.OnContactsLoadFinishedListener)

  fun loadConversations(
  listener: MainInteractor.OnConversationsLoadFinishedListener)

  fun logout(listener: MainInteractor.OnLogoutFinishedListener)
}
```

代码向 MainInteractor 接口整合添加了 OnConversationsLoadFinishedListener、OnContactsLoadFinishedListener 以及 OnLogoutFinishedListener 接口，且为须由 MainPresenter 实现的全部接口。另外，无论会话加载、联系方式加载或用户注销操作成功与否，presenter 中的回调函数均不可或缺，进而可执行相关动作。

包含了（实现后的）loadContacts()方法的 MainInteractorImpl 类如下所示：

```
package com.example.messenger.ui.main

import android.content.Context
import android.util.Log
import com.example.messenger.data.local.AppPreferences
import com.example.messenger.data.remote.repository.ConversationRepository
import com.example.messenger.data.remote.repository.ConversationRepositoryImpl
import com.example.messenger.data.remote.repository.UserRepository
import com.example.messenger.data.remote.repository.UserRepositoryImpl
import io.reactivex.android.schedulers.AndroidSchedulers
import io.reactivex.schedulers.Schedulers

class MainInteractorImpl(val context: Context) : MainInteractor {

    private val userRepository: UserRepository =
    UserRepositoryImpl(context)
    private val conversationRepository: ConversationRepository =
    ConversationRepositoryImpl(context)
    override fun loadContacts(listener:
    MainInteractor.OnContactsLoadFinishedListener) {
```

下面加载 Messenger API 平台中的全部注册用户，这一类用户可与当前登录用户进行通信，如下所示：

```
userRepository.all()
.subscribeOn(Schedulers.io())
.observeOn(AndroidSchedulers.mainThread())
.subscribe({ res ->
```

当前，联系方式被成功地载入。通过作为参数传递的 API 响应数据，onContactsLoadSuccess()将被调用，如下所示：

```
listener.onContactsLoadSuccess(res) },
{ error ->
```

如果联系方式加载失败，onContactsLoadError()将被调用，如下所示：

```
    listener.onContactsLoadError()
  error.printStackTrace()})
 }
}
```

　　loadContacts()利用 UserRepository 加载 Messenger 平台上现有的用户列表。如果检索成功，监听器的 onContactsLoadSuccess()将被打通，其中，载入的用户列表将作为参数予以传递。否则，onContactsLoadError()将被调用，同时输出错误消息。

　　当前，MainInteractorImpl 尚未实现，还需添加 loadConversations()和 logout()函数。下列代码片段定义了这两个函数，随后可将其添加至 MainInteractorImpl 中。

```
override fun loadConversations(
listener: MainInteractor.OnConversationsLoadFinishedListener) {
```

该函数通过会话存储库实例，检索当前登录用户的所有会话，如下所示：

```
 conversationRepository.all()
 .subscribeOn(Schedulers.io())
 .observeOn(AndroidSchedulers.mainThread())
 .subscribe({ res -> listener.onConversationsLoadSuccess(res) },
 { error ->
   listener.onConversationsLoadError()
 error.printStackTrace()})
}

override fun logout(
listener: MainInteractor.OnLogoutFinishedListener) {
```

注销操作将从共享预置文件中清空用户数据，并调用监听器的 onLogoutSuccess()回调函数，如下所示：

```
val preferences: AppPreferences = AppPreferences.create(context)
preferences.clear()
  listener.onLogoutSuccess()
}
```

　　loadConversations()的工作方式类似于loadContacts()，唯一差别在于，ConversationRepository用于检索当前用户包含的活动会话，而非联系方式列表。logout()只是简单地清空应用程序所用的预置文件，并移除当前登录用户的数据。随后，OnLogoutFinishedListener 提供的 onLogoutSuccess()方法将被调用。

MainInteractorImpl 类定义暂告一段落，下面将实现 MainPresenter。

6.1.3 创建 MainPresenter

像以往一样，首先需要定义一个 presenter 接口，进而定义 presenter 实现类所需的函数。MainPresenter 接口定义如下所示：

```
package com.example.messenger.ui.main

interface MainPresenter {
  fun loadConversations()
  fun loadContacts()
  fun executeLogout()
}
```

loadConversations()、loadContacts()和 executeLogout()函数通过 MainView 被调用，且需要由 MainPresenterImpl 类予以实现。下列代码显示了 MainPresenterImpl 类定义，其中包含了相关属性，以及 onConversationsLoadSuccess()和 onConversationsLoadError()方法。

```
package com.iyanuadelekan.messenger.ui.main

import com.iyanuadelekan.messenger.data.vo.ConversationListVO
import com.iyanuadelekan.messenger.data.vo.UserListVO

class MainPresenterImpl(val view: MainView) : MainPresenter,
        MainInteractor.OnConversationsLoadFinishedListener,
        MainInteractor.OnContactsLoadFinishedListener,
        MainInteractor.OnLogoutFinishedListener {

  private val interactor: MainInteractor = MainInteractorImpl
                                              (view.getContext())

  override fun onConversationsLoadSuccess(conversationsListVo:
                                              ConversationListVO) {
```

下列代码负责检测当前登录用户是否包含活动会话。

```
if (!conversationsListVo.conversations.isEmpty()) {
  val conversationsFragment = view.getConversationsFragment()
  val conversations = conversationsFragment.conversations
  val adapter = conversationsFragment.conversationsAdapter
```

```
conversations.clear()
adapter.notifyDataSetChanged()
```

在从 API 中检索到会话后,可将每个会话添加至 ConversationFragment 的会话列表中,在加入每项内容后,会话适配器将被通知,如下所示:

```
    conversationsListVo.conversations.forEach { contact ->
      conversations.add(contact)
      adapter.notifyItemInserted(conversations.size - 1)
    }
  } else {
    view.showNoConversations()
  }
}

override fun onConversationsLoadError() {
  view.showConversationsLoadError()
}
```

除此之外,还需要向 MainPresenterImpl 添加 onContactsLoadSuccess()、onContactsLoadError()、onLogoutSuccess()、loadConversations()、loadContacts() 和 executeLogout() 函数,如下所示:

```
override fun onContactsLoadSuccess(userListVO: UserListVO) {
  val contactsFragment = view.getContactsFragment()
  val contacts = contactsFragment.contacts
  val adapter = contactsFragment.contactsAdapter
```

下列代码负责清空联系方式列表中所加载的联系方式,并通知适配器数据已产生变化,如下所示:

```
contacts.clear()
adapter.notifyDataSetChanged()
```

下面将从 API 中检索到的每项联系方式添加至 ContactsFragment 的联系列表中;在每项内容被添加后,联系方式适配器将被通知,如下所示:

```
  userListVO.users.forEach { contact ->
    contacts.add(contact)
    contactsFragment.contactsAdapter.notifyItemInserted(contacts.size-1)
  }
}
```

```kotlin
override fun onContactsLoadError() {
  view.showContactsLoadError()
}

override fun onLogoutSuccess() {
  view.navigateToLogin()
}

override fun loadConversations() {
  interactor.loadConversations(this)
}

override fun loadContacts() {
  interactor.loadContacts(this)
}

override fun executeLogout() {
  interactor.logout(this)
}
```

至此，我们已成功地创建了 MainInteractor 和 MainPresenter，下面将完成 MainView 及其布局内容。

6.1.4 封装 MainView

首先需要对 activity_main.xml 布局文件进行适当调整，如下所示：

```xml
<?xml version="1.0" encoding="utf-8"?>
<android.support.design.widget.CoordinatorLayout
  xmlns:android="http://schemas.android.com/apk/res/android"
  xmlns:tools="http://schemas.android.com/tools"
  android:layout_width="match_parent"
  android:layout_height="match_parent"
  tools:context=".ui.main.MainActivity">
  <LinearLayout
    android:id="@+id/ll_container"
    android:layout_width="match_parent"
    android:layout_height="match_parent"
    android:orientation="vertical"/>
</android.support.design.widget.CoordinatorLayout>
```

在布局文件的根视图中，设置了单一的 LinearLayout。其中，ViewGroup 设置为会话和联系方式片段的容器。关于会话和联系方式片段，需要对其构建需要的布局，并在项目的布局resource目录中生成fragment_conversations.xml布局文件，如下所示：

```xml
<?xml version="1.0" encoding="utf-8"?>
<android.support.design.widget.CoordinatorLayout
xmlns:android="http://schemas.android.com/apk/res/android"
  android:layout_width="match_parent"
  android:layout_height="match_parent"
xmlns:app="http://schemas.android.com/apk/res-auto">
<android.support.v7.widget.RecyclerView
  android:id="@+id/rv_conversations"
  android:layout_width="match_parent"
android:layout_height="match_parent"/>
<android.support.design.widget.FloatingActionButton
  android:id="@+id/fab_contacts"
  android:layout_width="wrap_content"
  android:layout_height="wrap_content"
  android:layout_margin="@dimen/default_margin"
  android:src="@android:drawable/ic_menu_edit"
  app:layout_anchor="@id/rv_conversations"
  app:layout_anchorGravity="bottom|right|end"/>
</android.support.design.widget.CoordinatorLayout>
```

这里，在CoordinatorLayout根视图中使用了两个子视图，分别是RecyclerView和FloatingActionButton。RecyclerView是一个Android微件，它被用作显示大量数据的容器，通过维护有限的视图数量，可以有效地遍历这些数据。由于向项目中的build.gradle模块脚本中加入了RecyclerView依赖关系，因而此处需要使用到该微件，如下所示：

```
implementation 'com.android.support:recyclerview-v7:26.1.0'
```

鉴于使用了RecyclerView微件，因而需要针对每个RecyclerView微件创建相应的视图容器布局。针对于此，可在布局resource目录中生成vh_contacts.xml文件和vh_conversations.xml文件。

vh_contacts.xml布局文件如下所示：

```xml
<?xml version="1.0" encoding="utf-8"?>
<LinearLayout xmlns:android="http://schemas.android.com/apk/res/android"
  android:orientation="vertical" android:layout_width="match_parent"
  android:id="@+id/ll_container"
  android:layout_height="wrap_content">
  <LinearLayout
    android:layout_width="match_parent"
```

```xml
    android:layout_height="wrap_content"
    android:orientation="vertical"
    android:padding="@dimen/default_padding">
  <LinearLayout
    android:layout_width="match_parent"
    android:layout_height="wrap_content"
    android:orientation="horizontal">
    <TextView
      android:id="@+id/tv_username"
      android:layout_width="wrap_content"
      android:layout_height="wrap_content"
      android:textSize="18sp"
      android:textStyle="bold"/>
    <LinearLayout
      android:layout_width="0dp"
      android:layout_height="wrap_content"
      android:layout_weight="1"
      android:gravity="end">
      <TextView
        android:id="@+id/tv_phone"
        android:layout_width="wrap_content"
        android:layout_height="wrap_content"
        android:layout_marginLeft="@dimen/default_margin"
        android:layout_marginStart="@dimen/default_margin"/>
    </LinearLayout>
  </LinearLayout>
  <TextView
    android:id="@+id/tv_status"
    android:layout_width="wrap_content"
    android:layout_height="wrap_content"/>
  </LinearLayout>
  <View
    android:layout_width="match_parent"
    android:layout_height="1dp"
    android:background="#e8e8e8"/>
</LinearLayout>
```

vh_conversations.xml 布局文件包含了以下代码：

```xml
<?xml version="1.0" encoding="utf-8"?>
<LinearLayout xmlns:android="http://schemas.android.com/apk/res/android"
  android:orientation="vertical" android:layout_width="match_parent"
  android:id="@+id/ll_container"
```

```xml
      android:layout_height="wrap_content">
    <LinearLayout
      android:layout_width="match_parent"
      android:layout_height="wrap_content"
      android:orientation="vertical"
      android:padding="@dimen/default_padding">
      <TextView
          android:id="@+id/tv_username"
          android:layout_width="wrap_content"
          android:layout_height="wrap_content"
          android:textStyle="bold"
          android:textSize="18sp"/>
      <TextView
          android:id="@+id/tv_preview"
          android:layout_width="wrap_content"
          android:layout_height="wrap_content"/>
    </LinearLayout>
    <View
      android:layout_width="match_parent"
      android:layout_height="1dp"
      android:background="#e8e8e8"/>
</LinearLayout>
```

Android 开发参考中提到：浮动按钮多用于特定的提示类型，并通过浮动于 UI 上方的圆形图标加以区分，同时包含了与变形、启动以及移动锚点相关的特殊的运动行为。这里，我们可使用 FloatingActionButton 微件——之前曾向项目的 build.gradle 脚本中添加了 Android 设计库依赖关系，如下所示：

```
implementation 'com.android.support:design:26.1.0'
```

在包含了下列 XML 文件的 resource 布局目录中，创建 fragment_contacts.xml 布局文件。

```xml
<?xml version="1.0" encoding="utf-8"?>
<LinearLayout xmlns:android="http://schemas.android.com/apk/res/android"
  android:orientation="vertical" android:layout_width="match_parent"
  android:layout_height="match_parent">
  <android.support.v7.widget.RecyclerView
    android:id="@+id/rv_contacts"
    android:layout_width="match_parent"
    android:layout_height="match_parent"/>
</LinearLayout>
```

下面完成 MainActivity 类定义，其中涉及了较多内容。首先，需要声明所需的相关类；其次，还应实现下列方法：bindViews()、showConversationsLoadError()、showContactsLoadError()、showConversationsScreen()、showContactsScreen()、getContext()、getContactsFragment()、getConversationsFragment()、navigateToLogin()和navigateToSettings()。最后，还需要定义ConversationsFragment 和 ContactsFragment 类。

下面先向 MainActivity 中加入 ConversationsFragment 和 ContactsFragment。下列代码定义了 ConversationsFragment，将其添加至 MainActivity 中。

```kotlin
//ConversationsFragment class extending the Fragment class
 class ConversationsFragment : Fragment(), View.OnClickListener {

    private lateinit var activity: MainActivity
    private lateinit var rvConversations: RecyclerView
    private lateinit var fabContacts: FloatingActionButton
    var conversations: ArrayList<ConversationVO> = ArrayList()
    lateinit var conversationsAdapter: ConversationsAdapter
```

当 ConversationsFragment 的用户实例首次被绘制时，将调用下列方法：

```kotlin
override fun onCreateView(inflater: LayoutInflater, container:
ViewGroup, savedInstanceState: Bundle?): View? {
 // fragment layout inflation
 val baseLayout =
 inflater.inflate(R.layout.fragment_conversations,
 container, false)

 // Layout view bindings
 rvConversations = baseLayout.findViewById(R.id.rv_conversations)
 fabContacts = baseLayout.findViewById(R.id.fab_contacts)

 conversationsAdapter = ConversationsAdapter(
 getActivity(), conversations)

 // Setting the adapter of conversations recycler view to
 // created conversations adapter
 rvConversations.adapter = conversationsAdapter
```

下面设置会话回收布局管理器，并考察线性布局管理器的查看方式，如下所示：

```kotlin
    rvConversations.layoutManager =
    LinearLayoutManager(getActivity().baseContext)
    fabContacts.setOnClickListener(this)
    return baseLayout
 }
```

```kotlin
override fun onClick(view: View) {
  if (view.id == R.id.fab_contacts) {
    this.activity.showContactsScreen()
  }
}

fun setActivity(activity: MainActivity) {
  this.activity = activity
}
```

ConversationsFragment 包含了一个 RecyclerView 布局元素,并需要使用到适配器,以提供数据集到视图(显示于 RecyclerView 中)之间的绑定。简而言之,RecyclerView 使用一个 Adapter,并为所显示的视图提供数据。作为 ConversationsFragment 的嵌套类(内部类),下列代码显示了 ConversationsAdapter 类定义。

```kotlin
class ConversationsAdapter(private val context:
Context, private val dataSet: List<ConversationVO>) :
RecyclerView.Adapter<ConversationsAdapter.ViewHolder>(),
ChatView.ChatAdapter {

  val preferences: AppPreferences =
  AppPreferences.create(context)

  override fun onBindViewHolder(holder: ViewHolder, position:
  Int) {
    val item = dataSet[position] // get item at current position
    val itemLayout = holder.itemLayout // bind view holder layout
    // to local variable

    itemLayout.findViewById<TextView>(R.id.tv_username).text =
    item.secondPartyUsername
    itemLayout.findViewById<TextView>(R.id.tv_preview).text =
    item.messages[item.messages.size - 1].body
```

下列代码设置 itemLayout 的 View.OnClickListener。

```kotlin
itemLayout.setOnClickListener {
  val message = item.messages[0]
  val recipientId: Long

  recipientId = if (message.senderId ==
```

```
    preferences.userDetails.id) {
      message.recipientId
    } else {
      message.senderId
    }

    navigateToChat(item.secondPartyUsername,
    recipientId, item.conversationId)
  }
}

override fun onCreateViewHolder(parent: ViewGroup,
viewType: Int): ViewHolder {
```

下列代码创建 ViewHolder 布局。

```
  val itemLayout = LayoutInflater.from(parent.context)
    .inflate(R.layout.vh_conversations, null, false)
    .findViewById<LinearLayout>(R.id.ll_container)

  return ViewHolder(itemLayout)
}

override fun getItemCount(): Int {
  return dataSet.size
}

override fun navigateToChat(recipientName: String,
recipientId: Long, conversationId: Long?) {
  val intent = Intent(context, ChatActivity::class.java)
  intent.putExtra("CONVERSATION_ID", conversationId)
  intent.putExtra("RECIPIENT_ID", recipientId)
  intent.putExtra("RECIPIENT_NAME", recipientName)

  context.startActivity(intent)
}

class ViewHolder(val itemLayout: LinearLayout) :
RecyclerView.ViewHolder(itemLayout)
}
```

当创建回收视图适配器时，一些较为重要的方法需要提供相应的自定义实现，其中包括 onCreateViewHolder()、onBindViewHolder()和 getItemCount()方法。当回收视图需要一个新的视图容器实例时，onCreateViewHolder()将被调用。回收视图调用 onBindViewHolder()，进而显示特定位置数据集中的数据。调用 getItemCount()将得到数据集中的数据数量。ViewHolder 用于描述正在使用的项目视图，以及 RecyclerView 中与其位置相关的元数据。

> **注意**：
> 内部类表示为嵌套于另一个类中的类。

在理解了 ConversationsFragment 的含义后，下面讨论其实现过程。首先将 ContactsFragment 类添加至 MainActivity 中，如下所示：

```kotlin
class ContactsFragment : Fragment() {

    private lateinit var activity: MainActivity
    private lateinit var rvContacts: RecyclerView
    var contacts: ArrayList<UserVO> = ArrayList()
    lateinit var contactsAdapter: ContactsAdapter

    override fun onCreateView(inflater: LayoutInflater,
    container: ViewGroup, savedInstanceState: Bundle?): View? {
        val baseLayout = inflater.inflate(R.layout.fragment_contacts,
        container, false)
        rvContacts = baseLayout.findViewById(R.id.rv_contacts)
        contactsAdapter = ContactsAdapter(getActivity(), contacts)

        rvContacts.adapter = contactsAdapter
        rvContacts.layoutManager =
        LinearLayoutManager(getActivity().baseContext)

        return baseLayout
    }

    fun setActivity(activity: MainActivity) {
        this.activity = activity
    }
}
```

相信读者已经意识到，ContactsFragment 使用 RecyclerView 向应用程序用户显示联系方式视图元素，这一点与 ConversationsFragment 十分相似。ContactsAdapter 则表示为 RecyclerView 对应的适配器类。作为 ContactsFragment 的内部类，ContactsAdapter 类定义如下所示：

```kotlin
class ContactsAdapter(private val context: Context,
                      private val dataSet: List<UserVO>) :
                      RecyclerView.Adapter<ContactsAdapter.ViewHolder>(),
                      ChatView.ChatAdapter {

  override fun onCreateViewHolder(parent: ViewGroup,
                                  viewType: Int): ViewHolder {
    val itemLayout = LayoutInflater.from(parent.context)
                    .inflate(R.layout.vh_contacts, parent, false)
    val llContainer = itemLayout.findViewById<LinearLayout>
                    (R.id.ll_container)

    return ViewHolder(llContainer)
  }

  override fun onBindViewHolder(holder: ViewHolder, position: Int) {
    val item = dataSet[position]
    val itemLayout = holder.itemLayout

    itemLayout.findViewById<TextView>(R.id.tv_username).text =
item.username
    itemLayout.findViewById<TextView>(R.id.tv_phone).text =
item.phoneNumber
    itemLayout.findViewById<TextView>(R.id.tv_status).text = item.status

    itemLayout.setOnClickListener {
      navigateToChat(item.username, item.id)
    }
  }

  override fun getItemCount(): Int {
    return dataSet.size
  }

  override fun navigateToChat(recipientName: String,
                              recipientId: Long, conversationId: Long?) {
    val intent = Intent(context, ChatActivity::class.java)
    intent.putExtra("RECIPIENT_ID", recipientId)
    intent.putExtra("RECIPIENT_NAME", recipientName)

    context.startActivity(intent)
  }
```

```
class ViewHolder(val itemLayout: LinearLayout) :
  RecyclerView.ViewHolder(itemLayout)
}
```

截至目前，一切均工作正常。下面考察 MainActivity 中的相关属性和方法，并将属性定义添加至 MainActivity 类的开始处，如下所示：

```
private lateinit var llContainer: LinearLayout
private lateinit var presenter: MainPresenter

// Creation of fragment instances
private val contactsFragment = ContactsFragment()
private val conversationsFragment = ConversationsFragment()
```

接下来，onCreate()修改方法以体现相关变化，如下所示：

```
override fun onCreate(savedInstanceState: Bundle?) {
  super.onCreate(savedInstanceState)
  setContentView(R.layout.activity_main)
  presenter = MainPresenterImpl(this)

  conversationsFragment.setActivity(this)
  contactsFragment.setActivity(this)

  bindViews()
  showConversationsScreen()
}
```

随后，向 MainActivity 中添加 bindViews()、showConversationsLoadError()、showContactsLoadError()、showConversationsScreen()和 showContactsScreen()方法，如下所示：

```
override fun bindViews() {
  llContainer = findViewById(R.id.ll_container)
}

override fun onCreateOptionsMenu(menu: Menu?): Boolean {
  menuInflater.inflate(R.menu.main, menu)
  return super.onCreateOptionsMenu(menu)
}

override fun showConversationsLoadError() {
  Toast.makeText(this, "Unable to load conversations.
```

```
  Try again later.",
  Toast.LENGTH_LONG).show()
}

override fun showContactsLoadError() {
  Toast.makeText(this, "Unable to load contacts. Try again later.",
  Toast.LENGTH_LONG).show()
}
```

下面利用 ConversationsFragment 替换活动容器中的片段，如下所示：

```
override fun showConversationsScreen() {
  val fragmentTransaction = fragmentManager.beginTransaction()
  fragmentTransaction.replace(R.id.ll_container, conversationsFragment)
  fragmentTransaction.commit()

  // Begin conversation loading process
  presenter.loadConversations()

  supportActionBar?.title = "Messenger"
  supportActionBar?.setDisplayHomeAsUpEnabled(false)
}

override fun showContactsScreen() {
  val fragmentTransaction = fragmentManager.beginTransaction()
  fragmentTransaction.replace(R.id.ll_container, contactsFragment)
  fragmentTransaction.commit()
  presenter.loadContacts()

  supportActionBar?.title = "Contacts"
  supportActionBar?.setDisplayHomeAsUpEnabled(true)
}
```

最后，将 showNoConversations()、onOptionsItemSelected()、getContext()、getContactsFragment()、getConversationsFragment()、navigateToLogin()和 navigateToSettings()函数添加至 MainActivity 最后，如下所示：

```
override fun showNoConversations() {
  Toast.makeText(this, "You have no active conversations.",
  Toast.LENGTH_LONG).show()
}

override fun onOptionsItemSelected(item: MenuItem?): Boolean {
  when (item?.itemId) {
```

```
        android.R.id.home -> showConversationsScreen()
        R.id.action_settings -> navigateToSettings()
        R.id.action_logout -> presenter.executeLogout()
    }

    return super.onOptionsItemSelected(item)
}

override fun getContext(): Context {
    return this
}

override fun getContactsFragment(): ContactsFragment {
    return contactsFragment
}

override fun getConversationsFragment(): ConversationsFragment {
    return conversationsFragment
}

override fun navigateToLogin() {
    startActivity(Intent(this, LoginActivity::class.java))
    finish()
}

override fun navigateToSettings() {
    startActivity(Intent(this, SettingsActivity::class.java))
}
```

读者应注意上述代码片段中的注释内容,进而深入理解相关操作。

6.1.5 创建 MainActivity 菜单

在 MainActivity 的 onCreateOptionsMenu(Menu)函数中,包含了尚未实现的菜单内容。下面向应用程序 resource 目录下的 menu 包中添加 main.xml 文件,对应内容如下所示:

```
<?xml version="1.0" encoding="utf-8"?>
<menu xmlns:android="http://schemas.android.com/apk/res/android"
    xmlns:app="http://schemas.android.com/apk/res-auto">
<item
    android:id="@+id/action_settings"
    android:orderInCategory="100"
```

```xml
    android:title="@string/action_settings"
    app:showAsAction="never" />
<item
    android:id="@+id/action_logout"
    android:orderInCategory="100"
    android:title="@string/action_logout"
    app:showAsAction="never" />
</menu>
```

至此，我们距离项目的完成更近了一步。下面开始处理聊天用户方面的内容。相应地，showConversationLoadError()和 showMessageSendError()分别表示为对应的函数和接口（聊天机制所处的实际位置）。

6.2 创建聊天 UI

聊天 UI 需要显示活动会话的消息线程，并支持用户向其聊天对象发送消息。本节首先创建显示于用户的视图布局。

6.2.1 创建聊天布局

这里将使用到开源库 ChatKit 构建聊天视图布局。ChatKit 是一个 Android 库，针对 Android 项目中的聊天用户界面实现提供了灵活的组件，同时还包含了针对聊天用户界面数据管理和自定义操作的各种工具。

下列代码（位于 build.gradle 脚本文件中）将 ChatKit 添加至 Messenger 项目中。

```
implementation 'com.github.stfalcon:chatkit:0.2.2'
```

如前所述，ChatKit 针对聊天 UI 提供了有效的用户界面微件，例如 MessagesList 和 MessageInput。其中，微件 MessagesList 用于显示会话线程中的消息管理；MessageInput 则用于消息输入。除了支持多种风格的选项之外，MessageInput 还支持简单的输入验证处理。

下面考察如何在布局文件中使用 MessagesList 和 MessageInput。在 com.example.messenger.ui 中创建 chat 包，并将空活动 ChatActivity 添加至其中。打开 ChatActivity 活动布局文件（activity_chat.xml 文件）并添加下列 XML 内容：

```
<?xml version="1.0" encoding="utf-8"?>
<RelativeLayout xmlns:android="http://schemas.android.com/apk/res/android"
```

```xml
  xmlns:app="http://schemas.android.com/apk/res-auto"
  xmlns:tools="http://schemas.android.com/tools"
  android:layout_width="match_parent"
  android:layout_height="match_parent"
  tools:context="com.example.messenger.ui.chat.ChatActivity">
  <com.stfalcon.chatkit.messages.MessagesList
    android:id="@+id/messages_list"
    android:layout_width="match_parent"
    android:layout_height="match_parent"
    android:layout_above="@+id/message_input"/>
  <com.stfalcon.chatkit.messages.MessageInput
    android:id="@+id/message_input"
    android:layout_width="match_parent"
    android:layout_height="wrap_content"
    android:layout_alignParentBottom="true"
    app:inputHint="@string/hint_enter_a_message" />
</RelativeLayout>
```

在上述 XML 文件中可以看到，其中使用了 ChatKit 中的 MessagesList 和 MessageInput UI 微件。MessagesList 和 MessageInput 微件位于 com.stfalcon.chatkit.messages 包中。读者可打开布局设计窗口，并以可视化方式查看布局的外观。

下面考察 ChatView 类定义，如下所示：

```kotlin
package com.example.messenger.ui.chat

import com.example.messenger.ui.base.BaseView
import com.example.messenger.utils.message.Message
import com.stfalcon.chatkit.messages.MessagesListAdapter

interface ChatView : BaseView {

  interface ChatAdapter {
    fun navigateToChat(recipientName: String, recipientId: Long,
      conversationId: Long? = null)
  }

  fun showConversationLoadError()

  fun showMessageSendError()

  fun getMessageListAdapter(): MessagesListAdapter<Message>
}
```

在 ChatView 类中，我们定义了 ChatAdapter 接口，其中包含了唯一的 navigateToChat(String, Long, Long)函数。该接口须通过将用户导向至 ChatView 的适配器予以实现，之前定义的 ConversationsAdapter 和 ContactsAdapter 实现了该接口。

若会话和消息加载失败，应分别调用 showConversationLoadError() 和 showMessageSendError()函数，并显示相应的错误消息。

针对消息数据集的管理，ChatKit 的 MessagesList UI 微件需要包含一个 MessagesListAdapter。当通过 ChatView 实现了函数 getMessageListAdapter()时，该函数将返回 UI MessagesList 的 MessagesListAdapter。

6.2.2 准备聊天 UI 模型

在将消息添加至 MessageList 的 MessagesListAdapter 中时，需要实现对应模型中的、ChatKit 的 IMessage 接口。此处将实现该模型，对此，可创建 com.example.messenger.utils.message 包，并将下列 Message 类添加于其中。

```kotlin
package com.example.messenger.utils.message

import com.stfalcon.chatkit.commons.models.IMessage
import com.stfalcon.chatkit.commons.models.IUser
import java.util.*

data class Message(private val authorId: Long, private val body: String,
private val createdAt: Date) : IMessage {

  override fun getId(): String {
    return authorId.toString()
  }

  override fun getCreatedAt(): Date {
    return createdAt
  }

  override fun getUser(): IUser {
    return Author(authorId, "")
  }

  override fun getText(): String {
    return body
```

```
        }
    }
```

除此之外,还需要定义 Author 类以实现 ChatKit 中的 IUser 接口。该类的具体实现如下所示:

```
package com.example.messenger.utils.message

import com.stfalcon.chatkit.commons.models.IUser

data class Author(val id: Long, val username: String) : IUser {

    override fun getAvatar(): String? {
        return null
    }

    override fun getName(): String {
        return username
    }

    override fun getId(): String {
        return id.toString()
    }
}
```

Author 类对消息创建者的用户信息建模,例如消息创建者的名字、ID 以及昵称(如果存在)。下面将处理视图和布局方面的问题,并实现 ChatInteractor 和 ChatPresenter。

6.2.3 创建 ChatInteractor 和 ChatPresenter

前述内容讨论了 presenter 和交互器的具体含义,这里将直接对其进行编码。下列代码定义了 ChatInteractor 接口。相应地,该接口以及所有其他的 Chat_ 文件均位于 com.example.messenger.ui.chat 包中。

```
package com.example.messenger.ui.chat

import com.example.messenger.data.vo.ConversationVO

interface ChatInteractor {
```

```
interface OnMessageSendFinishedListener {
  fun onSendSuccess()

 fun onSendError()
 }

 interface onMessageLoadFinishedListener {
   fun onLoadSuccess(conversationVO: ConversationVO)
   fun onLoadError()
 }

 fun sendMessage(recipientId: Long, message: String, listener:
 OnMessageSendFinishedListener)

 fun loadMessages(conversationId: Long, listener:
 onMessageLoadFinishedListener)
}
```

下列代码显示了基于 ChatInteractor 接口的 ChatInteractorImpl 类定义。

```
package com.example.messenger.ui.chat

import android.content.Context
import com.example.messenger.data.local.AppPreferences
import com.example.messenger.data.remote.repository.ConversationRepository
Import com.example.messenger.data.remote.repository.ConversationRepositoryImpl
import com.example.messenger.data.remote.request.MessageRequestObject
import com.example.messenger.service.MessengerApiService
import io.reactivex.android.schedulers.AndroidSchedulers
import io.reactivex.schedulers.Schedulers

class ChatInteractorImpl(context: Context) : ChatInteractor {

 private val preferences: AppPreferences = AppPreferences.create(context)
 private val service: MessengerApiService = MessengerApiService
                                               .getInstance()
 private val conversationsRepository: ConversationRepository =
                        ConversationRepositoryImpl(context)
```

当调用下列方法时,将加载会话线程的消息内容。

```kotlin
override fun loadMessages(conversationId: Long, listener:
ChatInteractor.onMessageLoadFinishedListener) {
 conversationsRepository.findConversationById(conversationId)
 .subscribeOn(Schedulers.io())
 .observeOn(AndroidSchedulers.mainThread())
 .subscribe({ res -> listener.onLoadSuccess(res) },
 { error ->
  listener.onLoadError()
  error.printStackTrace() })
}
```

当调用下列方法时,将向用户发送一条消息。

```kotlin
override fun sendMessage(recipientId: Long, message: String,
listener: ChatInteractor.OnMessageSendFinishedListener) {
service.createMessage(MessageRequestObject(
recipientId, message), preferences.accessToken as String)
 .subscribeOn(Schedulers.io())
 .observeOn(AndroidSchedulers.mainThread())
 .subscribe({ _ -> listener.onSendSuccess() },
 { error ->
  listener.onSendError()
  error.printStackTrace() })
 }
}
```

接下来将处理与 ChatPresenter 和 ChatPresenterImpl 相关的代码。对于 ChatPresenter,需要定义一个接口,并声明两个函数:sendMessage(Long, String)和 loadMessages(Long)。ChatPresenter 接口定义如下所示:

```kotlin
package com.example.messenger.ui.chat

interface ChatPresenter {

 fun sendMessage(recipientId: Long, message: String)

 fun loadMessages(conversationId: Long)
}
```

ChatPresenter 接口的实现类如下所示:

```kotlin
package com.iyanuadelekan.messenger.ui.chat
```

```kotlin
import android.widget.Toast
import com.iyanuadelekan.messenger.data.vo.ConversationVO
import com.iyanuadelekan.messenger.utils.message.Message
import java.text.SimpleDateFormat

class ChatPresenterImpl(val view: ChatView) : ChatPresenter,
        ChatInteractor.OnMessageSendFinishedListener,
        ChatInteractor.onMessageLoadFinishedListener {

    private val interactor: ChatInteractor = ChatInteractorImpl
                                                (view.getContext())

    override fun onLoadSuccess(conversationVO: ConversationVO) {
      val adapter = view.getMessageListAdapter()

      // create date formatter to format createdAt dates
      // received from Messenger API
      val dateFormatter = SimpleDateFormat("yyyy-MM-dd HH:mm:ss")
```

下面遍历从 API 中加载的会话消息，对当前遍历的消息创建新的 IMessage 对象，并将 IMessage 添加至 MessagesListAdapter 的开始处，如下所示：

```kotlin
      conversationVO.messages.forEach { message ->
        adapter.addToStart(Message(message.senderId, message.body,
            dateFormatter.parse(message.createdAt.split(".")[0])), true)
      }
    }

    override fun onLoadError() {
      view.showConversationLoadError()
    }

    override fun onSendSuccess() {
      Toast.makeText(view.getContext(), "Message sent",
Toast.LENGTH_LONG).show()
    }

    override fun onSendError() {
      view.showMessageSendError()
    }

    override fun sendMessage(recipientId: Long, message: String) {
      interactor.sendMessage(recipientId, message,this)
```

```
    }

    override fun loadMessages(conversationId: Long) {
        interactor.loadMessages(conversationId, this)
    }
}
```

读者应留意上述代码中的注释内容，从而深入理解当前操作的具体含义。

下面讨论 ChatActivity，声明所需的属性并考察 onCreate()函数。

对此，调整 ChatActivity 方法并包含以下内容：

```
package com.example.messenger.ui.chat

import android.content.Context
import android.content.Intent
import android.support.v7.app.AppCompatActivity
import android.os.Bundle
import android.view.MenuItem
import android.widget.Toast
import com.example.messenger.R
import com.example.messenger.data.local.AppPreferences
import com.example.messenger.ui.main.MainActivity
import com.example.messenger.utils.message.Message
import com.stfalcon.chatkit.messages.MessageInput
import com.stfalcon.chatkit.messages.MessagesList
import com.stfalcon.chatkit.messages.MessagesListAdapter
import java.util.*

class ChatActivity : AppCompatActivity(), ChatView,
MessageInput.InputListener {

    private var recipientId: Long = -1
    private lateinit var messageList: MessagesList
    private lateinit var messageInput: MessageInput
    private lateinit var preferences: AppPreferences
    private lateinit var presenter: ChatPresenter
    private lateinit var messageListAdapter: MessagesListAdapter<Message>

    override fun onCreate(savedInstanceState: Bundle?) {
        super.onCreate(savedInstanceState)

        setContentView(R.layout.activity_chat)
        supportActionBar?.setDisplayHomeAsUpEnabled(true)
```

```
supportActionBar?.title =
intent.getStringExtra("RECIPIENT_NAME")

preferences = AppPreferences.create(this)
messageListAdapter = MessagesListAdapter(
preferences.userDetails.id.toString(), null)
presenter = ChatPresenterImpl(this)
bindViews()
```

下面解析源自 intent 的额外包（用于启动 ChatActivity）。若不存在由 CONVERSATION_ID 和 RECIPIENT_ID 所标识的包，则返回默认值-1，如下所示：

```
val conversationId = intent.getLongExtra("CONVERSATION_ID", -1)
recipientId = intent.getLongExtra("RECIPIENT_ID", -1)
```

如果 conversationId 不等于-1，那么，该 conversationId 视为有效，并加载会话中的消息，如下所示：

```
if (conversationId != -1L) {
    presenter.loadMessages(conversationId)
  }
 }
}
```

上述代码创建了 recipientId、messageList、messageInput、preferences、presenter 以及 messageListAdapter 属性，对应类型分别为 Long、MessageList、MessageInput、AppPreferences、ChatPresenter 和 MessageListAdapter。其中，视图 messageList 用于显示 messageListAdapter 向其提供的消息视图。onCreate()中包含的全部逻辑将处理当前活动中的视图初始化操作，读者应留意其中的注释内容。ChatActivity 实现了 MessageInput.InputListener，实现了该接口的相关类须提供相应的 onSubmit()方法。

当用户提交了包含 MessageInput 的微件后，MessageInput.InputListener 中被覆写的函数将被调用，如下所示：

```
override fun onSubmit(input: CharSequence?): Boolean {
    // create a new Message object and add it to the
    // start of the MessagesListAdapter
    messageListAdapter.addToStart(Message(
    preferences.userDetails.id, input.toString(), Date()), true)

    // start message sending procedure with the ChatPresenter
    presenter.sendMessage(recipientId, input.toString())
```

```
    return true
}
```

onSubmit()接收 MessageInput 所提交消息的 CharSequence,并对此创建相应的 Message 实例。随后,该实例添加至 MessageList 的开始处,即调用 messageListAdapter. addToStart()并包含作为参数传递的 Message 实例。在向 MessageList 添加了生成的 Message 后,ChatPresenter 实例用于初始化针对服务器的发送程序。

下面考察其他需要覆写的方法,并将 showConversationLoadError()、showMessageSendError()、getContext()和 getMessageListAdapter()方法添加至 ChatActivity 中,如下所示:

```
override fun showConversationLoadError() {
  Toast.makeText(this, "Unable to load thread.
  Please try again later.",
    Toast.LENGTH_LONG).show()
}

override fun showMessageSendError() {
  Toast.makeText(this, "Unable to send message.
  Please try again later.",
    Toast.LENGTH_LONG).show()
}

override fun getContext(): Context {
  return this
}

override fun getMessageListAdapter(): MessagesListAdapter<Message> {
  return messageListAdapter
}
```

最后,还需要覆写 bindViews()、onOptionsItemSelected()和 onBackPressed()方法,如下所示:

```
override fun bindViews() {
  messageList = findViewById(R.id.messages_list)
  messageInput = findViewById(R.id.message_input)
  messageList.setAdapter(messageListAdapter)
  messageInput.setInputListener(this)
}

override fun onOptionsItemSelected(item: MenuItem?): Boolean {
```

```
  if (item?.itemId == android.R.id.home) {
    onBackPressed()
  }
  return super.onOptionsItemSelected(item)
}

override fun onBackPressed() {
  super.onBackPressed()
  finish()
}
```

当前，一切均故作正常，我们已成功地构建了 Messenger 应用程序中的大部分内容。剩下的工作则是构建设置活动，以使用户能够更新其配置状态。

6.3 应用程序设置

本节将开发一个简单的应用程序设置活动，据此，用户可更新其配置状态。对此，可在 com.example.messenger.ui 下生成新的数据包，同时创建新的设置活动，并将该活动命名为 SettingsActivity。当创建设置活动时，可在 settings 上单击鼠标右键，在弹出的快捷菜单中选择 New | Activity | Settings Activity 命令，输入新设置活动必要的信息，例如活动名称和活动标题，随后单击 Finish 按钮。

在 SettingsActivity 的创建过程中，Android Studio 会向当前项目中添加多个文件。除此之外，新的资源目录（app | res | xml）也将添加至项目中。该目录中包含了以下文件：

- ❏ pref_data_sync.xml 文件
- ❏ pref_general.xml 文件。
- ❏ pref_headers.xml 文件。
- ❏ pref_notification.xml 文件。

读者也可选择删除 pref_notification.xml 和 pref_data_sync.xml 文件，当前项目暂不会使用到这些文件。下面考察 pref_general.xml 文件，如下所示：

```
<PreferenceScreen
xmlns:android="http://schemas.android.com/apk/res/android">
  <SwitchPreference
  android:defaultValue="true"
  android:key="example_switch"
  android:summary=
  "@string/pref_description_social_recommendations"
```

```xml
    android:title="@string/pref_title_social_recommendations" />

<!-- NOTE: EditTextPreference accepts EditText attributes. -->
<!-- NOTE: EditTextPreference's summary should be set to
     its value by the activity code. -->
<EditTextPreference
    android:capitalize="words"
    android:defaultValue="@string/pref_default_display_name"
    android:inputType="textCapWords"
    android:key="example_text"
    android:maxLines="1"
    android:selectAllOnFocus="true"
    android:singleLine="true"
    android:title="@string/pref_title_display_name" />

<!-- NOTE: Hide buttons to simplify the UI.
     Users can touch outside the dialog to
     dismiss it. -->
<!-- NOTE: ListPreference's summary should be set to
     its value by the activity code. -->
<ListPreference
    android:defaultValue="-1"
    android:entries="@array/pref_example_list_titles"
    android:entryValues="@array/pref_example_list_values"
    android:key="example_list"
    android:negativeButtonText="@null"
    android:positiveButtonText="@null"
    android:title="@string/pref_title_add_friends_to_messages" />

</PreferenceScreen>
```

上述 xml 布局文件的根视图为 PreferenceScreen。PreferenceScreen 表示为 Preference 层次结构的根。PreferenceScreen 自身表示为一个顶级（top-level）Preference。这里，术语 Preference 将被多次使用，下面对此加以定义。Preference 表示为基本的 Preference 用户界面构造块，并以 PreferenceActivity 列表显示加以显示。

对于显示于 PreferenceActivity 中的首选项及其所关联的 SharedPreferences（用于首选项的存储和检索），Preference 类提供了相应的视图。上述代码片段中的 SwitchPreference、EditTextPreference 以及 ListPreference 均为 DialogPreference 的子类，同时也是 Preference 类的超类。PreferenceActivity 表示为某个活动所需的基类，进而向用户显示首选项的层次结构。

当前并不需要使用到 pref_general.xml 文件中的 SwitchPreference、EditTextPreference 和 ListPreference，因而可将其从 XML 文件中移除。此处需要一个首选项，以使用户可在 Messenger 平台上更新其状态。作为一个特例，目前尚不存在首选项微件可提供这一功能。但读者也不必过分担心，下面将实现一个自定义首选项以完成这项任务。该首选项可命名为 ProfileStatusPreference。接下来在 settings 包中定义对应的 ProfileStatusPreference 类，如下所示：

```
package com.example.messenger.ui.settings

import android.content.Context
import android.preference.EditTextPreference
import android.text.TextUtils
import android.util.AttributeSet
import android.widget.Toast
import com.example.messenger.data.local.AppPreferences
import com.example.messenger.data.remote.request.
StatusUpdateRequestObject
import com.example.messenger.service.MessengerApiService
import io.reactivex.android.schedulers.AndroidSchedulers
import io.reactivex.schedulers.Schedulers

class ProfileStatusPreference(context: Context, attributeSet:
AttributeSet) : EditTextPreference(context, attributeSet) {

  private val service: MessengerApiService = MessengerApiService
                                              .getInstance()
  private val preferences: AppPreferences = AppPreferences
                                              .create(context)

  override fun onDialogClosed(positiveResult: Boolean) {
    if (positiveResult) {
```

下列代码片段将 ProfileStatusPreference 的 EditText 绑定至 etStatus 变量上：

```
  val etStatus = editText

  if (TextUtils.isEmpty(etStatus.text)) {
    // Display error message when user tries
    // to submit an empty status.
  Toast.makeText(context, "Status cannot be empty.",
  Toast.LENGTH_LONG).show()
```

```
} else {
 val requestObject =
 StatusUpdateRequestObject(etStatus.text.toString())
```

下面利用 MessengerApiService 更新用户的状态,如下所示:

```
service.updateUserStatus(requestObject,
preferences.accessToken as String)
.subscribeOn(Schedulers.io())
.observeOn(AndroidSchedulers.mainThread())
.subscribe({ res ->
```

如果状态关系成功,则存储更新后的用户信息,如下所示:

```
    preferences.storeUserDetails(res) },
    { error ->
    Toast.makeText(context, "Unable to update status at the " +
    "moment. Try again later.", Toast.LENGTH_LONG).show()
    error.printStackTrace()})
  }
 }

 super.onDialogClosed(positiveResult)
 }
}
```

ProfileStatusPreference 类扩展了 EditTextPreference。相应地,EditTextPreference 表示为支持 EditText 中字符串输入的 Preference。EditTextPreference 表示为一个 DialogPreference,因此,当单击 Preference 时,可向用户显示一个包含 Preference 视图的对话框。当关闭 DialogPreference 的对话框时,其 onDialogClosed(Boolean)方法将被调用。其中,当对话框利用正值被关闭时,正布尔值参数 true 将被传递至 onDialogClosed()中;相应地,当对话框利用负值被关闭时,false 将传递至 onDialogClosed()中,例如单击对话框的取消按钮。

ProfileStatusPreference 覆写了 EditTextPreference 的 onDialogClosed()函数。如果该对话框利用正值被关闭,将对 ProfileStatusPreference 的 EditText 函数中的状态有效性进行检测。如果状态消息有效,该状态将通过 API 更新;否则将显示一条错误消息。

在创建了 ProfileStatusPreference 后,下面返回至 pref_general.xml 文件,并对其进行更新,如下所示:

```
<PreferenceScreen
xmlns:android="http://schemas.android.com/apk/res/android">
 <com.example.messenger.ui.settings.ProfileStatusPreference
```

```xml
    android:key="profile_status"
    android:singleLine="true"
    android:inputType="text"
    android:maxLines="1"
    android:selectAllOnFocus="true"
    android:title="Profile status"
    android:defaultValue="Available"
    android:summary="Set profile status (visible to contacts)."/>
</PreferenceScreen>
```

不难发现,上述代码中使用了 ProfileStatusPreference,就像在 Android 应用程序框架中绑定的任何其他首选项一样。

接下来查看 pref_headers.xml 文件,如下所示:

```xml
<preference-headers
xmlns:android="http://schemas.android.com/apk/res/android">
  <!-- These settings headers are only used on tablets. -->
  <header
    android:fragment=
    "com.example.messenger.ui.settings.SettingsActivity
    $GeneralPreferenceFragment"
    android:icon="@drawable/ic_info_black_24dp"
    android:title="@string/pref_header_general" />
  <header
    android:fragment=
    "com.example.messenger.ui.settings.SettingsActivity
    $NotificationPreferenceFragment"
    android:icon="@drawable/ic_notifications_black_24dp"
    android:title="@string/pref_header_notifications" />
  <header
    android:fragment=
    "com.example.messenger.ui.settings.SettingsActivity
    $DataSyncPreferenceFragment"
    android:icon="@drawable/ic_sync_black_24dp"
    android:title="@string/pref_header_data_sync" />
</preference-headers>
```

针对 SettingsActivity 中的各种首选项,首选项头文件定义了相应的数据头,如下所示:

```xml
<preference-headers
xmlns:android="http://schemas.android.com/apk/res/android">
<header
  android:fragment=
```

```xml
    "com.example.messenger.ui.settings.SettingsActivity
    $GeneralPreferenceFragment"
    android:icon="@drawable/ic_info_black_24dp"
    android:title="@string/pref_header_account" />
</preference-headers>
```

下面考察 SettingsActivity,如下所示:

```kotlin
package com.example.messenger.ui.settings

import android.content.Intent
import android.os.Bundle
import android.preference.PreferenceActivity
import android.preference.PreferenceFragment
import android.view.MenuItem
import android.support.v4.app.NavUtils
import com.example.messenger.R
```

其中,PreferenceActivity 表示为应用程序设置集,如下所示:

```kotlin
class SettingsActivity : AppCompatPreferenceActivity() {

  override fun onCreate(savedInstanceState: Bundle?) {
    super.onCreate(savedInstanceState)
    supportActionBar?.setDisplayHomeAsUpEnabled(true)
  }

  override fun onMenuItemSelected(featureId: Int, item: MenuItem):
Boolean {
    val id = item.itemId

    if (id == android.R.id.home) {
      if (!super.onMenuItemSelected(featureId, item)) {
        NavUtils.navigateUpFromSameTask(this)
      }
      return true
    }
    return super.onMenuItemSelected(featureId, item)
  }
```

当相关活动需要使用到数据头列表时,下列 onBuildHeaders()函数将被调用:

```kotlin
override fun onBuildHeaders(target: List<PreferenceActivity.Header>)
{
```

```
    loadHeadersFromResource(R.xml.pref_headers, target)
}
```

下面的方法可以防止恶意应用程序的片段注入,且所有未知的片段均在此处被拒绝。

```
override fun isValidFragment(fragmentName: String): Boolean {
  return PreferenceFragment::class.java.name == fragmentName
  || GeneralPreferenceFragment::class.java.name == fragmentName
}
```

下列片段显示了通用首选项:

```
class GeneralPreferenceFragment : PreferenceFragment() {

  override fun onCreate(savedInstanceState: Bundle?) {
    super.onCreate(savedInstanceState)

    addPreferencesFromResource(R.xml.pref_general)
    setHasOptionsMenu(true)
  }

  override fun onOptionsItemSelected(item: MenuItem): Boolean {
   val id = item.itemId

   if (id == android.R.id.home) {
    startActivity(Intent(activity, SettingsActivity::class.java))
     return true
   }
   return super.onOptionsItemSelected(item)
  }
 }
}
```

SettingsActivity 扩展了 AppCompatPreferenceActivity——该活动实现了与 AppCompat 连同使用的多个调用。SettingsActivity 表示为一个 PreferenceActivity,并显示了一组应用程序设置集。若当前活动需要使用到数据头列表时,SettingsActivity 的 onBuildHeaders() 函数将被调用。isValidFragment()则用于阻止恶意应用程序向 SettingsActivity 中注入片段。若某个片段有效,那么,isValidFragment()将返回 true;否则返回 false。

SettingsActivity 中声明了一个 GeneralPreferenceFragment 类,并对 PreferenceFragment 进行扩展。PreferenceFragment 表示为定义于 Android 应用程序框架中的抽象类,并以列表形式显示了 Preference 实例的层次结构。

通过调用 addPreferencesFromResource(R.xml.pref_general),pref_general.xml 文件中的

首选项将被添加至 onCreate()方法的 GeneralPreferenceFragment 中。

在对 SettingsActivity 进行了上述调整后，读者即成功地完成了 Messenger 应用程序的设置工作。

在结束了 SettingsActivity 后，下面将尝试运行 Messenger 应用程序。相应地，读者可在虚拟或物理设备上构建、运行当前 Messenger 应用程序。当应用程序启动后，读者将直接转至 LoginActivity。

首先需要在 Messenger 平台上注册新用户，我们可在 SignUpActivity 上完成此项操作。单击"DON'T HAVE AN ACCOUNT? SIGN UP!"按钮，用户将被转至 SignUpActivity，如图 6.2 所示。

在当前活动中创建新用户，并输入 popeye 作为用户名，以及电话号码和密码，随后单击 SIGN UP 按钮。此时，新用户将在 Messenger 平台上注册，对应的用户名为 popeye。当注册结束后，用户将被转至 MainActivity，且会话视图将被即刻显示，如图 6.3 所示。

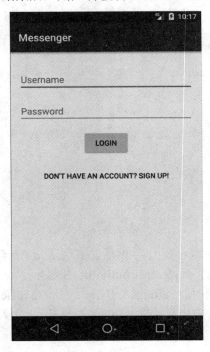
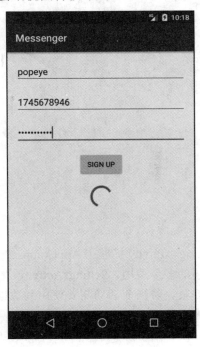

图 6.2　注册页面　　　　　　　　图 6.3　会话视图

由于新注册用户尚不包含任何活动对话，因而此处将会显示一条提示信息，如图 6.4 所示。对此，需要在 Messenger 平台上创建另一个用户，以展示聊天功能。单击屏幕右上角处的3个竖点图标，选择logout命令并注销popeye账户。

当注销用户后，可利用用户名dexter创建一个新的Messenger账号。在以dexter登录后，单击会话视图右下方的新消息创建浮动动作按钮，随后将显示联系方式视图，如图6.5所示。

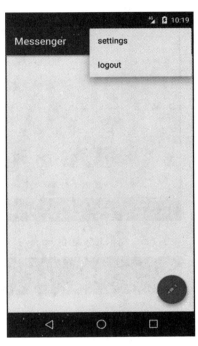

图6.4　新注册用户尚不包含任何活动对话　　　　图6.5　联系方式视图

单击popeye联系方式将打开ChatActivity，下面向popeye发送消息。在屏幕右下方的消息输入框中输入"Hey Popeye!"，随后单击Send按钮，该消息将发送至popeye处，如图6.6所示。

当返回至MainActivity的会话视图，读者将注意到一个对话项现在存在于popeye启动的对话中，如图6.7所示。

下面检测消息是否真正传送至popeye处。注销平台，随后以popeye身份登录。登录后，读者将会看到dexter启动的对话，如图6.8所示。

恭喜！消息已被正确地发送。下面尝试回复dexter。打开会话，并向dexter发送如图6.9所示的消息。

图 6.6　将消息发送至 popeye

图 6.7　对话项存在于 popeye 启动的对话中

图 6.8　dexter 启动的对话

图 6.9　回复 dexter

在图 6.9 中，我们发送了一条简单的消息"How are you Dexter?"。当前，需要相应地更新 popeye 的配置状态。对此，返回至主活动并访问设置活动（单击操作栏上的 3 个竖点形状的图标）。单击启动设置显示中的 Account 将显示通用首选项片段。单击 Profile status 首选项，如图 6.10 所示。

随后将显示一个包含 EditText 的对话框，用户可输入新的配置状态。输入所选的状态消息并单击 OK 按钮，如图 6.11 所示。

图 6.10　更新配置状态

图 6.11　输入新的配置状态

当前配置状态将即刻被更新。

至此，我们完整地实现了 Messenger 应用程序，读者可对其进行调整并尝试添加相关代码。本章剩余部分还将涉及两个话题，即应用程序测试和执行后台任务。

6.4　Android 应用程序测试

应用程序测试是指测试开发完毕的程序，并对其质量进行检测，诸多因素均会对软件质量产生影响，包括应用程序可用性、功能、可靠性以及一致性。Android 应用程序测试包含许多优点，包含但不仅限于以下内容：
- 错误检测。
- 软件的可靠性。

Android 应用程序集成测试涵盖了大量内容，因而也超出了本书的讨论范围，但读者应留意下列 Android 测试资源：
- Espresso，对应网址为 https://developer.android.com/training/testing/espresso/index.html。
- Roboelectric，对应网址为 http://robolectric.org。
- Mockito，对应网址为 http://site.mockito.org。
- Calabash，对应网址为 https://github.com/calabash/calabash-android。

6.5 执行后台操作

在开发 Messenger 应用程序时，我们曾使用了 RxAndroid 并执行异步操作。在许多场合下，当使用 RxAndroid 时，可在 Android 应用程序的主线程上查看后台运行结果。某些时候，读者可能不希望使用第三方库实现这一功能，例如 RxAndroid。相反，读者更愿意采用 Android 应用程序框架中的解决方案。针对于此，Android 提供了多种选择方案，AsyncTask 便是其中之一。

6.5.1 AsyncTask

AsyncTask 类可查看后台操作的性能，并在应用程序 UI 线程中显示操作结果，且无须管理处理程序和线程。AsyncTask 适用于运行小型操作任务。其间，AsyncTask 的计算部分运行于后台线程，对应结果将显示于 UI 线程中。关于 AsyncTask 的更多信息，读者可访问 https://developer.android.com/reference/android/os/AsyncTask.html。

6.5.2 IntentService

当在后台中运行调度任务时，IntentService 是一类较好的候选方案。Android 开发参考中指出，IntentService 定义为服务基类，用于即时处理异步请求（表示为 Intent）。客户端通过 startService（Intent）调用发送请求；服务在必要时启用，并通过工作线程处理每个 Intent，并在任务结束时执行终止。关于 IntentService 的更多内容，读者可访问 https://developer.android.com/reference/android/app/IntentService.html。

6.6 本章小结

本章结束了 Messenger Android 应用程序开发之旅。其间,我们学习了用于创建聊天用户界面的第三方库 ChatKit。除此之外,还进一步讨论了 Android 应用程序框架提供的各种工具。本章首先介绍了 Android 中的设置活动开发过程,并了解了 PreferenceScreen、PreferenceActivity、DialogPreference、Preference 和 PreferenceFragment。最后,本章还简要介绍了 Android 应用程序测试机制,以及执行后台操作任务。

第 7 章将讨论 Android 应用程序框架中的各种存储选择方案。

第 7 章　在数据库中存储信息

在第 6 章中，我们介绍了一些较为重要的话题，包括第三方库的使用、Android 应用程序测试机制，以及如何在 Android 平台上运行后台任务。本章主要讨论数据的存储。前述内容曾针对不同实例对持久化应用程序数据进行存储，例如通过 SharedPreferences 满足数据存储需求，但该方案并不是 Android 应用程序框架中的唯一数据存储方式。本章将深度探讨 Android 中的数据存储方式，且主要涉及以下内容：

- 内部存储。
- 外部存储。
- 网络存储。
- 内容提供商。

除此之外，我们还将针对不同应用选取最佳存储方案。下面首先讨论内部存储。

> **提示：**
> 代码中省略号所表示的内容位于对应的代码文件中。

7.1　与内部存储协同工作

Android 应用程序框架中的现有存储媒介可使开发人员存储设备内存中的私人数据。这里，"私人数据"是指其他应用程序无法通过内部存储访问 App 存储的数据。除此之外，当应用程序被卸载时，此类文件将从存储中移除。

7.1.1　向内部存储中写入文件

当在内部存储中创建私有文件时，可调用 openFileOutput()函数。openFileOutput()函数接收两个参数。其中，第一个参数表示为文件名（以 String 方式表示）；第二个参数为操作模式。需要注意的是，openFileOutput()函数须在 Context 实例中被调用，例如 Activity。

openFileOutput()函数返回一个 FileOutputStream，随后用于通知当前文件可利用 write()方法完成操作。当写入操作结束后，FileOutputStream 通过调用 close()方法被关闭。下列代码显示了这一处理过程。

```kotlin
private fun writeFile(fileName: String) {
    val content: String = "Hello world"
    val stream: FileOutputStream = openFileOutput(fileName,
                                        Context.MODE_PRIVATE)

    stream.write(content.toByteArray())
    stream.close()
}
```

7.1.2 从内部存储中读取私有文件

当读取私有文件时，可通过调用 openFileInput() 获取 FileInputStream。该方法包含单一参数，即所读取的文件名。同样，openFileInput() 须在 Context 实例中被调用。当获得 FileInputStream 之后，即可调用 read() 函数从当前文件中读取字节。当文件读取完毕后，可调用 close() 执行关闭操作。

考察下列代码：

```kotlin
private fun readFile(fileName: String) {
    val stream: FileInputStream = openFileInput(fileName)
    val data = ByteArray(1024)

    stream.read(data)
    stream.close()
}
```

7.1.3 基于内部存储的示例程序

适宜的示例程序可帮助我们进一步理解相关概念，本节将利用内部存储机制实现一个文件更新程序。文件的更新操作较为简单，即收集用户文本数据（源自输入框），并更新内部存储中的文件。随后，用户可通过应用程序中的视图查看文件文本内容。下面创建一个新的 Android 项目,其名称应适当反映当前应用程序的功能。在项目创建完毕后，在 src 项目包中分别创建 base 包和 main 包。

相应地，base 包中的 BaseView 接口包含下列代码：

```kotlin
package com.example.storageexamples.base

interface BaseView {
```

```kotlin
  fun bindViews()

  fun setupInstances()
}
```

如果读者阅读了前述章节，相信已对视图界面有所了解。在 main 包中，可定义一个 MainView 并扩展 BaseView，如下所示：

```kotlin
package com.example.storageexamples.main

import com.example.storageexamples.base.BaseView

interface MainView : BaseView {

  fun navigateToHome()
  fun navigateToContent()
}
```

文件更新应用程序包含了两个片段形式的视图，其中，第一个视图表示为主视图，用户可以此更新文件内容；第二个视图则表示为内容视图，用户可据此读取更新文件的内容。

下面在 main 中创建空活动，并将其命名为 MainActivity，同时确保 MainActivity 为启动活动。当 MainActivity 创建完毕后，应保证其在执行前扩展了 MainView。打开 activity_main.xml 并添加下列内容：

```xml
<?xml version="1.0" encoding="utf-8"?>
<android.support.constraint.ConstraintLayout

xmlns:android="http://schemas.android.com/apk/res/android"
 xmlns:tools="http://schemas.android.com/tools"
 android:layout_width="match_parent"
 android:layout_height="match_parent"
 tools:context="com.example.storageexamples.main.MainActivity">

 <LinearLayout
      android:id="@+id/ll_container"
      android:layout_width="match_parent"
      android:layout_height="match_parent"
      android:orientation="vertical"/>
</android.support.constraint.ConstraintLayout>
```

上述代码将 LinearLayout 视图组添加至布局文件的根视图中，该布局可视作应用程序主片段和内容片段的容器。下列代码表示为主片段布局（位于 fragment_home.xml 文件中）：

```xml
<?xml version="1.0" encoding="utf-8"?>
<LinearLayout
xmlns:android="http://schemas.android.com/apk/res/android"
  android:orientation="vertical"
  android:layout_width="match_parent"
  android:paddingTop="@dimen/padding_default"
  android:paddingBottom="@dimen/padding_default"
  android:paddingStart="@dimen/padding_default"
  android:paddingEnd="@dimen/padding_default"
  android:gravity="center_horizontal"
  android:layout_height="match_parent">
<TextView
      android:id="@+id/tv_header"
      android:layout_width="wrap_content"
      android:layout_height="wrap_content"
      android:text="@string/header_title"
      android:textSize="45sp"
      android:textStyle="bold"/>
<EditText
      android:id="@+id/et_input"
      android:layout_width="match_parent"
      android:layout_height="wrap_content"
      android:layout_marginTop="@dimen/margin_top_large"
      android:hint="@string/hint_enter_text"/>
<Button
      android:id="@+id/btn_submit"
      android:layout_width="match_parent"
      android:layout_height="wrap_content"
      android:layout_marginTop="@dimen/margin_default"
      android:text="@string/submit"/>
<Button
      android:id="@+id/btn_view_file"
      android:layout_width="match_parent"
      android:layout_height="wrap_content"
      android:text="@string/view_file"
      android:background="@android:color/transparent"/>
</LinearLayout>
```

在打开设计窗口以查看布局变化之前,还需要添加一些数值资源。当前项目的 strings.xml 文件应涵盖以下内容(除 app_name 字符串资源之外):

```
<resources>
  <string name="app_name">Storage Examples</string>
  <string name="hint_enter_text">Enter text here…</string>
  <string name="submit">Update file</string>
  <string name="view_file">View file</string>
  <string name="header_title">FILE UPDATER</string>
</resources>
```

除此之外,当前项目还应包含 dimens.xml 文件,并涵盖以下内容:

```
<?xml version="1.0" encoding="utf-8"?>
<resources>
  <dimen name="padding_default">16dp</dimen>
  <dimen name="margin_default">16dp</dimen>
  <dimen name="margin_top_large">64dp</dimen>
</resources>
```

当添加了上述资源后,即可看到如图 7.1 所示的、fragment_home.xml 中的布局设计窗口。

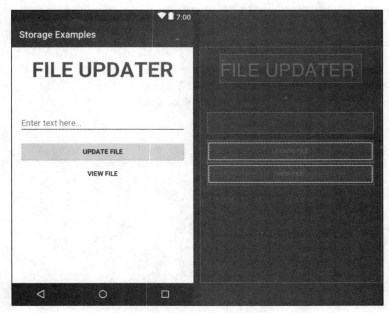

图 7.1 布局设计窗口

对于内容片段布局，向布局资源目录的 fragment_content.xml 布局文件中添加下列代码：

```xml
<?xml version="1.0" encoding="utf-8"?>
<LinearLayout
xmlns:android="http://schemas.android.com/apk/res/android"
  android:orientation="vertical"
  android:layout_width="match_parent"
  android:padding="@dimen/padding_default"
  android:layout_height="match_parent">
  <TextView
      android:id="@+id/tv_content"
      android:layout_width="match_parent"
      android:layout_height="wrap_content"
      android:textSize="20sp"
      android:textStyle="bold"
      android:layout_marginTop="@dimen/margin_default"/>
</LinearLayout>
```

该布局包含了单一的 TextView，并向应用程序用户显示内部存储文件中的文本。如果读者感兴趣的话，可查看当前布局设计窗口。

此时，需要定义相应的片段类，以显示刚刚创建的片段布局。对此，可将 HomeFragment 类添加至 MainActivity 中，如下所示：

```kotlin
class HomeFragment : Fragment(), BaseView, View.OnClickListener {

  private lateinit var layout: LinearLayout
  private lateinit var tvHeader: TextView
  private lateinit var etInput: EditText
  private lateinit var btnSubmit: Button
  private lateinit var btnViewFile: Button

  private var outputStream: FileOutputStream? = null

  override fun onCreateView(inflater: LayoutInflater,
                            container: ViewGroup?,
                            savedInstanceState: Bundle?): View {

    // inflate the fragment_home.xml layout
    layout = inflater.inflate(R.layout.fragment_home,
                              container, false) as LinearLayout
    setupInstances()
    bindViews()
```

```
    return layout
}

override fun bindViews() {
    tvHeader = layout.findViewById(R.id.tv_header)
    etInput = layout.findViewById(R.id.et_input)
    btnSubmit = layout.findViewById(R.id.btn_submit)
    btnViewFile = layout.findViewById(R.id.btn_view_file)

    btnSubmit.setOnClickListener(this)
    btnViewFile.setOnClickListener(this)
}
```

下列方法用于实例属性的实例化操作。

```
override fun setupInstances() {
```

下面向一个名为 content_file 的文件启用 FileOutputStream,该文件表示为内部存储的私有文件,因而只可被当前应用程序所访问,如下所示:

```
    outputStream = activity?.openFileOutput("content_file",
                                Context.MODE_PRIVATE)
}
```

如果出现无效输入,下列函数将被调用,并向用户显示一条错误消息。

```
private fun showInputError() {
    etInput.error = "File input cannot be empty."
    etInput.requestFocus()
}
```

下面通过 FileOutputStream 写入字符串内容。

```
private fun writeFile(content: String) {
    outputStream?.write(content.toByteArray())
    outputStream?.close()
}
```

下列函数用于清空输入框中的输入内容。

```
private fun clearInput() {
    etInput.setText("")
}
```

调用下列函数将向用户显示一条成功消息。

第 7 章 在数据库中存储信息

```kotlin
private fun showSaveSuccess() {
  Toast.makeText(activity, "File updated successfully.",
            Toast.LENGTH_LONG).show()
}

override fun onClick(view: View?) {
  val id = view?.id

  if (id == R.id.btn_submit) {
    if (TextUtils.isEmpty(etInput.text)) {
```

如果用户提供了一个空值作为文件输入内容,那么,函数将显示一条错误消息,如下所示:

```kotlin
  showInputError()
} else {
```

下面向文件中写入内容,清空 EditText 并显示文件更新成功消息。

```kotlin
    writeFile(etInput.text.toString())
    clearInput()
    showSaveSuccess()
  }
} else if (id == R.id.btn_view_file) {
  // retrieve a reference to MainActivity
  val mainActivity = activity as MainActivity
```

下列代码将用户转至内容片段,并在动作栏上显示主按钮,以使用户可返回至上一个片段。

```kotlin
    mainActivity.navigateToContent()
    mainActivity.showHomeNavigation()
    }
  }
}
```

读者应留意 HomeFragment 中的注释内容,以确保理解具体的操作含义。在将 HomeFragment 添加至 MainActivity 后,还需要添加一个片段,并向用户显示 fragment_content.xml 布局。对此,可将 ContentFragment 类添加至 MainActivity 中,如下所示:

```kotlin
class ContentFragment : Fragment(), BaseView {

  private lateinit var layout: LinearLayout
```

```
    private lateinit var tvContent: TextView

    private lateinit var inputStream: FileInputStream

    override fun onCreateView(inflater: LayoutInflater?,
                              container: ViewGroup?,
                              savedInstanceState: Bundle?): View {

      layout = inflater?.inflate(R.layout.fragment_content,
                                 container, false) as LinearLayout
      setupInstances()
      bindViews()

      return layout
    }

    override fun onResume() {
```

下列代码在恢复片段时更新 TextView 中呈现的内容。

```
      updateContent()
      super.onResume()
}

private fun updateContent() {
  tvContent.text = readFile()
}

override fun bindViews() {
  tvContent = layout.findViewById(R.id.tv_content)
}

override fun setupInstances() {
  inputStream = activity.openFileInput("content_file")
}
```

下列代码将读取内部存储中的文件内容,并以字符串显示返回对应内容,如下所示:

```
    private fun readFile(): String {
      var c: Int
      var content = ""

      c = inputStream.read()
```

```
    while (c != -1) {
      content += Character.toString(c.toChar())
      c = inputStream.read()
    }

    inputStream.close()

    return content
  }
}
```

当恢复 ContentFragment 时，tvContent 实例（向用户显示文件内容的 TextView）将被更新。TextView 的更新过程可描述为：将 TextView 的内容设置为文件的读取内容（利用 readFile()函数）。最后一项任务是完成 MainActivity。完整的 MainActivity 类定义如下所示：

```
package com.example.storageexamples.main

import android.aupport.v4.app.Fragment
import android.content.Context
import android.support.v7.app.AppCompatActivity
import android.os.Bundle
import android.text.TextUtils
import android.view.LayoutInflater
import android.view.MenuItem
import android.view.View
import android.view.ViewGroup
import android.widget.*
import com.example.storageexamples.R
import com.example.storageexamples.base.BaseView
import java.io.FileInputStream
import java.io.FileOutputStream

class MainActivity : AppCompatActivity(), MainView {

  private lateinit var llContainer: LinearLayout
```

下面设置片段实例，如下所示：

```
private lateinit var homeFragment: HomeFragment
private lateinit var contentFragment: ContentFragment

override fun onCreate(savedInstanceState: Bundle?) {
```

```kotlin
    super.onCreate(savedInstanceState)
    setContentView(R.layout.activity_main)
    setupInstances()
    bindViews()
    navigateToHome()
}

override fun bindViews() {
    llContainer = findViewById(R.id.ll_container)
}

override fun setupInstances() {
    homeFragment = HomeFragment()
    contentFragment = ContentFragment()
}

private fun hideHomeNavigation() {
    supportActionBar?.setDisplayHomeAsUpEnabled(false)
}

private fun showHomeNavigation() {
    supportActionBar?.setDisplayHomeAsUpEnabled(true)
}

override fun navigateToHome() {
    val transaction = supportFragmentManager.beginTransaction()
    transaction.replace(R.id.ll_container, homeFragment)
    transaction.commit()

    supportActionBar?.title = "Home"
}

override fun navigateToContent() {
    val transaction = supportFragmentManager.beginTransaction()
    transaction.replace(R.id.ll_container, contentFragment)
    transaction.commit()

    supportActionBar?.title = "File content"
}

override fun onOptionsItemSelected(item: MenuItem?): Boolean {
    val id = item?.itemId
```

```kotlin
  if (id == android.R.id.home) {
    navigateToHome()
    hideHomeNavigation()
  }

  return super.onOptionsItemSelected(item)
}

class HomeFragment : Fragment(), BaseView, View.OnClickListener {

  private lateinit var layout: LinearLayout
  private lateinit var tvHeader: TextView
  private lateinit var etInput: EditText
  private lateinit var btnSubmit: Button
  private lateinit var btnViewFile: Button

  private lateinit var outputStream: FileOutputStream

  override fun onCreateView(inflater: LayoutInflater?,
                            container: ViewGroup?,
                            savedInstanceState: Bundle?): View {
```

下列代码用于创建 fragment_home.xml 布局：

```kotlin
    layout = inflater .inflate(R.layout.fragment_home,
                            container, false) as LinearLayout
    setupInstances()
    bindViews()
    return layout
  }

override fun bindViews() {
  tvHeader = layout.findViewById(R.id.tv_header)
  etInput = layout.findViewById(R.id.et_input)
  btnSubmit = layout.findViewById(R.id.btn_submit)
  btnViewFile = layout.findViewById(R.id.btn_view_file)

  btnSubmit.setOnClickListener(this)
  btnViewFile.setOnClickListener(this)
}

//Method for the instantiation of instance properties
override fun setupInstances() {
```

下面针对 content_file 打开 FileOutputstream，该文件为内部存储中的私有文件，因而仅可被当前应用程序所访问，如下所示：

```kotlin
  outputStream = activity.openFileOutput("content_file",
                                Context.MODE_PRIVATE)
}

//Called to display an error to the user if an invalid input is given
private fun showInputError() {
  etInput.error = "File input cannot be empty."
  etInput.requestFocus()
}

// Writes string content to a file via a [FileOutputStream]
private fun writeFile(content: String) {
  outputStream.write(content.toByteArray())
}

//Called to clear the input in the input field
private fun clearInput() {
  etInput.setText("")
}

//Shows a success message to the user when invoked.
private fun showSaveSuccess() {
  Toast.makeText(activity, "File updated successfully.",
              Toast.LENGTH_LONG).show()
}

override fun onClick(view: View?) {
  val id = view?.id

  if (id == R.id.btn_submit) {
```

如果用户提交了空值作为文件输入内容，下列代码片段将显示一条错误消息。

```kotlin
  if (TextUtils.isEmpty(etInput.text)) {
    showInputError()
  } else {
    //Write content to the file, clear the input
    //EditText and show a file update success message
```

```
        writeFile(etInput.text.toString())
        clearInput()
        showSaveSuccess()
    }
} else if (id == R.id.btn_view_file) {
    // retrieve a reference to MainActivity
    val mainActivity = activity as MainActivity
```

下面将用户转至内容片段,并在动作栏上显示主按钮,以使用户能够返回至上一个片段,如下所示:

```
        mainActivity.navigateToContent()
        mainActivity.showHomeNavigation()
    }
  }
}

class ContentFragment : Fragment(), BaseView {

    private lateinit var layout: LinearLayout
    private lateinit var tvContent: TextView

    private lateinit var inputStream: FileInputStream

    override fun onCreateView(inflater: LayoutInflater?,
                              container: ViewGroup?,
                              savedInstanceState: Bundle?): View {

        layout = inflater?.inflate(R.layout.fragment_content,
                              container, false) as LinearLayout
        setupInstances()
        bindViews()

        return layout
    }
```

当恢复片段时,下列代码将更新 TextView 中的内容。

```
override fun onResume() {
    updateContent()
    super.onResume()
}
```

```kotlin
private fun updateContent() {
  tvContent.text = readFile()
}

override fun bindViews() {
  tvContent = layout.findViewById(R.id.tv_content)
}

override fun setupInstances() {
  inputStream = activity.openFileInput("content_file")
}
```

下列代码读取内部存储中的文件内容，并以字符串形式返回该内容。

```kotlin
  private fun readFile(): String {
    var c: Int
    var content = ""

    c = inputStream.read()

    while (c != -1) {
      content += Character.toString(c.toChar())
      c = inputStream.read()
    }

    inputStream.close()

    return content
  }
}
```

MainActivity 已经全部完成，下面准备运行应用程序。

在所选取的设备上构建并运行当前项目。但启动应用程序后，MainActivity 的主片段将显示在设备上，读者可在 EditText 输入框中输入任何内容，如图 7.2 所示。

在向 EditText 中输入相关内容后，单击 UPDATE FILE 按钮。相应地，内部存储文件通过所提供的内容予以更新，随后将显示一条信息。在文件更新完毕后，可单击 VIEW FILE 按钮，如图 7.3 所示。

内容片段显示于 TextView 中，取值包含了更新后的文件内容。尽管该程序较为简单，但却展示了 Android 应用程序与内部存储之间的工作方式。

图 7.2　在 EditText 输入框中输入内容　　　　图 7.3　文件更新完毕

7.1.4　保存缓存文件

如果用户并不希望永久存储数据，而是在存储中缓存数据，则可使用 cacheDir 打开内部存储中的目录，其中临时保存了缓存文件。

cacheDir 返回一个 File。因此，可使用 File 类提供的全部方法，例如 outputStream() 方法，该方法将返回一个 FileOutputStream。

7.2　与外部存储协同工作

外部存储用于创建和访问非私有、共享的公有文件。Android 设备均支持共享外部存储机制。对此，首先需要获得外部存储许可。

7.2.1　获得外部存储许可

应用程序在使用外部存储 API 之前，须获取 READ_EXTERNAL_STORAGE 和

WRITE_EXTERNAL_STORAGE 许可。当仅使用外部存储的读取功能时，READ_EXTERNAL_STORAGE 许可不可或缺；相应地，WRITE_EXTERNAL_STORAGE 许可则表示外部存储的写入操作。

上述两项许可可方便地添加至清单文件中，如下所示：

```
<uses-permission android:name="android.permission.WRITE_EXTERNAL_STORAGE"/>
<uses-permission android:name="android.permission.READ_EXTERNAL_STORAGE" />
```

需要注意的是，WRITE_EXTERNAL_STORAGE 隐式地包含了 READ_EXTERNAL_STORAGE。因此，若需要使用到两项许可，那么，请求 WRITE_EXTERNAL_STORAGE 即可，如下所示：

```
<uses-permission android:name="android.permission.WRITE_EXTERNAL_STORAGE"/>
```

7.2.2 媒介的有效性

某些时候，由于各种原因，存储设备往往会意外丢失——外部存储媒介无法被正常访问。相应地，在使用外部存储媒介之前，需要对其进行检测。

getExternalStorageState()方法可用于检测媒介的有效性，下列代码片段将检测外部存储是否可写入应用程序中。

```kotlin
private fun isExternalStorageWritable(): Boolean {
  val state = Environment.getExternalStorageState()

  return Environment.MEDIA_MOUNTED == state
}
```

首先应检索外部存储的当前状态，并于随后检测其是否处于 MEDIA_MOUNTED 状态。若是，那么，应用程序可向其写入。对此，isExternalStorageWritable()将返回 true。

外部存储的读取检测同样十分简单，如下所示：

```kotlin
private fun isExternalStorageReadable(): Boolean {
  val state = Environment.getExternalStorageState()

  return Environment.MEDIA_MOUNTED == state ||
         Environment.MEDIA_MOUNTED_READ_ONLY == state
}
```

若处于 MEDIA_MOUNTED 或 MEDIA_MOUNTED_READ_ONLY 状态，应用程序则可从外部存储中读取数据。

7.2.3 存储共享文件

供用户或其他应用程序访问的文件应存储于共享公共目录中,例如 Pictures/和 Music/目录。

当检索表示为公共目录的 File 时,应用程序可调用 getExternalStoragePulicDirectory() 方法。其中,检索的目录类型则作为唯一参数传递至该方法中。

下列方法针对存储的音乐素材创建了一个目录。

```
private fun getMusicStorageDir(collectionName: String): File {
 val file = File(Environment.getExternalStoragePublicDirectory(
     Environment.DIRECTORY_MUSIC), collectionName)

 if (!file.mkdir()) {
   Log.d("DIR_CREATION_STATUS", "Directory creation failed.")
 }

 return file
}
```

在生成公共目录过程中若出现错误,相关消息将输出至控制台中。

7.2.4 利用外部存储缓存文件

某些时候,可能会出现需要用外部存储缓存文件的场景。对此,可以打开一个表示外部存储目录的文件,其中应用程序应该使用 externalCacheDir 保存缓存的文件。

7.3 网络存储

网络存储是指远程服务器上的数据存储。与之前讨论的其他存储方式不同,此类存储方式使用了网络连接存储、检索远程服务器上的数据。当构建第 6 章中的 Messenger Android 应用程序时,即可采用这一类存储媒介。其中,Messenger 应用程序依赖于远程服务器存储、检索信息。当远程服务器用于客户端应用程序的数据源时,常会使用到客户端-服务器架构。客户端通过 HTTP 向服务器发送数据请求(一般采用 GET 请求);作为响应,服务器将发送所需数据,进而完成了 HTTP 事务处理周期。

SQLite 是一种较为流行的关系型数据库管理系统(RDBMS),与其他一些 RDBMS

不同，SQLite并未采用客户机-服务器数据库引擎，而是直接嵌入应用程序中。

Android完全支持SQLite，并可通过类这一Android项目形式访问SQLite数据库。需要注意的是，在Android中，仅可通过构建数据库的应用程序对SQLite进行访问。

当在Android中与SQLite协同工作时，建议使用Room持久库。在Android中，处理Room的第一个步骤在项目的build.gradle脚本中包含下列依赖关系：

```
implementation "android.arch.persistence.room:runtime:1.0.0-alpha9-1"
implementation "android.arch.persistence.room:rxjava2:1.0.0-alpha9-1"
implementation "io.reactivex.rxjava2:rxandroid:2.0.1"
kapt "android.arch.persistence.room:compiler:1.0.0-alpha9-1"
```

借助于Room，实体访问变得更加简单。这里，所有实体须添加@Entity注解。下列代码显示了简单的User实体。

```
package com.example.roomexample.data

import android.arch.persistence.room.ColumnInfo
import android.arch.persistence.room.Entity
import android.arch.persistence.room.PrimaryKey

@Entity
data class User(
  @ColumnInfo(name = "first_name")
  var firstName: String = "",
  @ColumnInfo(name = "surname")
  var surname: String = "",
  @ColumnInfo(name = "phone_number")
  var phoneNumber: String = "",
  @PrimaryKey(autoGenerate = true)
  var id: Long = 0
)
```

Room针对所定义的User实体生成所需的SQLite表，该表中包含了名称、user以及4个属性，即id、first_name、surname和phone_number。其中，id属性是创建的用户表的主属性，我们通过@PrimaryKey注解对此加以指定。用户表每条记录中的主键可通过Room生成，即在@PrimaryKey注解中设置autoGenerate = true。注解@ColumnInfo则用于指定与表中某个列相关的附加信息。例如，考察下列代码片段：

```
@ColumnInfo(name = "first_name")
var firstName: String = ""
```

其中，User 中定义了一个 firstName 属性。@ColumnInfo(name ="first_name")将用户表中的列名（针对 firstName 属性）设置为 first_name。

当在数据库中读取、写入记录时，需要使用到数据访问对象（DAO）。DAO 利用标注方法执行数据库操作。下列代码显示了 User 实体中的 DAO。

```
package com.example.roomexample.data

import android.arch.persistence.room.Dao
import android.arch.persistence.room.Insert
import android.arch.persistence.room.OnConflictStrategy
import android.arch.persistence.room.Query
import io.reactivex.Flowable

@Dao
interface UserDao {

  @Query("SELECT * FROM user")
  fun all(): Flowable<List<User>>

  @Query("SELECT * FROM user WHERE id = :id")
  fun findById(id: Long): Flowable<User>

  @Insert(onConflict = OnConflictStrategy.REPLACE)
  fun insert(user: User)
}
```

@Query 注解将 DAO 中的方法标识为查询方法。其中，方法被调用时执行的查询操作将作为数值传递至注解中。自然情况下，传递至@Query 的查询表示为 SQL 查询。SQL 查询涉及大量的内容，读者应了解其编写方式。

@Insert 注解用于向表中插入数据。相应地，其他较为重要的注解还包括@Update 和@Delete，分别用于更新、删除数据库表中的数据。

最后，在创建了实体和 DAO 后，还需要定义应用程序的数据库。对此，可定义 RoomDatabase 的子类，并利用@Database 对其进行注解。最小限度下，该注解应包括实体类引用集合以及数据库版本号。下列代码显示了简单的 AppDatabase 抽象类定义。

```
@Database(entities = [User::class], version = 1)
public abstract class AppDatabase : RoomDatabase() {
  abstract fun userDao(): UserDao
}Now that we have our DAO and entity created, we must create an AppDatabase class. Add
```

在定义了应用程序数据库类之后,通过调用 databaseBuilder() 可得到数据库实例,如下所示:

```
val db = Room.databaseBuilder(<context>, AppDatabase::class.java,
                    "app-database").build()
```

一旦获得 RoomDatabase 实例,即可以此检索数据访问对象,进而读取、写入、更新、查询以及删除数据库中的数据。

下面构建一个简单的应用程序,并展示如何借助于 Android 中的 Room 使用 SQLite。在该应用程序中,用户可手动输入与人员相关的信息,并于随后查看用户输入信息。

利用空 MainActivity 构建新的 Android 项目,并将其作为启动活动;随后将下列依赖关系添加至应用程序的 build.gradle 脚本中。

```
implementation 'com.android.support:design:26.1.0'
implementation "android.arch.persistence.room:runtime:1.0.0"
implementation "android.arch.persistence.room:rxjava2:1.0.0"
implementation "io.reactivex.rxjava2:rxandroid:2.0.1"
kapt "android.arch.persistence.room:compiler:1.0.0"
```

此外,还需要将 kotlin-kapt 独立插件添加至 build.gradle 脚本中,如下所示:

```
apply plugin: 'kotlin-kapt'
```

在加入了上述项目依赖关系后,在项目的 source 包中创建 data 和 ui 包。在 ui 包中,添加 MainView,如下所示:

```
package com.example.roomexample.ui

interface MainView {

  fun bindViews()
  fun setupInstances()
}
```

在向 ui 包中加入了 MainView 后,可将 MainActivity 重置到 ui 包中。下面处理应用程序化的数据库问题。由于应用程序将存储用户信息,因此需要定义 User 实体。相应地,可向 data 包中添加下列 User 实体。

```
package com.example.roomexample.data

import android.arch.persistence.room.ColumnInfo
import android.arch.persistence.room.Entity
import android.arch.persistence.room.PrimaryKey
```

```kotlin
@Entity
data class User(
  @ColumnInfo(name = "first_name")
  var firstName: String = "",
  @ColumnInfo(name = "surname")
  var surname: String = "",
  @ColumnInfo(name = "phone_number")
  var phoneNumber: String = "",
  @PrimaryKey(autoGenerate = true)
  var id: Long = 0
)
```

下列代码在 data 包中创建 UserDao。

```kotlin
package com.example.roomexample.data

import android.arch.persistence.room.Dao
import android.arch.persistence.room.Insert
import android.arch.persistence.room.OnConflictStrategy
import android.arch.persistence.room.Query
import io.reactivex.Flowable

@Dao
interface UserDao {

  @Query("SELECT * FROM user")
  fun all(): Flowable<List<User>>

  @Query("SELECT * FROM user WHERE id = :id")
  fun findById(id: Long): Flowable<User>

  @Insert(onConflict = OnConflictStrategy.REPLACE)
  fun insert(user: User)
}
```

UserDao 接口定义了 3 个方法，分别是 all()、findById() 和 insert()。其中，all() 方法返回包含所有用户列表的 Flowable；findById() 方法搜索与传递至该方法中 id 相匹配的 User。若存在，该方法则在 Flowable 中返回 User。insert() 方法则用于将用户作为一条记录插入 User 表中。

在创建了 DAO 和实体后，还需要定义 AppDatabase 类。对此，可将下列代码添加至 data 包中。

```kotlin
package com.example.roomexample.data

import android.arch.persistence.room.Database
import android.arch.persistence.room.Room
import android.arch.persistence.room.RoomDatabase
import android.content.Context

@Database(entities = arrayOf(User::class), version = 1, exportSchema = false)
internal abstract class AppDatabase : RoomDatabase() {

  abstract fun userDao(): UserDao

  companion object Factory {
    private var appDatabase: AppDatabase? = null

    fun create(ctx: Context): AppDatabase {
      if (appDatabase == null) {
        appDatabase = Room.databaseBuilder(ctx.applicationContext,
                                  AppDatabase::class.java,
                                  "app-database").build()

      }
      return appDatabase as AppDatabase
    }
  }
}
```

上述代码创建了包含单一 create() 函数的 Factory 伴生对象，其唯一任务是生成 AppDatabase 实例（若不存在），并返回该实例以供使用。

最后一项与数据相关的任务是构建 AppDatabase。当前，需要针对应用程序视图生成相应的布局。对此，可在 MainActivity 中使用两个片段。第一个片段针对新创建用户收集输入信息；第二个片段则在 RecyclerView 中显示全部用户的信息。这里，可修改 activity_main.xml 布局并包含下列代码：

```xml
<?xml version="1.0" encoding="utf-8"?>
<android.support.constraint.ConstraintLayout
xmlns:android="http://schemas.android.com/apk/res/android"
xmlns:tools="http://schemas.android.com/tools"
android:layout_width="match_parent"
android:layout_height="match_parent"
```

```xml
    tools:context="com.example.roomexample.ui.MainActivity">

    <LinearLayout
        android:id="@+id/ll_container"
        android:layout_width="match_parent"
        android:layout_height="match_parent"
        android:orientation="vertical"/>
</android.support.constraint.ConstraintLayout>
```

activity_main.xml 中的 LinearLayout 包含了 MainActivity 中的片段。利用下列代码，可将 fragment_create_user.xml 文件加入 resource 布局目录中。

```xml
<?xml version="1.0" encoding="utf-8"?>
<LinearLayout
xmlns:android="http://schemas.android.com/apk/res/android"
android:orientation="vertical"
android:layout_width="match_parent"
android:layout_height="match_parent"
android:gravity="center_horizontal"
android:padding="@dimen/padding_default">
 <TextView
        android:layout_width="wrap_content"
        android:layout_height="wrap_content"
        android:textSize="32sp"
        android:text="@string/create_user"/>
 <EditText
        android:id="@+id/et_first_name"
        android:layout_width="match_parent"
        android:layout_height="wrap_content"
        android:layout_marginTop="@dimen/margin_default"
        android:hint="@string/first_name"
        android:inputType="text"/>
 <EditText
        android:id="@+id/et_surname"
        android:layout_width="match_parent"
        android:layout_height="wrap_content"
        android:layout_marginTop="@dimen/margin_default"
        android:hint="@string/surname"
        android:inputType="text"/>
 <EditText
        android:id="@+id/et_phone_number"
        android:layout_width="match_parent"
        android:layout_height="wrap_content"
```

```xml
        android:layout_marginTop="@dimen/margin_default"
        android:hint="@string/phone_number"
        android:inputType="phone"/>
<Button
        android:id="@+id/btn_submit"
        android:layout_width="match_parent"
        android:layout_height="wrap_content"
        android:layout_marginTop="@dimen/margin_default"
        android:text="@string/submit"/>
<Button
        android:id="@+id/btn_view_users"
        android:layout_width="match_parent"
        android:layout_height="wrap_content"
        android:layout_marginTop="@dimen/margin_default"
        android:text="@string/view_users"/>
</LinearLayout>
```

随后将下列代码添加至 fragment_list_users.xml 布局资源中。

```xml
<?xml version="1.0" encoding="utf-8"?>
<LinearLayout xmlns:android="http://schemas.android.com/apk/res/android"
 android:orientation="vertical"
 android:layout_width="match_parent"
 android:layout_height="match_parent">
 <android.support.v7.widget.RecyclerView
        android:id="@+id/rv_users"
        android:layout_width="match_parent"
        android:layout_height="match_parent"/>
</LinearLayout>
```

fragment_list_users.xml 文件包含了一个 RecyclerView，进而显示保存至数据库中的每个用户的信息。对此，须针对 RecyclerView 创建一个视图容器布局资源项，对应布局文件称作 vh_user.xml，并将添加下列代码：

```xml
<?xml version="1.0" encoding="utf-8"?>
<LinearLayout xmlns:android="http://schemas.android.com/apk/res/android"
 android:orientation="vertical"
 android:layout_width="match_parent"
 android:padding="@dimen/padding_default"
 android:layout_height="wrap_content">
 <TextView
        android:id="@+id/tv_first_name"
        android:layout_width="wrap_content"
```

```xml
        android:layout_height="wrap_content"/>
    <TextView
        android:id="@+id/tv_surname"
        android:layout_width="wrap_content"
        android:layout_height="wrap_content"
        android:layout_marginTop="@dimen/margin_default"/>
    <TextView
        android:id="@+id/tv_phone_number"
        android:layout_width="wrap_content"
        android:layout_height="wrap_content"
        android:layout_marginTop="@dimen/margin_default"/>
    <View
        android:layout_width="match_parent"
        android:layout_height="1dp"
        android:layout_marginTop="@dimen/margin_default"
        android:background="#e8e8e8"/>
</LinearLayout>
```

相信读者已经猜测到，这里还需要向当前项目中添加一些字符串和尺寸资源。相应地，打开应用程序的 strings.xml 布局文件，并向其中加入下列字符串资源：

```xml
<resources>
...
  <string name="first_name">First name</string>
  <string name="surname">Surname</string>
  <string name="phone_number">Phone number</string>
  <string name="submit">Submit</string>
  <string name="create_user">Create User</string>
  <string name="view_users">View users</string>
</resources>
```

随后在项目中生成下列尺寸资源：

```xml
<?xml version="1.0" encoding="utf-8"?>
<resources>
  <dimen name="padding_default">16dp</dimen>
  <dimen name="margin_default">16dp</dimen>
</resources>
```

下面开始着手处理 MainActivity。如前所述，我们将在 MainActivity 中使用到两个独立的片段。其中，第一个片段可将个人数据保存至 SQL 数据库中；第二个片段则允许用户查看已保存至数据库中的人员信息。

首先定义 CreateUserFragment，并将下列片段类添加至 MainActivity 中（位于文件

MainActivity.kt 中）。

```kotlin
class CreateUserFragment : Fragment(), MainView, View.OnClickListener {

    private lateinit var btnSubmit: Button
    private lateinit var etSurname: EditText
    private lateinit var btnViewUsers: Button
    private lateinit var layout: LinearLayout
    private lateinit var etFirstName: EditText
    private lateinit var etPhoneNumber: EditText

    private lateinit var userDao: UserDao
    private lateinit var appDatabase: AppDatabase

    override fun onCreateView(inflater: LayoutInflater,
                container: ViewGroup?, savedInstanceState: Bundle?): View {
        layout = inflater.inflate(R.layout.fragment_create_user,
                        container, false) as LinearLayout
        bindViews()
        setupInstances()
        return layout
    }

    override fun bindViews() {
        btnSubmit = layout.findViewById(R.id.btn_submit)
        btnViewUsers = layout.findViewById(R.id.btn_view_users)
        etSurname = layout.findViewById(R.id.et_surname)
        etFirstName = layout.findViewById(R.id.et_first_name)
        etPhoneNumber = layout.findViewById(R.id.et_phone_number)

        btnSubmit.setOnClickListener(this)
        btnViewUsers.setOnClickListener(this)
    }

    override fun setupInstances() {
        appDatabase = AppDatabase.create(activity)
            // getting an instance of AppDatabase
        userDao = appDatabase.userDao() // getting an instance of UserDao
    }
```

下列方法验证以用户表单形式提交的输入内容：

```kotlin
private fun inputsValid(): Boolean {
    var inputValid = true
```

```kotlin
val firstName = etFirstName.text
val surname = etSurname.text
val phoneNumber = etPhoneNumber.text

if (TextUtils.isEmpty(firstName)) {
  etFirstName.error = "First name cannot be empty"
  etFirstName.requestFocus()
  inputValid = false

} else if (TextUtils.isEmpty(surname)) {
  etSurname.error = "Surname cannot be empty"
  etSurname.requestFocus()
  inputValid = false

} else if (TextUtils.isEmpty(phoneNumber)) {
  etPhoneNumber.error = "Phone number cannot be empty"
  etPhoneNumber.requestFocus()
  inputValid = false

} else if (!android.util.Patterns.PHONE
                 .matcher(phoneNumber).matches()) {
  etPhoneNumber.error = "Valid phone number required"
  etPhoneNumber.requestFocus()
  inputValid = false
}

return inputValid
}
```

下列函数用于显示相关消息，以表明用户已成功创建。

```kotlin
private fun showCreationSuccess() {
  Toast.makeText(activity, "User successfully created.",
             Toast.LENGTH_LONG).show()
}

override fun onClick(view: View?) {
  val id = view?.id

  if (id == R.id.btn_submit) {
    if (inputsValid()) {
      val user = User(
        etFirstName.text.toString(),
```

```
            etSurname.text.toString(),
            etPhoneNumber.text.toString())

         Observable.just(userDao)
                 .subscribeOn(Schedulers.io())
                 .subscribe( { dao ->
            dao.insert(user) // using UserDao to save user to database.
            activity?.runOnUiThread { showCreationSuccess() }
         }, Throwable::printStackTrace)
      }
   } else if (id == R.id.btn_view_users) {
     val mainActivity = activity as MainActivity

     mainActivity.navigateToList()
     mainActivity.showHomeButton()
   }
 }
}
```

前述内容已经讨论过片段多次，因而此处重点在于与 AppDatabase 协同工作的片段部分。在 setupInstances()中，将设置指向 AppDatabase 和 UserDao 的引用。通过调用 AppDatabase 的 Factory 伴生对象的 create()函数，可得到 AppDatabase 实例。UserDao 则可通过调用 appDatabase.userDao()得到。

下面考察片段类中的 onClick()方法。当单击提交按钮时，所提交的用户信息将执行有效性检测。如果输入内容无效，则会显示相应的错误信息；否则，将创建包含所提交的用户信息的新 User 对象，并保存至数据库中。下列代码实现了上述功能。

```
if (inputsValid()) {
 val user = User(
   etFirstName.text.toString(),
   etSurname.text.toString(),
   etPhoneNumber.text.toString())

 Observable.just(userDao)
          .subscribeOn(Schedulers.io())
          .subscribe( { dao ->
   dao.insert(user) // using UserDao to save user to database.
   activity?.runOnUiThread { showCreationSuccess() }
 }, Throwable::printStackTrace)
}
```

同样，ListUsersFragment 的构建过程也较为简单。对此，将下列 ListUsersFragment

添加至 MainActivity 中，如下所示：

```kotlin
class ListUsersFragment : Fragment(), MainView {

  private lateinit var layout: LinearLayout
  private lateinit var rvUsers: RecyclerView

  private lateinit var appDatabase: AppDatabase

  override fun onCreateView(inflater: LayoutInflater,
          container: ViewGroup?, savedInstanceState: Bundle?): View {

    layout = inflater.inflate(R.layout.fragment_list_users,
    container, false) as LinearLayout
    bindViews()
    setupInstances()

    return layout
  }
```

下面将用户回收视图绑定至其布局元素上，如下所示：

```kotlin
override fun bindViews() {
    rvUsers = layout.findViewById(R.id.rv_users)
}

override fun setupInstances() {
    appDatabase = AppDatabase.create(activity)
    rvUsers.layoutManager = LinearLayoutManager(activity)
    rvUsers.adapter = UsersAdapter(appDatabase)
}

private class UsersAdapter(appDatabase: AppDatabase) :
        RecyclerView.Adapter<UsersAdapter.ViewHolder>() {

    private val users: ArrayList<User> = ArrayList()
    private val userDao: UserDao = appDatabase.userDao()

    init {
      populateUsers()
    }

    override fun onCreateViewHolder(parent: ViewGroup?, viewType: Int):
```

```kotlin
                ViewHolder {
    val layout = LayoutInflater.from(parent?.context)
                        .inflate(R.layout.vh_user, parent, false)
    return ViewHolder(layout)
}

override fun onBindViewHolder(holder: ViewHolder?, position: Int) {
    val layout = holder?.itemView
    val user = users[position]

    val tvFirstName = layout?.findViewById<TextView>(R.id.tv_first_name)
    val tvSurname = layout?.findViewById<TextView>(R.id.tv_surname)
    val tvPhoneNumber = layout?.findViewById<TextView>
                    (R.id.tv_phone_number)

    tvFirstName?.text = "First name: ${user.firstName}"
    tvSurname?.text = "Surname: ${user.surname}"
    tvPhoneNumber?.text = "Phone number: ${user.phoneNumber}"
}

//Populates users ArrayList with User objects
    private fun populateUsers() {
        users.clear()
```

下面获取数据库用户表中的所有用户。当成功获取该列表后，可将列表中的全部用户对象添加至用户 ArrayList 中，如下所示：

```kotlin
        userDao.all()
                .subscribeOn(Schedulers.io())
                .observeOn(AndroidSchedulers.mainThread())
                .subscribe({ res ->
            users.addAll(res)
            notifyDataSetChanged()
        }, Throwable::printStackTrace)
    }

    override fun getItemCount(): Int {
        return users.size
    }

    class ViewHolder(itemView: View) : RecyclerView.ViewHolder(itemView)
}
```

ListUsersFragment 中的 UsersAdapter 使用了 UserDao 实例设置其用户列表，这一操作在 populateUsers() 中完成。当调用 populateUsers() 时，应用程序保存的用户列表通过调用 userDao.all() 被检索。当成功检索到全部用户后，所有的 User 对象将添加至 UserAdapter 的用户 ArrayList 中去。随后，通过调用 notifyDataSetChanged()，适配器将被通知其数据集中的数据发生变化。

MainActivity 自身也需要稍作修改。完整的 MainActivity 类定义如下所示：

```kotlin
package com.example.roomexample.ui

import android.app.Fragment
import android.support.v7.app.AppCompatActivity
import android.os.Bundle
import android.support.v7.widget.LinearLayoutManager
import android.support.v7.widget.RecyclerView
import android.text.TextUtils
import android.view.LayoutInflater
import android.view.MenuItem
import android.view.View
import android.view.ViewGroup
import android.widget.*
import com.example.roomexample.R
import com.example.roomexample.data.AppDatabase
import com.example.roomexample.data.User
import com.example.roomexample.data.UserDao
import io.reactivex.Observable
import io.reactivex.android.schedulers.AndroidSchedulers
import io.reactivex.schedulers.Schedulers

class MainActivity : AppCompatActivity() {

  override fun onCreate(savedInstanceState: Bundle?) {
    super.onCreate(savedInstanceState)
    setContentView(R.layout.activity_main)
    navigateToForm()
  }

  private fun showHomeButton() {
    supportActionBar?.setDisplayHomeAsUpEnabled(true)
  }

  private fun hideHomeButton() {
```

```
    supportActionBar?.setDisplayHomeAsUpEnabled(false)
}

private fun navigateToForm() {
    val transaction = fragmentManager.beginTransaction()
    transaction.add(R.id.ll_container, CreateUserFragment())
    transaction.commit()
}
```

当用户单击返回按钮时，将调用下列函数。如果片段回退栈包含一个或多个片段，片段过滤器将弹出片段，并向用户予以显示，如下所示：

```
override fun onBackPressed() {
    if (fragmentManager.backStackEntryCount > 0) {
        fragmentManager.popBackStack()
        hideHomeButton()
    } else {
        super.onBackPressed()
    }
}

private fun navigateToList() {
    val transaction = fragmentManager.beginTransaction()
    transaction.replace(R.id.ll_container, ListUsersFragment())
    transaction.addToBackStack(null)
    transaction.commit()
}

override fun onOptionsItemSelected(item: MenuItem?): Boolean {
    val id = item?.itemId

    if (id == android.R.id.home) {
        onBackPressed()
        hideHomeButton()
    }

    return super.onOptionsItemSelected(item)
}

class CreateUserFragment : Fragment(), MainView, View.OnClickListener {
    ...
}
```

```
class ListUsersFragment : Fragment(), MainView {
    ...
    }
}
```

当前,需要运行该应用程序,以查看是否按照期望的方式运行。在所选取的设备上构建并运行该项目。当项目启动时,即会看到用户表单,读者可在其中输入用户信息,如图 7.4 所示。

在向用户表单中输入有效信息后,单击 SUBMIT 按钮,并将该用户保存至应用程序的 SQLite 数据中。当保存成功后,用户将会看到一条提示信息。随后,单击 VIEW USERS 按钮可查看刚刚保存的用户信息,如图 7.5 所示。

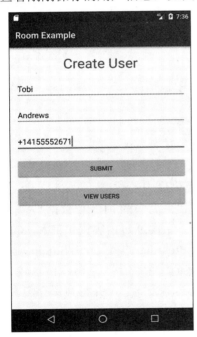

图 7.4　用户表单

图 7.5　查看刚刚保存的用户信息

读者可查看多个用户的信息,就数据库的容量而言,此类信息数量并不存在上限。

7.4　与内容提供商协同工作

第 2 章曾简要介绍了内容提供商方面的内容,并帮助应用程序控制访问数据(此类数据存储于当前应用程序中或者是另一个 App 中）。除此之外，我们还通过应用程序编程接口使得内容提供商实现了与另一个应用程序之间的数据共享。

内容提供商的行为类似于数据库，支持内容的插入、删除、编辑、更新以及查询操作，对应方法包括 insert()、update()、delete()以及 query()。多数时候，由内容提供商控制的数据存在于 SQLite 数据库中。

基于应用程序的内容提供商可通过以下 5 个步骤创建：

（1）定义扩展了 ContentProvider 的内容提供商类。
（2）定义内容 URI 地址。
（3）创建与内容提供商交互的数据源，该数据源往往以 SQLite 数据库这一形式出现。当 SQLite 作为数据源时，需要定义 SQLiteOpenHelper，并覆写其 onCreate()方法，进而构建内容提供商可控制的数据库。
（4）实现所需的内容提供商方法。
（5）在项目的清单文件中注册内容提供商。

综上所述，内容提供商需要实现 6 个方法，其中包括：

- onCreate()：该方法调用后将初始化数据库。
- query()：该方法通过 Cursor 向调用方返回数据。
- insert()：该方法被调用后，将向内容提供商插入新的数据。
- delete()：该方法被调用后将从内容提供商中删除数据。
- update()：该方法被调用后，将更新内容提供商中的数据。
- getType()：该方法被调用后，将返回内容提供商中的 MIME 数据类型。

为了确保读者完全理解内容提供商的工作机制，下面通过内容提供商和 SQLite 数据库创建一个示例项目。对此，创建一个名为 ContentProvider 的 Android Studio 项目，并将空 MainActivity 加入其中。类似于本章所构建的其他应用程序，当前示例也较为简单：用户可在文本框中输入产品信息（包括产品名称及其制造商），并将其保存至 SQLite 数据库中。随后，用户可查看保存后的产品信息。

修改 activity_main.xml 文件，并使其包含以下 XML 内容。

```xml
<?xml version="1.0" encoding="utf-8"?>
<LinearLayout xmlns:android="http://schemas.android.com/apk/res/android"
 xmlns:tools="http://schemas.android.com/tools"
 android:layout_width="match_parent"
 android:layout_height="match_parent"
 android:orientation="vertical"
 android:gravity="center_horizontal"
 android:padding="16dp"
```

```xml
    tools:context="com.example.contentproviderexample.MainActivity">
    <TextView
        android:layout_width="wrap_content"
        android:layout_height="wrap_content"
        android:gravity="center"
        android:text="@string/content_provider_example"
        android:textColor="@color/colorAccent"
        android:textSize="32sp"/>
    <EditText
        android:id="@+id/et_product_name"
        android:layout_width="match_parent"
        android:layout_height="wrap_content"
        android:layout_marginTop="16dp"
        android:hint="Product Name"/>
    <EditText
        android:id="@+id/et_product_manufacturer"
        android:layout_width="match_parent"
        android:layout_height="wrap_content"
        android:layout_marginTop="16dp"
        android:hint="Product Manufacturer"/>
    <Button
        android:id="@+id/btn_add_product"
        android:layout_width="match_parent"
        android:layout_height="wrap_content"
        android:layout_marginTop="16dp"
        android:text="Add product"/>
    <Button
        android:id="@+id/btn_show_products"
        android:layout_width="match_parent"
        android:layout_height="wrap_content"
        android:layout_marginTop="16dp"
        android:text="Show products"/>
</LinearLayout>
```

在经过上述调整后,可将下列字符串资源添加至项目的 strings.xml 文件中。

```xml
<string name="content_provider_example">Content Provider Example</string>
```

下面在 com.example.contentproviderexample 包中创建文件,并添加下列代码:

```
package com.example.contentproviderexample

import android.content.*
import android.database.Cursor
```

```kotlin
import android.database.SQLException
import android.database.sqlite.SQLiteDatabase
import android.database.sqlite.SQLiteOpenHelper
import android.database.sqlite.SQLiteQueryBuilder
import android.net.Uri
import android.text.TextUtils

internal class ProductProvider : ContentProvider() {

  companion object {

    val PROVIDER_NAME: String = "com.example.contentproviderexample
                                .ProductProvider"
    val URL: String = "content://$PROVIDER_NAME/products"
    val CONTENT_URI: Uri = Uri.parse(URL)

    val PRODUCTS = 1
    val PRODUCT_ID = 2

    // Database and table property declarations
    val DATABASE_VERSION = 1
    val DATABASE_NAME = "Depot"
    val PRODUCTS_TABLE_NAME = "products"

    // 'products' table column name declarations
    val ID: String = "id"
    val NAME: String = "name"
    val MANUFACTURER: String = "manufacturer"
    val uriMatcher: UriMatcher = UriMatcher(UriMatcher.NO_MATCH)
    val PRODUCTS_PROJECTION_MAP: HashMap<String, String> = HashMap()
```

SQLiteOpenHelper 类负责创建内容提供商的数据库，如下所示：

```kotlin
    private class DatabaseHelper(context: Context) :
      SQLiteOpenHelper(context, DATABASE_NAME, null,
DATABASE_VERSION) {

      override fun onCreate(db: SQLiteDatabase) {
        val query = " CREATE TABLE " + PRODUCTS_TABLE_NAME +
                    " (id INTEGER PRIMARY KEY AUTOINCREMENT, " +
                    " name VARCHAR(255) NOT NULL, " +
                    " manufacturer VARCHAR(255) NOT NULL);"
```

```
    db.execSQL(query)
  }

  override fun onUpgrade(db: SQLiteDatabase, oldVersion: Int,
                         newVersion: Int) {
    val query = "DROP TABLE IF EXISTS $PRODUCTS_TABLE_NAME"

    db.execSQL(query)
    onCreate(db)
  }
 }
}
private lateinit var db: SQLiteDatabase

override fun onCreate(): Boolean {
  uriMatcher.addURI(PROVIDER_NAME, "products", PRODUCTS)
  uriMatcher.addURI(PROVIDER_NAME, "products/#", PRODUCT_ID)

  val helper = DatabaseHelper(context)
```

下面利用 SQLiteOpenHelper 获取可写入的数据库。如果数据库尚未创建，下列代码将创建一个新的数据库。

```
    db = helper.writableDatabase

    return true
}

override fun insert(uri: Uri, values: ContentValues): Uri {
  //Insert a new product record into the products table
  val rowId = db.insert(PRODUCTS_TABLE_NAME, "", values)

  //If rowId is greater than 0 then the product record was added
  //successfully.
  if (rowId > 0) {
    val _uri = ContentUris.withAppendedId(CONTENT_URI, rowId)
    context.contentResolver.notifyChange(_uri, null)

    return _uri
  }

  // throws an exception if the product was not successfully added.
```

```kotlin
    throw SQLException("Failed to add product into " + uri)
}

override fun query(uri: Uri, projection: Array<String>?,
                   selection: String?, selectionArgs: Array<String>?,
                   sortOrder: String): Cursor {

    val queryBuilder = SQLiteQueryBuilder()
    queryBuilder.tables = PRODUCTS_TABLE_NAME

    when (uriMatcher.match(uri)) {
      PRODUCTS ->
      queryBuilder.setProjectionMap(PRODUCTS_PROJECTION_MAP)
        PRODUCT_ID -> queryBuilder.appendWhere(
       "$ID = ${uri.pathSegments[1]}"
      )
    }

    val cursor: Cursor = queryBuilder.query(db, projection, selection,
                    selectionArgs, null, null, sortOrder)

      cursor.setNotificationUri(context.contentResolver, uri)
      return cursor
}

override fun delete(uri: Uri, selection: String,
                    selectionArgs: Array<String>): Int {

  val count = when(uriMatcher.match(uri)) {

    PRODUCTS -> db.delete(PRODUCTS_TABLE_NAME, selection, selectionArgs)
    PRODUCT_ID -> {
      val id = uri.pathSegments[1]
      db.delete(PRODUCTS_TABLE_NAME, "$ID = $id " +
        if (!TextUtils.isEmpty(selection)) "AND
          ($selection)" else "", selectionArgs)
     }
    else -> throw IllegalArgumentException("Unknown URI: $uri")
  }

  context.contentResolver.notifyChange(uri, null)
  return count
```

```kotlin
}
override fun update(uri: Uri, values: ContentValues, selection: String,
                    selectionArgs: Array<String>): Int {

    val count = when(uriMatcher.match(uri)) {
        PRODUCTS -> db.update(PRODUCTS_TABLE_NAME, values,
                              selection, selectionArgs)
        PRODUCT_ID -> {
            db.update(PRODUCTS_TABLE_NAME, values,
                  "$ID = ${uri.pathSegments[1]} " +
                  if (!TextUtils.isEmpty(selection)) " AND
                  ($selection)" else "", selectionArgs)
        }
        else -> throw IllegalArgumentException("Unknown URI: $uri")
    }

    context.contentResolver.notifyChange(uri, null)
    return count
}

override fun getType(uri: Uri): String {
    //Returns the appropriate MIME type of records

    return when (uriMatcher.match(uri)){
        PRODUCTS -> "vnd.android.cursor.dir/vnd.example.products"
        PRODUCT_ID -> "vnd.android.cursor.item/vnd.example.products"
        else -> throw IllegalArgumentException("Unpermitted URI: " + uri)
    }
}
}
```

再加入相应的 ProductProvider，并提供与保存后的产品相关的内容后，还需要在 AndroidManifest 文件中注册新的组件。下列代码片段向清单文件中添加了当前提供商。

```xml
<?xml version="1.0" encoding="utf-8"?>
<manifest xmlns:android="http://schemas.android.com/apk/res/android"
    package="com.example.contentproviderexample">

<application
      android:allowBackup="true"
      android:icon="@mipmap/ic_launcher"
      android:label="@string/app_name"
```

```xml
            android:roundIcon="@mipmap/ic_launcher_round"
            android:supportsRtl="true"
            android:theme="@style/AppTheme">
    <activity android:name=".MainActivity">
      <intent-filter>
        <action android:name="android.intent.action.MAIN" />

        <category android:name="android.intent.category.LAUNCHER" />
      </intent-filter>
    </activity>
    <provider android:authorities="com.example.contentproviderexample
                    .ProductProvider" android:name="ProductProvider"/>
  </application>

</manifest>
```

下面修改 MainActivity 并使用新注册的提供商。调整 MainActivity.kt 文件，并包含下列代码：

```kotlin
package com.example.contentproviderexample

import android.content.ContentValues
import android.net.Uri
import android.support.v7.app.AppCompatActivity
import android.os.Bundle
import android.text.TextUtils
import android.view.View
import android.widget.Button
import android.widget.EditText
import android.widget.Toast

class MainActivity : AppCompatActivity(), View.OnClickListener {

  private lateinit var etProductName: EditText
  private lateinit var etProductManufacturer: EditText
  private lateinit var btnAddProduct: Button
  private lateinit var btnShowProduct: Button

  override fun onCreate(savedInstanceState: Bundle?) {
    super.onCreate(savedInstanceState)
    setContentView(R.layout.activity_main)
    bindViews()
    setupInstances()
```

```kotlin
}

private fun bindViews() {
    etProductName = findViewById(R.id.et_product_name)
    etProductManufacturer = findViewById(R.id.et_product_manufacturer)
    btnAddProduct = findViewById(R.id.btn_add_product)
    btnShowProduct = findViewById(R.id.btn_show_products)
}

private fun setupInstances() {
    btnAddProduct.setOnClickListener(this)
    btnShowProduct.setOnClickListener(this)
    supportActionBar?.hide()
}

private fun inputsValid(): Boolean {
    var inputsValid = true
    if (TextUtils.isEmpty(etProductName.text)) {
        etProductName.error = "Field required."
        etProductName.requestFocus()
        inputsValid = false

    } else if (TextUtils.isEmpty(etProductManufacturer.text)) {
        etProductManufacturer.error = "Field required."
        etProductManufacturer.requestFocus()
        inputsValid = false
    }

    return inputsValid
}

private fun addProduct() {
    val contentValues = ContentValues()

    contentValues.put(ProductProvider.NAME, etProductName.text.toString())
    contentValues.put(ProductProvider.MANUFACTURER,
                etProductManufacturer.text.toString())
    contentResolver.insert(ProductProvider.CONTENT_URI, contentValues)

    showSaveSuccess()
}
```

当调用下列函数时，将显示数据库中的对应产品。

```kotlin
private fun showProducts() {
  val uri = Uri.parse(ProductProvider.URL)
  val cursor = managedQuery(uri, null, null, null, "name")

  if (cursor != null) {
    if (cursor.moveToFirst()) {

      do {
        val res = "ID: ${cursor.getString(cursor.getColumnIndex
                (ProductProvider.ID))}" + ",
          \nPRODUCT NAME: ${cursor.getString(cursor.getColumnIndex
                ( ProductProvider.NAME))}" + ",
          \nPRODUCT MANUFACTURER: ${cursor.getString(cursor.getColumnIndex
                (ProductProvider.MANUFACTURER))}"

          Toast.makeText(this, res, Toast.LENGTH_LONG).show()
        } while (cursor.moveToNext())
      }
    } else {
      Toast.makeText(this, "Oops, something went wrong.",
                Toast.LENGTH_LONG).show()
    }
  }
}

private fun showSaveSuccess() {
  Toast.makeText(this, "Product successfully saved.",
            Toast.LENGTH_LONG).show()
}

override fun onClick(view: View) {
  val id = view.id

  if (id == R.id.btn_add_product) {
    if (inputsValid()) {
      addProduct()
    }
  } else if (id == R.id.btn_show_products) {
    showProducts()
  }
}
```

在上述代码中,应注意 addProduct()和 showProducts()这两个方法。其中,addProduct() 将产品数据存储于 contentValues 实例中,并于随后借助于 ProductProvider 将该信息插入 SQLite 数据库中,即调用 contentResolver.insert(ProductProvider.CONTENT_URI,contentValues) 方法。showProducts()则使用 Cursor 显示于存储于数据库中的产品信息。

下面构建并运行该应用程序。启动程序后,用户将被转至 MainActivity,并显示一个表单,等待用户输入产品名称及其制造商,如图 7.6 所示。

当输入有效的产品信息后,单击 ADD PRODUCT 按钮。该产品将作为一条新的记录插入应用程序 SQLite 数据库的产品表中。此外,还可在当前表单中添加新的产品项,并单击 SHOW PRODUCTS 按钮,如图 7.7 所示。

 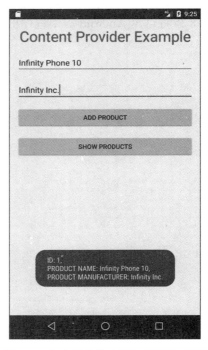

图 7.6 用户输入产品名称及其制造商　　　　图 7.7 在当前表单中添加新的产品项

上述操作将调用 MainActivity 中的 showProducts()。随后,全部产品记录将被检索并逐一显示。

本章讨论了内容提供商的工作机制,读者可尝试添加、更新和删除产品记录等功能,进而完善该应用程序。对于读者来说,这也是一次极好的实践机会。

7.5 本章小结

本章介绍了 Android 应用程序框架所支持的各种数据存储方式,包括内部存储和外部存储,进而实现私有和公共文件中的数据存储。除此之外,我们还学习了如何借助于内部存储和外部存储与缓存文件协同工作。

同时,本章还深入讨论了 SQLite RDBMS 及其在 Android 应用程序中的应用方式。例如,利用 Room 检索和存储 SQLite 数据库中的数据。接下来还探讨了如何通过 SQLite 数据库(作为底层数据存储)并使用内容提供商控制数据的访问。

第 8 章将讨论 Android 应用程序框架中 Android 应用程序安全和部署方面的问题。

第 8 章　Android App 的安全和部署

前述章节讨论了 Kotlin 语言在移动 App 领域中的应用，以及 Android 应用程序框架中所支持的各类存储媒介，包括内部存储、外部存储、网络存储以及 SQLite，并构建了大量的应用示例。其中涉及基于 Room 和内容提供商的 SQLite 数据库中的数据存储和检索方法。

本章主要涉及以下两项较为重要的内容：
- Android 应用程序安全。
- Android 应用程序部署。

8.1　Android 应用程序安全

在软件的构建过程中，安全一直是一类重要的问题。除了 Android 操作系统中的安全措施之外，开发人员还应确保应用程序符合既定的安全标准。在本节中，我们将对一些重要的安全性考虑事项和最佳实践进行分析，以供读者深入理解安全性问题。据此，可确保应用程序避免遭受到客户端设备上的恶意程序的攻击。

8.1.1　内部存储

在条件均等的情况下，对于应用程序保存到设备上的数据，隐私性是开发 Android 应用程序时最常见的安全问题。对此，可以遵循一些简单的规则，以使应用程序数据更加安全。

1. 使用内部存储

如前所述，内部存储是在设备上保存私有数据的一种较好方式。每个 Android 应用程序都设置了一个相应的内部存储目录，可以在其中创建和写入私有文件。这些文件是构建应用程序的私有文件，因此客户端设备上的其他应用程序不能访问这些文件。

根据经验，如果应用程序只能访问数据，并且可以合理地将数据存储在内部存储中，那么就秉承这一原则。读者可参考第 7 章以了解如何使用内部存储。

2. 使用外部存储

外部存储文件并不是应用程序的私有文件，因此，可以由同一客户端设备上的其他应用程序轻松地加以访问。因此，在将应用程序数据存储到外部存储之前，应考虑对其进行加密。在将数据保存到外部存储之前，可以使用许多库和包对数据进行加密。例如，对于外部存储数据加密，Facebook 的 Conceal 库（http://facebook.github.io/conceal/）是一个较好的选择方案。

此外，作为另一个经验法则，不要在外部存储中存储敏感数据，以防止外部存储文件被他人操控。同时，还应该对外部存储检索的输入内容进行验证。这种验证是针对外部存储中非受信数据的一种防范措施。

3. 内容提供商

如前所述，内容提供商可阻止或启用应用程序数据的外部访问。当在清单文件中注册内容提供商时，可使用 android:exported 属性确定是否允许对内容提供商进行外部访问。若是，则将 android:exported 设置为 true；否则将其设置为 false。

除此之外，内容提供商中的查询方法还可用于防止 SQL 注入（即攻击者在输入字段中执行恶意 SQL 语句的代码注入技术），例如 query()、update()和 delete()方法。

8.1.2 网络安全

在通过 Android 应用程序执行网络事务时，应该遵循一些最佳实践方案，相关方案可划分为不同的类别。本节将讨论互联网协议（IP）网络和电话网络。

1. IP 网络

当通过 IP 与远程计算机通信时，应确保应用程序尽可能地使用 HTTP（因此，服务器均支持 HTTP）。这样做的一个主要原因是，设备经常会连接到不安全的网络，例如公共无线连接。HTTP 可确保客户端和服务器之间的加密通信，而不考虑所连接的网络。在 Java 中，HttpsURLConnection 可以用于网络上的安全数据传输。值得注意的是，通过不安全的网络连接接收的数据不应予以信任。

2. 电话网络

当需要在服务器和客户端应用程序之间自由传输数据的情况下，应该使用 Firebase 云消息传递（FCM）以及 IP 网络，而不是使用其他方案，例如短消息传递服务（SMS）协议。FCM 是一种多平台的消息传递解决方案，并有助于在应用程序之间无缝、可靠地传输消息。

对于数据消息传输，SMA 并不是一种较好的候选方案，其原因包括：
- 未采用加密操作。
- 未经过强认证。
- 通过 SMS 发送的消息可能受到欺骗。
- SMS 消息可能被拦截。

8.1.3 输入验证

用户的输入验证十分重要，进而可避免可能出现的安全风险，SQL 注入便是其中之一。对此，可通过参数化查询以及在原始 SQL 查询中使用输入过滤来防止恶意的 SQL 脚本注入。

除此之外，源自外部存储的输入还需执行验证操作——外部存储并非是受信数据源。

8.1.4 与用户凭证协同工作

通过减少对应用程序中用户凭证输入的要求，可以降低钓鱼攻击的风险。与其不断地请求用户凭证，不如考虑使用授权令牌。其中，用户不需要在设备上存储用户名和密码。相反，可使用更新的授权令牌。

8.1.5 代码混淆技术

在发布 Android 应用程序之前，必须使用代码混淆工具（如 ProGuard），以防止他人利用各种方法（如反编译）不受限制地访问源代码。ProGuard 是在 Android SDK 中预置打包的工具，因此不需要包含依赖项。如果用户将构建类型指定为发布，那么它将自动包含在构建过程中。关于 ProGuard 的更多内容，读者可访问 https://www.guardsquare.com/en/proguard。

8.1.6 广播接收器的安全性

默认情况下，广播接收器组件将被导出，因此可以在同一设备上被其他应用程序调用。通过安全权限，可控制应用程序对广播接收器的访问。对此，可在应用程序清单文件中利用<receiver>元素设置广播接收器的权限。

8.1.7 动态加载代码

当应用程序需要动态加载代码时，必须确保所加载的代码来自受信任的源代码。除此之外，还必须确保不惜任何代价降低篡改代码的风险。加载和执行被篡改的代码是一种巨大的安全威胁。当从远程服务器加载代码时，应确保它通过安全、加密的网络传输。请记住，动态加载的代码运行时具有与应用程序相同的安全权限（应用程序清单文件中定义的权限）。

8.1.8 服务的安全性

与广播接收器不同，Android 系统默认状态下不导出服务。仅当 Intent 过滤器添加至清单文件中的服务声明时，才会出现默认的服务导出行为。对此，应使用 android:exported 属性确保只在需要服务时才导出服务。当需要导出服务时，可将 android:exported 设置为 true，否则设置为 false。

8.2 启用和发布 Android 应用程序

截至目前，本章前述内容介绍了 Android 中的系统、应用程序、开发、安全特性等内容。下面探讨 Android 应用程序的启动和发布。

读者可能会对此处的术语启用和发布有所疑惑。启用表示为一个活动，并向公众（终端用户）引入新产品；而发布 Android 应用程序仅是简单地令该程序对用户可用。相应地，需要执行各种活动和处理过程以确保成功地启用 Android 应用程序。此处共计 15 种活动，如下所示：

- ❑ 理解 Android 开发者程序策略。
- ❑ 设置 Android 开发者账户。
- ❑ 本地化规划。
- ❑ 规划同步版本。
- ❑ 根据质量标准进行测试。
- ❑ 构建可发布的 APK。
- ❑ 规划应用程序的 Play Store 列表。
- ❑ 将应用程序包上传至 alpha 或 beta 测试。
- ❑ 设备兼容性定义。
- ❑ 启用前报告评估。
- ❑ 定价和应用程序分发配置。
- ❑ 分发选项的选取。
- ❑ 应用程序内产品和订阅设置。

- ❏ 制定应用程序内容评级。
- ❏ 发布应用程序。

这是一份长长的名单。如果读者尚不理解清单中的所有内容，不要担心，下面让我们详细地查看每一项内容。

8.2.1 理解 Android 开发者程序策略

制定开发者程序策略的唯一宗旨是确保 Play Store 仍然是用户受信的软件源。违反这些政策会带来一定的后果。因此，在启动应用程序之前，仔细阅读并充分理解这些开发人员策略（其目的和结果）是非常重要的。

8.2.2 设置 Android 开发者账号

读者需要设置一个 Android 开发人员账号，进而在 Play Store 上启用应用程序，同时还应确保账号细节的准确性。另外，如果需要在 Android 应用程序上销售产品，还需要设置一个商业账号。

8.2.3 本地化规划

某些时候，出于本地化的目的，用户可能持有多个应用程序副本，每个副本都本地化为不同的语言。在这种情况下，需要尽早地计划本地化操作，并遵循 Android 开发人员推荐的本地化检查表。关于本地化检查表，读者可访问 https://developer.android.com/distribute/best-practices/launch/localization-checklist.html 获取更多内容。

8.2.4 规划同步版本

用户可能希望在多个平台上发布产品，其中包含了诸多优点，如增加产品的潜在市场规模，减少对产品的访问壁垒，以及最大化应用程序的潜在安装数量。同时在多个平台上发布产品通常是一种较好的做法。对此，应确保提前做好计划。如果无法在多个平台上启用应用程序，应确保提供某种方式，以使潜在用户可提交他们的联系方式。一旦产品在平台上被选用，可确保与其进行联系。

8.2.5 根据质量标准进行测试

测试标准提供了一种测试模板,以确认应用程序满足 Android 用户期望的基本功能和非功能需求。在启用之前，应确保参照这些质量标准运行应用程序。

8.2.6　构建可发布的 APK

可发布的 APK 是一个 Android 应用程序，它经过优化打包，然后使用发布密钥进行构建和签名。构建一个可发布的 APK 是 Android 应用程序发布的一个重要步骤，读者应对此予以足够的重视。

8.2.7　规划应用程序的 Play Store 列表

对于 Play Store 清单，这一步骤涉及产品资源列表。这些资源包括但不限于应用程序的日志、屏幕截图、描述、推广图形和视频（若存在）。另外，确保将应用程序的隐私策略与应用程序的 Play Store 列表连接在一起。同时，将应用程序的产品列表本地化为应用程序支持的所有语言也很重要。

8.2.8　将应用程序包上传至 alpha 或 beta 测试

测试是检测软件缺陷和提高软件质量的一种有效方法，一种较好的做法是将应用程序包上传到 alpha 和 beta 测试，以便针对产品进行 alpha 和 beta 软件测试。另外，alpha 测试和 beta 测试都是验收型测试。

8.2.9　设备兼容性定义

这一步骤涉及应用程序开发过程中 Android 版本和屏幕尺寸的声明。在这一步骤中，应尽可能准确地定义 Android 版本和屏幕尺寸，这一点非常重要；否则将导致用户体验方面的问题。

8.2.10　启用前报告评估

启用前报告评估用于识别 Android 设备上的应用程序在自动测试后所产生的问题。在将一个应用程序包上传到 alpha 或 beta 测试时，如果选择该项功能，用户将会得到启用前的报告评估结果。

8.2.11　定价和应用程序分发配置

首先，需要确定应用程序的定价方法。此后，可将应用程序设置为免费安装或付费下载。在应用程序价格制定完毕后，可选择应用程序分发的国家。

8.2.12 分发选项的选取

这一步涉及应用程序发布时设备和平台的选择——例如，Android TV 和 Android Wear。在此之后，谷歌 Play 团队将对应用程序进行审查。若审核通过，Google Play 将会使该应用程序排位更加靠前。

8.2.13 应用程序内产品和订阅设置

如果希望在应用程序内销售产品，则需要设置应用程序中的产品和订阅功能。其中，可指定所销售的国家，并处理与货币相关的各种问题，例如税金问题。在该步骤中，用户还将建立个人商业账户。

8.2.14 制定应用程序内容评级

读者需要针对 Play Store 发布的应用程序提供准确的评级。这一步骤是由 Android 开发者程序策略所规定的，以帮助目标群体更容易地发现读者的应用程序。

8.2.15 发布应用程序

当完成了上述各步骤后，即可向 Play Store 发布应用程序。首先，读者需要发布一个软件版本，并可上传应用程序的 APK 文件，同时跟踪该应用程序的定位。在发布过程的结束阶段，读者可单击 Confirm rollout 按钮发布应用程序。

至此，读者已经了解了 Play Store 中应用程序的发布过程。多数时候，我们并不需要执行全部操作，仅需关注与所发布的应用程序相关的类型即可。下面尝试发布第 7 章中开发的一款应用程序，即 Messenger 应用程序。如果读者愿意的话，也可发布前述章节中的任意一款应用程序。

在将 Messenger 应用程序发布至 Play Store 时，首先需要注册一个 Google Play 开发人员账号。对此，可打开相应的浏览器，并访问 https://play.google.com/apps/publish/signup。

在开启相关网页后，读者将被提示输入 Google 账号，随后可接受开发人员程序协议，如图 8.1 所示。

读者可移至页面下方，并接受 Google Play 开发人员协议，随后选中 I agree and I am willing to associate my account registration with the Google Play Developer distribution agreement 复选框，如图 8.2 所示。

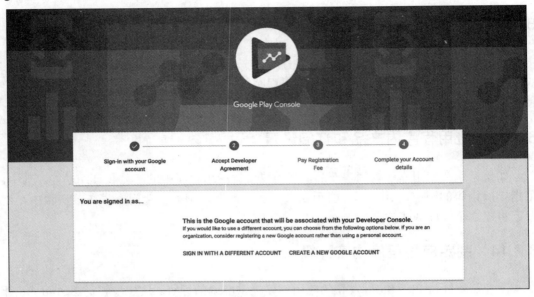

图 8.1　接受开发人员程序协议

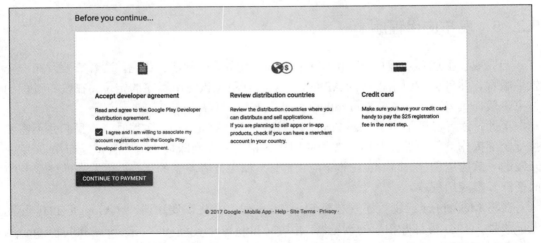

图 8.2　选中 I agree and I am willing to associate my account registration with the Google Play Developer distribution agreement 复选框

在接受了协议后，单击 CONTINUE TO PAYMENT 按钮，并创建开发人员账号。相应地，用户需要一次性地交付 25 美元的注册费用。待付费成功后，将提示如图 8.3 所示的信息。

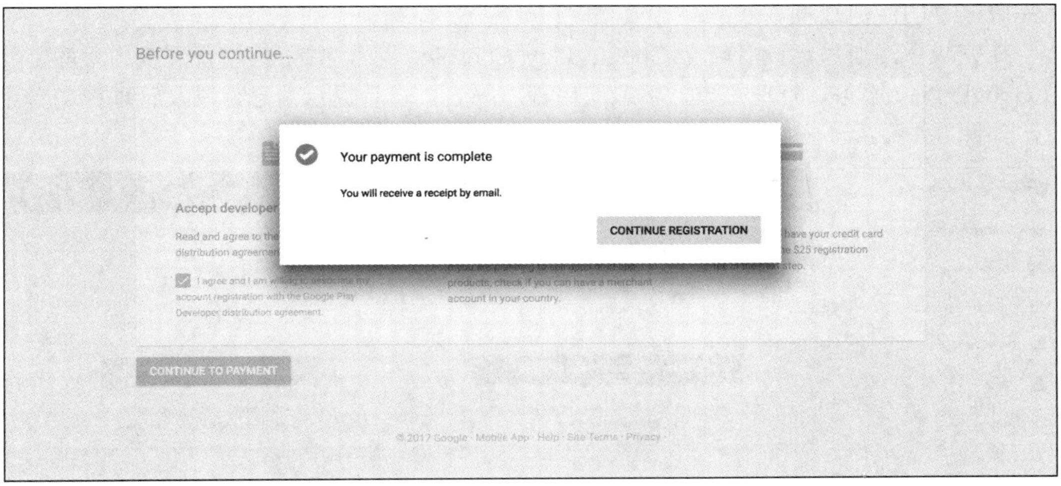

图 8.3　付费成功信息

接下来，单击 CONTINUE REGISTRATION 按钮将来到注册过程中的最后一步。其中，用户需要完善一些个人信息，如图 8.4 所示。

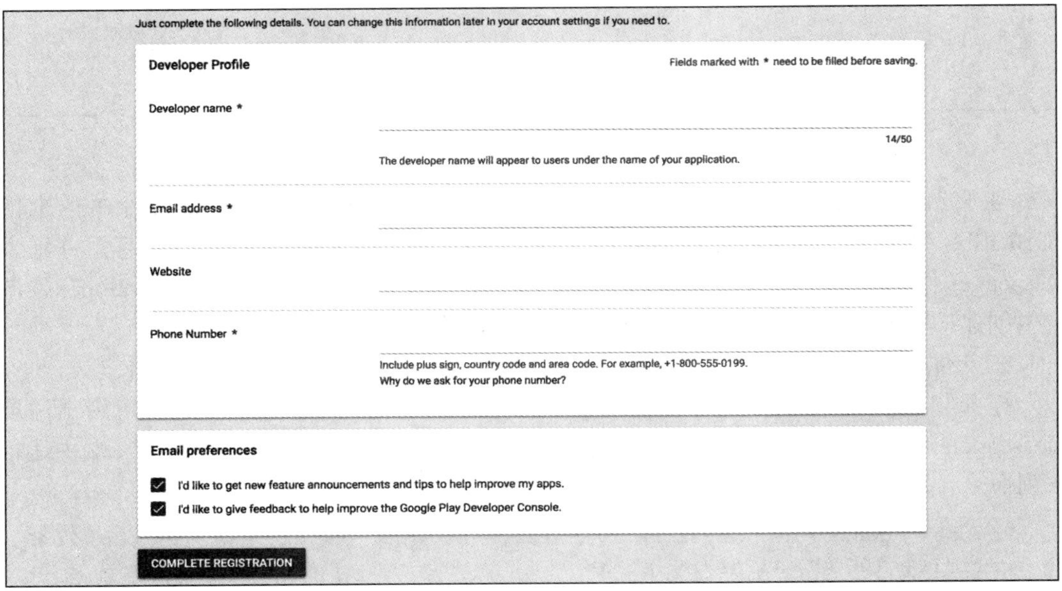

图 8.4　完善个人信息

读者需要输入账号所需的一些细节信息，随后单击 COMPLETE REGISTRATION 按钮完成账号的注册过程。

在完成了注册后，读者转至 Google Play Console 页面，并于其中管理应用程序、使用 Google Play 服务、管理账单、下载应用程序报告、查看提示消息，以及管理控制台设置，如图 8.5 所示。

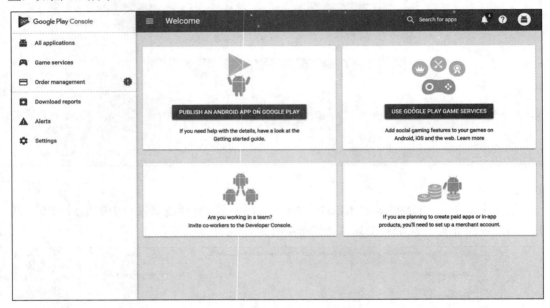

图 8.5　Google Play Console 页面

鉴于当前仅关注 Android 应用程序的发布流程，因而可单击控制台面板上的 PUBLISH AN ANDROID APP ON GOOGLE PLAY 按钮，选取默认的语言，并输入 Messenger 作为应用程序的标题，随后单击 CREATE 按钮。接下来，将在 Developer 控制台中创建如图 8.6 所示的应用程序。

在发布应用程序之前，还需要针对 Messenger 应用程序签署所发布的 APK。

打开 Android Studio 中的 Messenger 应用程序项目，Android Studio 可用于注册 Messenger App，虽然这并非是唯一的签署方法。首先，可在 Android Studio 终端中运行下列命令，以生成签署私钥。

```
keytool -genkey -v -keystore my-release-key.jks -keyalg RSA -keysize 2048 -validity 10000 -alias my-alias
```

执行上述命令将提示用户输入 keystore 密码，并提供与密钥相关的一些附加信息。

随后，keystore 将生成为一个 my-release-key.jks 文件，同时并存于当前目录中，keystone 中所保留的密钥其有效期为 10000 天。

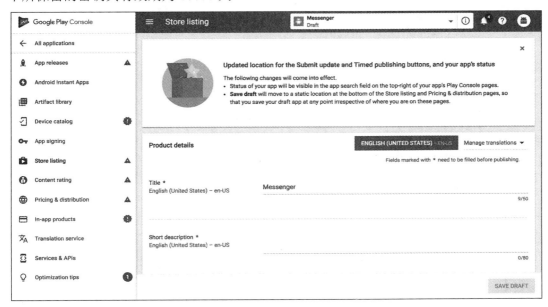

图 8.6　创建新的应用程序

在生成了私钥后，接下来将配置 Gradle 并签署 APK。打开 build.gradle 模块文件，并在 android {}代码块中添加 signingConfigs {}代码块，其中包含了 storeFile、storePassword、keyAlias 和 keyPassword 项。在操作完成后，可将该对象传递至应用程序发布类型的 signingConfig 属性中。考察下列示例：

```
android {
  compileSdkVersion 26
  buildToolsVersion "26.0.2"
  defaultConfig {
      applicationId "com.example.messenger"
      minSdkVersion 16
      targetSdkVersion 26
      versionCode 1
      versionName "1.0"
      testInstrumentationRunner
"android.support.test.runner.AndroidJUnitRunner"
      vectorDrawables.useSupportLibrary = true
  }
```

```
signingConfigs {
  release {
    storeFile file("../my-release-key.jks")
    storePassword "password"
    keyAlias "my-alias"
    keyPassword "password"
  }
}
buildTypes {
  release {
    minifyEnabled false
    proguardFiles getDefaultProguardFile('proguard-android.txt'),
'proguard-rules.pro'
    signingConfig signingConfigs.release
  }
}
}
```

在执行了上述操作后,即可签署 APK。在此之前,还需要调整包名。这里,已被 Google 所限用,因而在将应用程序发布至 Play Store 之前应对包名进行修改。在 Android Studio 中修改应用程序根包名十分简单。首先,确保已经设置了 Android Studio 并可显示目录结构。对此,可单击 IDE 窗口左上方的下拉菜单,并设置 Project,如图 8.7 所示。

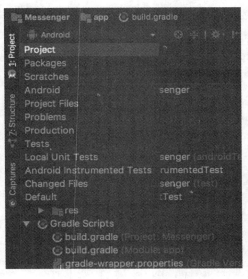

图 8.7　显示目录结构

完成上述操作后，通过取消选择项目结构设置菜单中的 Hide Empty Middle Packages，可显示项目结构视图中的全部空中间包，如图 8.8 所示。

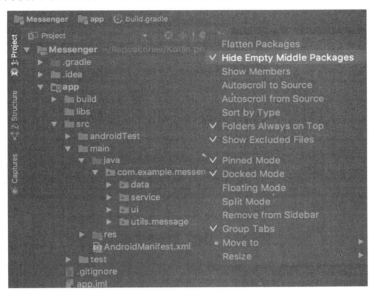

图 8.8　显示项目结构视图中的全部空中间包

在取消选择上述选项后，空中间包将不再处于隐藏状态，因此，com.example.messenger 分为 3 个可见包，即 com、example 和 messenger。下面将 example 重命名为其他名称。对此，可将 example 修改为姓和名的组合。例如，对于 Kevin Fakande 这一姓名，包名将从 example 重命名为 kevinfakande。在包上单击鼠标右键，在弹出的快捷菜单中选择 Refactor | Rename 命令，即可对其执行重命名操作。在包名修改完毕后，检查清单文件和 build.gradle 文件，以体现项目包名的变化结果。因此，若在 build.gradle 或清单文件中看到 com.example.messenger 字符串，可将其修改为 com.{full_name}.messenger。

在经过上述修改后，即可签署对应的应用程序。在 Android Studio 中输入下列命令：

```
./gradlew assembleRelease
```

运行上述命令将生成发布版本的 APK，并利用<project_name>/<module_name>/build/outputs/apk/release 路径中的私钥进行签署。鉴于当前项目中的模块命名为 app，因而 APK 将命名为 app-release.apk。基于私钥签署的 APK 可供发布使用。在 APK 签署完毕后，下面将完成 Messenger 应用程序的发布过程。

8.2.16 发布 Android 应用程序

当 Messenger 应用程序签署完毕后，可以继续完成所需的应用程序细节，以实现发布应用程序这一目标。首先，需要针对当前应用程序创建相应的存储列表。相应地，打开 Google Play Console 中的 Messenger 应用程序，并移至存储列表页（可选取导航栏上的 Store Listing）。

在执行后续操作之前，此处需要填写列表中的全部信息，相关信息包括标题、简短描述、完整描述、图像数据资源以及分类信息（包括应用程序类型、类别和内容评级、联系方式以及隐私策略）。图 8.9 显示了 Google Play Console 存储列表页面。

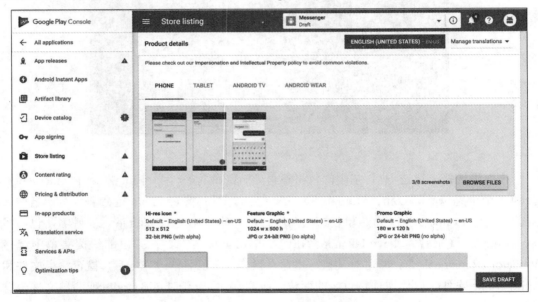

图 8.9　Google Play Console 存储列表页面

当存储列表信息填写完毕后，接下来需要完善价格和分发信息。对此，可选取左侧导航栏的 Pricing & distribution 选项，并打开偏好设置项。出于显示目的，此处可将当前 App 的价格设置为 FREE。除此之外，还需要选取 5 个随机国家以分发该 App，即 Nigeria、India、the United States of America、the United Kingdom 和 Australia，如图 8.10 所示。

除了选取价格类型和产品分发的国家之外，还需要提供额外的偏好设置信息，包括设备类别信息、用户程序信息以及知情信息。

下面向 Google Play Console 应用程序中添加签署后的 APK。对此，可访问 App

releases | MANAGE BETA | EDIT RELEASE。在对应的显示页面中,读者将被提示是否选择 Google Play 应用程序签署操作,如图 8.11 所示。

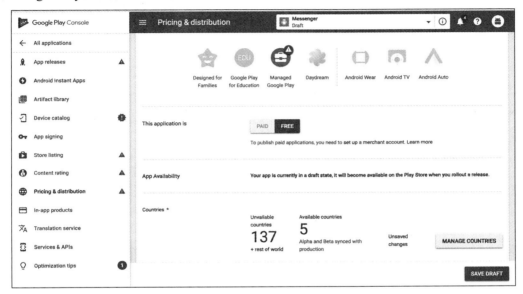

图 8.10　App 的价格和分发操作

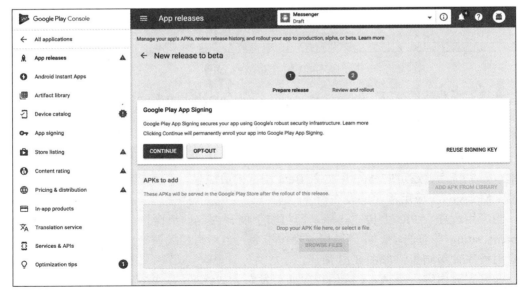

图 8.11　是否选择 Google Play 应用程序签署操作

出于演示目的，此处可选取 OPT-OUT。当选择 OPT-OUT 之后，即可选取从计算机文件系统中上传的 APK 文件。单击 BROWSE FILES 按钮，选择所上传的 APK，如图 8.12 所示。

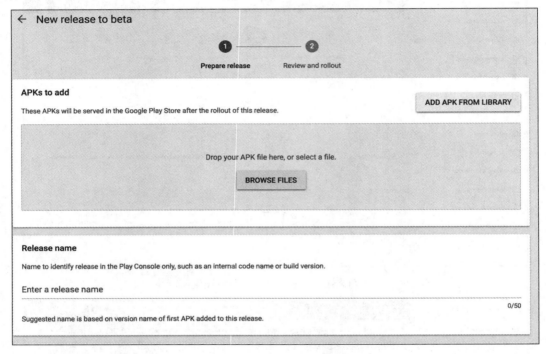

图 8.12　上传 APK 文件

在选取了相应的 APK 文件后，可将其上传至 Google Play Console 中。待上传完毕后，Google Play Console 将针对 Beat 版本自动添加所建议的发布名称，该名称基于上传的 APK 版本名。当然，读者也可自行更改版本名称。随后，可在文本框中填写相应的版本注释。在相关数据得以完善后，可单击页面下方的 Review 按钮，保存并执行后续操作。在 Beta 版本审查完毕后，还可向 App 中加入 Beta 测试人员信息。下面返回至我们所讨论的重点内容：Messenger 应用程序的发布。

当应用程序 APK 上传完毕后，即可着手展开内容评级操作。对此，单击侧栏上的 Content rating 导航选项，并遵循相关操作指令。当填写相关调查内容后，即可生成相应的应用程序评级结果，如图 8.13 所示。

一旦完成内容评级调查，应用程序即可发布为最终产品，并对 Google Play Store 上的所有用户可见。在 Google Play Console 上，可访问 App releases | Manage Production | Create releases 予以查看。当提示生成 APK 时，单击屏幕右侧的 ADD APK FROM LIBRARY 按钮，并选取之前上传的 APK（包含版本号 1.0 的 APK 文件），并完成必要的发布信息。单击 REVIEW 按钮完成后续操作，此时可以看到如图 8.14 所示的简要的发布摘要。

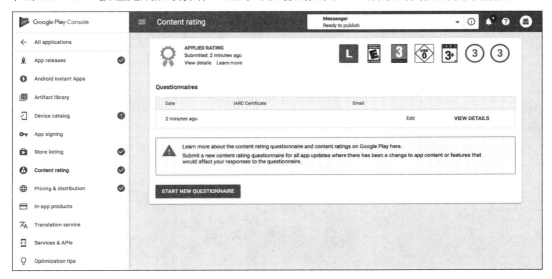

图 8.13　内容评级

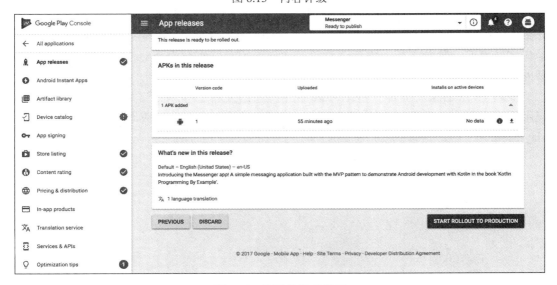

图 8.14　简要的发布摘要

读者应仔细阅读其中的信息。随后即进入产品阶段。此时，读者将被提示：当前 App 将发布 Play Store 中，以供广大用户所用，如图 8.15 所示。

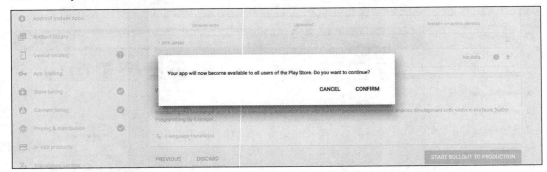

图 8.15　发布至 Play Store

单击 CONFIRM 按钮后，Messenger 应用程序即发布至 Google Play Store 中。

8.3　本章小结

本章讨论了 Android 应用程序框架，其中涉及了 Android 应用程序的安全性以及 Google Play Store 的发布流程。其中阐述了 Android 应用程序所面临的安全威胁，并详细介绍了缓解此类威胁的相关方法，还指出了 Android 生态系统开发应用程序时需要遵循的最佳实践案例。除此之外，本章还学习了如何使用存储媒介安全地执行网络操作过程。另外，我们还学习了如何保护 Android 组件，如服务和广播接收器。

最后，本章还深入研究了 Play Store 应用程序的发布过程。除了介绍成功发布 Android 应用程序所需的各项步骤之外，我们还进一步将之前的 Messenger 应用程序发布到 Google Play Store 中。

第 9 章将继续介绍 KotlinWeb 应用程序开发，并着手开发一个定位查看应用程序。

第 9 章 创建 Place Reviewer 后台应用程序

本书前 4 章讨论了 Android 平台上应用程序开发中 Kotlin 语言的应用。第 8 章则详细介绍了 Android 应用程序安全和部署方面的问题。另外，当与数据存储协同工作以及实现网络通信时，还展示了与安全相关的最佳实践方案。当处理用户输入和用户证书时，我们还考察了相关的安全措施。

进一步讲，第 8 章 Android 应用程序组件安全方面的各种方案，例如服务和广播接收器。最后，还逐步讲解了如何向 Google Play Store 中部署应用程序。本章将深入讨论基于 Web 的 Kotlin 解决方案，特别是如何使用 Spring 开发一款 Place Reviewer 应用程序，并主要集中于前端开发。本章主要涉及以下内容：

- 模型-视图-控制器设计模式。
- Logstash 及其在数据集中化、数据转换以及数据存储方面的应用。
- 基于 Spring Security 的网站安全问题。

9.1 MVC 设计模式

MVC 模式也称作模型-视图-控制器模式，主要用于应用程序中内容的分离。特别地，这种用户接口设计模式将应用程序分为 3 个独立的组件，将应用程序模块划分为多个独立部分包含多种原因。其中之一是将显示逻辑从核心业务逻辑中分离出来。下面考察 MVC 模式中这 3 类应用组件。

9.1.1 模型

模型主要负责数据管理以及 MVC 应用逻辑。由于模型视为全部数据和业务逻辑的管理者，因而可认为是 MVC 应用程序的动力源泉。

9.1.2 视图

视图表示为数据的视觉表达，并通过应用程序生成；同时，这也是用户与应用程序的主要交互点。

9.1.3 控制器

控制器是视图和模型之间的中介体，主要负责从视图中获取输入，并将输入数据的适当转换形式传递至模型中。除此之外，必要时，控制器还将通过数据更新视图，如图 9.1 所示。

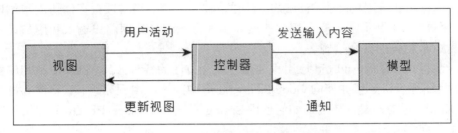

图 9.1 MVC 设计模式

9.2 设计并实现 Place Reviewer 后台程序

本章将对系统市场制定多个用例规范，针对系统数据库实现标识所需的实体，并深入考察系统的开发过程。下面首先讨论 Place Reviewer 系统的用例。

9.2.1 用例标识

本节首先标识系统中的 actor。在此之前，我们需要整体理解 Place Reviewer 应用程序的功能。

一如读者所了解到的，Place Reviewer 应用程序是一种基于互联网的应用程序，并支持平台用户的评论功能。其中，注册用户可借助于地图，并对相关地点发表评论。

在了解了 Place Reviewer 应用程序的功能后，下面将确定 Place Reviewer 应用程序的 actor。相信读者已经猜到，Place Reviewer 应用程序目前仅包含单一 actor，即用户。用户的用例包含以下内容：

- ❑ 用户利用 Place Reviewer 发表评论。
- ❑ 用户通过 Place Reviewer 应用程序查看其他用户的评论。
- ❑ 用户可在交互式地图上查看其他用户评论的地点。
- ❑ 用户可注册 Place Reviewer 平台。

❑ 用户可注销 Place Reviewer 账号。

截至目前，前述内容讨论了 Place Reviewer 系统的功能，标识了系统的 actor，并通过唯一的 actor——用户——标识了系统的用例。下面将标识系统中所需的数据。

9.2.2 标识数据

作为上述用例定义的结果，可通过创建相应的模型，方便地标识 Place Reviewer 应用程序所需的数据类型。其中，第一种数据类型为用户数据；第二种类型则是评论数据。顾名思义，用户数据表示与平台注册用户相关的数据，而评论数据则是针对平台中生成的评论所需的数据。

用户需要使用到以下数据：用户的电子邮件地址、用户名、密码和账号状态。除此之外，还需要针对每个平台用户使用唯一的标识符，即用户 ID，以及用户的注册日期。对于评论对应的数据，需要使用到评论标题、评论内容、评论地址、地名、额外的一些信息（例如经纬度坐标）以及标定地址的地名 ID。除此之外，还需要使用到评论的唯一标识符，以及与评论时间相关的信息。

此处，读者可能会产生疑问：为何需要使用到与地名（具体地址以及经纬度）外加评论相关的信息？为何不分离出此类信息，并将其视为独立的类型？这种思考方式是存在一定道理的，进而可据此定义一个数据库表，并包含与评论相关的地名信息。然而，当前暂未定义这一类表项。

接下来的问题是，如何使用户能够查看未包含任何信息的地址？答案十分简单：对此，可使用 Google 提供的 Places API，第 10 章将对此加以讨论，当前主要关注 Place Reviewer 后台程序的实现过程。

9.2.3 设置数据库

考虑到系统信息存储的必要性，我们需要针对当前应用程序配置数据库，以实现数据的持久化。前述应用程序利用 Postgres 作为数据库，此处也将继续沿袭这一方式。打开 Terminal 并运行下列命令：

```
createdb -h localhost —username=<username> place-reviewer
```

当预习上述命令时，系统中将创建名为 place-reviewer 的数据库。这里，所输入的、用以替代<username>参数的用户名将作为连接数据库的用户名。针对当前应用程序配置了数据库后，即可着手开始实现后台程序，此处将采用 Spring 框架。

9.2.4 实现后台应用程序

在定义了应用程序用例，并配置了应用程序所连接的数据库后，下面直接进入实现过程。打开 IntelliJ IDEA 并利用 Spring 初始化器创建新的项目。在单击 Next 按钮后，IntelliJ 将检索 Spring 初始化器，随后用户将被提示提供与应用程序相关的特定信息。在完成后续步骤之前，需要执行下列操作：

（1）输入 com.example 作为项目组织 ID。
（2）输入 place-reviewer 作为项目 ID。
（3）选择 Maven Project 作为项目类型（若未选取）。
（4）保留包选项以及 Java 版本。
（5）选择 Kotlin 作为当前语言。鉴于后续操作将使用 Kotlin 语言，因而该步骤十分重要。
（6）将版本属性修改为 1.0.0。
（7）输入选择描述，此处为 Ours is A nifty web application for the creation of location reviews。
（8）输入 com.example.placereviewer 作为包名。

在完善了所需的项目信息后，可单击 Next 按钮执行后续操作。在接下来的画面中，需要些当前项目的依赖关系。

> **注意：**
> Spring 初始化器与 Spring 插件一起出现。在本书编写时，仅 IntelliJ IDEA 旗舰版本对此予以支持，且需要提供付费证书。如果读者安装了 IntelliJ IDEA 共享版，可简单地使用 Spring 初始化器工具（对应网址为 https://start.spring.io）生成当前项目，并将该项目导入 IntelliJ IDEA 中。

随后可选择 Spring 中的 Security、Session、Cache 和 Web 依赖关系。除此之外，还需要在 template engine 分类项中选择 Thymeleaf。在 SQL 分类项下方，选取 PostgreSQL。此外，在屏幕上方的 Spring Boot Version 下拉菜单中，选择 2.0.0 M7 作为版本号。在依赖关系选择完毕后，最终结果如图 9.2 所示。

在指定了相应的依赖关系后，单击 Next 按钮执行后续的配置操作。此时需要提供项目名称和项目的保存位置。相应地，可填写 place-reviewer 作为项目名称，并选择项目的保存位置，如图 9.3 所示。

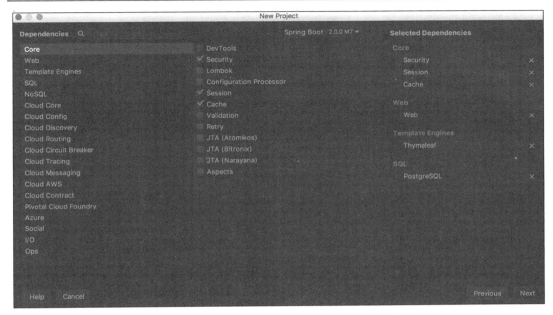

图 9.2 选取依赖关系项

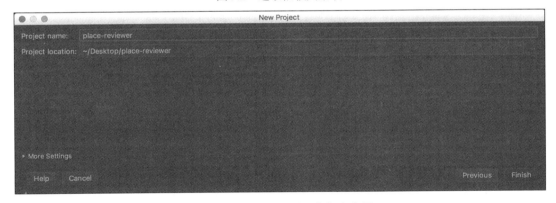

图 9.3 项目名称和项目的保存位置

随后，可单击 Finish 按钮，并等待项目配置完毕。接下来将显示包含初始化项目文件的新窗口。鉴于之前已经输入了 Spring 项目结构简述，此处无须重复操作。在进入下一阶段之前，还需要向当前项目的 pom 文件中添加下列依赖关系。

```
<dependency>
  <groupId>org.springframework.boot</groupId>
  <artifactId>spring-boot-starter-data-jpa</artifactId>
```

```xml
    </dependency>
    <dependency>
      <groupId>org.webjars</groupId>
      <artifactId>bootstrap</artifactId>
      <version>4.0.0-beta.3</version>
    </dependency>
    <dependency>
      <groupId>org.webjars</groupId>
      <artifactId>jquery</artifactId>
      <version>3.2.1</version>
    </dependency>
```

下面将当前应用程序连接至数据库中。

9.2.5 将后台程序连接至 Postgres

在将 Place Reviewer 后台应用程序连接至 PostgreSQL 数据库时，需要修改项目的 application.properties 文件，并添加连接 PostgreSQL 数据库所需的相关属性。打开项目的 application.properties 文件，并添加下列属性：

```
spring.jpa.hibernate.ddl-auto=create-drop
spring.jpa.generate-ddl=true
spring.datasource.url=jdbc:postgresql://localhost:5432/
place-reviewer
spring.datasource.driver.class-name=org.postgresql.Driver
spring.datasource.username=<username>
```

插入相应的用户名，其中，<username>属性位于上述代码片段中，在添加了数据库连接属性后，Spring Boot 将在应用程序启动时，将其连接至所指定的数据库。在项目的数据库连接属性设置完毕后，下面针对之前标识的 User 和 Review 实体构建对应的模型。

9.2.6 创建模型

前述内容标识了系统所需的两种实体类型，即 User 实体和 Review 实体。下面针对此类实体生成相应的模型。首先考察 User 实体。对此，可在 com.example.placereviewer 包中生成 data 包，并在新创建的 data 包中添加 model 包。随后，在新创建的 com.example.placereviewer.data.model 包中添加 User.kt 文件，并加入以下内容：

```
package com.example.placereviewer.data.model

import com.example.placereviewer.listener.UserListener
```

```
import org.springframework.format.annotation.DateTimeFormat
import java.time.Instant
import java.util.*
import javax.persistence.*
import javax.validation.constraints.Pattern
import javax.validation.constraints.Size

@Entity
@Table(name = "`user`")
@EntityListeners(UserListener::class)
data class User(
@Column(unique = true)
@Size(min = 2)
@Pattern(regexp = "^[A-Z0-9._%+-]+@[A-Z0-9.-]+\\\\.[A-Z]{2,6}\$")
var email: String = "",
  @Column(unique = true)
var username: String = "",
  @Size(min = 60, max = 60)
var password: String = "",
  @Column(name = "account_status")
  @Pattern(regexp = "\\A(activated|deactivated)\\z")
var accountStatus: String = "activated",
  @Id
  @GeneratedValue(strategy = GenerationType.AUTO)
var id: Long = 0,
  @DateTimeFormat
  @Column(name = "created_at")
var createdAt: Date = Date.from(Instant.now())
) {
@OneToMany(mappedBy = "reviewer", targetEntity = Review::class)
private var reviews: Collection<Review>? = null
}
```

根据之前 Spring 中实体的操作经验，上述代码无须做过多解释，其中定义了包含 email、username、password、accountStatus、id 以及 createdAt 属性的 User 实体。除此之外，User 中还包含了多个 Review 实体。针对包含@EntityListener 注解的实体，我们还对其设置了实体监听器。对此，将新的 listener 包添加至 com.example.placereviewer 中，并将 UserListener.kt 文件加入其中，对应代码如下所示：

```
package com.example.placereviewer.listener

import com.example.placereviewer.data.model.User
```

```
import org.springframework.security.crypto.bcrypt.BCryptPasswordEncoder
import javax.persistence.PrePersist
import javax.persistence.PreUpdate

class UserListener {

  @PrePersist
   @PreUpdate
     fun hashPassword(user: User) {
       user.password = BCryptPasswordEncoder().encode(user.password)
    }
}
```

UserListener 中定义了单一的 hashPassword 函数,该函数在 User 实体持久化和更新之前被调用,其唯一工作是在数据库持久化之前,将 password 编码为密文。

在针对 User 实体创建了相应的监听器后,下面考察 Review 实体。在 com.example.placereviewer.data.models 中设置 Review.kt 文件,并添加下列代码:

```
package com.example.placereviewer.data.model

import org.springframework.format.annotation.DateTimeFormat
import java.time.Instant
import java.util.*
import javax.persistence.*
import javax.validation.constraints.Size

@Entity
@Table(name = "`review`")
data class Review(
  @ManyToOne(optional = false)
  @JoinColumn(name = "user_id", referencedColumnName = "id")
  var reviewer: User? = null,
    @Size(min = 5)
  var title: String = "",
    @Size(min = 10)
  var body: String = "",
    @Column(name = "place_address")
    @Size(min = 2)
  var placeAddress: String = "",
    @Column(name = "place_name")
  var placeName: String = "",
    @Column(name = "place_id")
```

```
  var placeId: String = "",
  var latitude: Double = 0.0,
  var longitude: Double = 0.0,
    @Id
    @GeneratedValue(strategy = GenerationType.AUTO)
  var id: Long = 0,
    @DateTimeFormat
    @Column(name = "created_at")
  var createdAt: Date = Date.from(Instant.now())
)
```

其中定义了 Review 类，并包含 reviewer、title、body、placeAddress、placeName、placeId、latitude、longitude、id 和 createdAt 属性。其中，reviewer 属性表示为 User 类型，并引用了 review 的生成器——每条评论由用户生成。另外，一位用户可生成多条评论。相应地，可利用@ManyToOne 注解声明 Review 和 User 实体之间的关系。

9.2.7 创建数据存储库

前述内容设置了所需的实体，因而还应构建与实体数据访问相关的存储库。对此，可在 com.example.placereviewer 包中创建存储库包。鉴于当前包含两种实体，因而须创建两个存储库（其中之一用于访问与每个实体相关的数据），分别为 UserRepository 和 ReviewRepository。随后，可在 com.example.placereviewer.data.repository 中定义 UserRepository 接口，如下所示：

```
package com.example.placereviewer.data.repository

import com.example.placereviewer.data.model.User
import org.springframework.data.repository.CrudRepository

interface UserRepository : CrudRepository<User, Long> {

  fun findByUsername(username: String): User?
}
```

findByUsername(String)方法从数据库中检索 User，其中包含了作为参数传递至该方法中的用户名。ReviewRepository 接口定义如下：

```
package com.example.placereviewer.data.repository

import com.example.placereviewer.data.model.Review
import org.springframework.data.repository.CrudRepository
```

```kotlin
interface ReviewRepository : CrudRepository<Review, Long> {
  fun findByPlaceId(placeId: String)
}
```

在设置了查询实体的实体以及存储库后，即可以服务和服务实现的方式完成 Place Reviewer 应用程序的核心业务逻辑。

9.2.8 Place Reviewer 业务逻辑实现

如前所述，在基于 MVC 设计模式的应用程序中，存在 3 种主要的组件，即模型、视图和控制器。相应地，模型组件负责数据管理和业务逻辑的执行。在 Place Reviewer 应用程序中，模型将以服务的形式实现模型，并用于后台程序中。对此，可创建两个基本的服务，分别用于管理与应用程序用户相关的数据以及评论数据。

首先，需要定义 UserService 接口，其中定义了 UserServiceImpl 类需要实现的各种行为。之前在 Place Reviewer 应用程序用例中曾谈到，用户需要在平台上进行注册（因而需要创建账户）。因此，当前模型中须对此予以处理。相应地，可在项目的根数据包中创建 service 包，并将 UserService 接口添加于其中，如下所示：

```kotlin
package com.example.placereviewer.service

interface UserService {

  fun register(username: String, email: String, password: String): Boolean
}
```

其中仅声明了一个方法，且需要通过有效的 UserService 加以实现。register (String, String, String)方法接收 3 个参数：第一个参数表示为注册用户的用户名；第二个参数表示用户有效的电子邮件地址；第三个参数表示为密码。当利用相应的参数调用 register()方法时，该方法将通过用户提供的证书注册用户。如果用户注册成功，则该方法返回 true，否则返回 false。

下列代码表示为 UserService 实现，可将其添加至 service 包中。

```kotlin
package com.example.placereviewer.service

import com.example.placereviewer.data.model.User
import com.example.placereviewer.data.repository.UserRepository
import org.springframework.stereotype.Service
```

```
@Service
class UserServiceImpl(val userRepository: UserRepository) :
UserService {

  override fun register(username: String, email: String,
                        password: String): Boolean {
    val user = User(email, username, password)
    userRepository.save(user)

    return true
  }
}
```

UserServiceImpl 类实现的 register()其工作机制较为直观。当传递了有效的用户名、电子邮件地址以及密码参数后,将生成一个新的用户对象,同时向其构造函数传递相关参数。在用户对象创建完毕后,该用户将通过下列代码保存至数据库中。

```
userRepository.save(user)
```

userRepository 表示为之前生成的 UserRepository 实例,该实例通过 Spring 框架自动置入 UserServiceImpl 的构造函数中。当用户保存至数据库后,将返回 true。

下面实现评论服务接口,其中涉及平台用户生成的评论和评论列表。最终,用户服务接口中需要实现 createReview()和 listReview()方法。

下面向当前项目的 service 包中添加 ReviewService 接口,如下所示:

```
package com.example.placereviewer.service

import com.example.placereviewer.data.model.Review

interface ReviewService {

  fun createReview(reviewerUsername: String, reviewData: Review): Boolean

  fun listReviews(): Iterable<Review>
}
```

针对我们所创建的服务,下列代码定义了 ReviewServiceImpl 类,将该类连同后续章节创建的全部服务添加至 com.example.placereviewer.service 中,如下所示:

```
package com.example.placereviewer.service

import com.example.placereviewer.data.model.Review
import com.example.placereviewer.data.model.User
```

```kotlin
import com.example.placereviewer.data.repository.ReviewRepository
import com.example.placereviewer.data.repository.UserRepository
import org.springframework.stereotype.Service

@Service
class ReviewServiceImpl(val reviewRepository: ReviewRepository, val
userRepository: UserRepository) : ReviewService {

  override fun listReviews(): Iterable<Review> {
    return reviewRepository.findAll()
  }

  override fun createReview(reviewerUsername: String,
                            reviewData: Review): Boolean {
    val reviewer: User? = userRepository.findByUsername(reviewerUsername)

    if (reviewer != null) {
      reviewData.reviewer = reviewer
      reviewRepository.save(reviewData)
      return true
    }

    return false
  }
}
```

listReviews()返回包含存储于应用程序数据库中的、全部评论的 Iterable。另外一方面，createReview()接收一个字符串，其值表示为评论用户的用户名，以及一个 Review 实例，其中包含了所创建的评论数据。通过调用 UserRepository 的 findByUsername()方法，createReview()首先获得包含特定用户名的用户，对应的用户即为发表评论的用户。

如果 UserRepository 返回一个非 null 对象，说明存在相关用户；同时，该用户被赋予已保存的评论中的 reviewer 属性中。在赋值完毕后，评论将被保存至数据库中且函数返回 true，表明操作成功。如果为发现包含当前用户名的用户，那么，createReview()将返回 false。

在以服务形式构建了相关模型后，下面考察 Place Reviewer 的安全问题。考虑到非授权用户不可访问应用程序资源，因而这一问题较为重要。

9.2.9 Place Reviewer 后台应用程序的安全问题

与第 4 章讨论的 Messenger API 安全类似，此处也将使用 Spring Security 处理 Place

Reviewer 后台应用程序的安全问题。除了 Spring Security 之外，当前应用程序的安全处理稍有不同。在第 4 章，对于客户端应用程序的验证操作，我们配置了 Spring Security，并显式地依靠于 JSON Web 令牌；而此处的处理方式仅使用 Spring Security，且不再涉及 JSON Web 令牌。

首先需要针对应用程序构建一个自定义 Web 安全配置，并实现 Spring 框架的 WebSecurityConfigurerAdapter。在 com.example.placereviewer 中设置一个 config 包，并定义 WebSecurityConfig 类，如下所示：

```
package com.example.placereviewer.config

import com.example.placereviewer.service.AppUserDetailsService
import org.springframework.context.annotation.Bean
import org.springframework.context.annotation.Configuration
import org.springframework.http.HttpMethod
import org.springframework.security.authentication.AuthenticationManager
import org.springframework.security.config.BeanIds
import org.springframework.security.config.annotation
        .authentication.builders.AuthenticationManagerBuilder
import org.springframework.security.config.annotation
        .web.builders.HttpSecurity
import org.springframework.security.config.annotation
        .web.configuration.EnableWebSecurity
import org.springframework.security.config.annotation
        .web.configuration.WebSecurityConfigurerAdapter
import org.springframework.security.core.userdetails
        .UserDetailsService
import org.springframework.security.crypto.bcrypt
        .BCryptPasswordEncoder
import org.springframework.security.web
        .DefaultRedirectStrategy
import org.springframework.security.web.RedirectStrategy

@Configuration
@EnableWebSecurity
class WebSecurityConfig(val userDetailsService:AppUserDetailsService):
WebSecurityConfigurerAdapter() {

  private val redirectStrategy: RedirectStrategy =
                        DefaultRedirectStrategy()
```

```kotlin
@Throws(Exception::class)
override fun configure(http: HttpSecurity) {
  http.authorizeRequests()
      .antMatchers(HttpMethod.GET,"/register").permitAll()
      .antMatchers(HttpMethod.POST,"/users/registrations").permitAll()
      .antMatchers(HttpMethod.GET,"/css/**").permitAll()
      .antMatchers(HttpMethod.GET,"/webjars/**").permitAll()
      .anyRequest().authenticated()
      .and()
      .formLogin()
      .loginPage("/login")
      .successHandler { request, response, _ ->
        redirectStrategy.sendRedirect(request, response, "/home")
      }
      .permitAll()
      .and()
      .logout()
      .permitAll()
}

@Throws(Exception::class)
override fun configure(auth: AuthenticationManagerBuilder) {
  auth.userDetailsService<UserDetailsService>(userDetailsService)
      .passwordEncoder(BCryptPasswordEncoder())
}

@Bean(name = [BeanIds.AUTHENTICATION_MANAGER])
override fun authenticationManagerBean(): AuthenticationManager {
  return super.authenticationManagerBean()
}
}
```

如前所述，WebSecurityConfig 的 configure(HttpSecurity)方法其任务是配置 HTTP URL 安全路径。利用 configure(HttpSecurity)方法，我们配置了 Spring Security，并允许所有用户访问/users/registrations 和 GET 请求（其路径匹配于/register、/css 和/webjars/**）。除此之外，我们还允许所有的 HTTP 请求可进入从/login 路径访问的登录页面。

我们向登录动作中成功地添加了一个处理程序，并使用了 WebSecurityConfig 类定义的 redirectStrategy 属性；当用户成功登录后，可将客户端重定向至/home。最后，我们还应支持后端应用程序的全部登录请求。

configure(AuthenticationManagerBuilder)负责设置所用的 UserDetailsService，并指定了密码编码器。此处使用了 BcryptPasswordEncoder。读者可能已注意到，当前项目中还未实现 UserDetailsService。对此，将 AppUserDetailsService 添加至 com.example.placereviewer.service 包中，如下所示：

```
package com.example.placereviewer.service

import com.example.placereviewer.data.repository.UserRepository
import org.springframework.security.core.GrantedAuthority
import org.springframework.security.core.userdetails.User
import org.springframework.security.core.userdetails.UserDetails
import org.springframework.security.core.userdetails.UserDetailsService
import org.springframework.security.core.userdetails
        .UsernameNotFoundException
import org.springframework.stereotype.Service
import java.util.ArrayList

@Service
class AppUserDetailsService(private val userRepository:
UserRepository) : UserDetailsService {

  @Throws(UsernameNotFoundException::class)
  override fun loadUserByUsername(username: String): UserDetails {
    val user = userRepository.findByUsername(username) ?:
        throw UsernameNotFoundException("A user with the username
                                    $username doesn't exist")

    return User(user.username, user.password,
            ArrayList<GrantedAuthority>())
  }
}
```

loadUsername(String)加载用户的 UserDetails（与传递至函数中的用户名匹配的）。如果未发现匹配用户，那么将抛出 UsernameNotFoundException。

待全部工作就绪后，即成功地针对后台应用程序设置了 Spring Security。

下面将结束实体、存储库、服务、服务实现以及 Spring Security 安全配置方面的内容，并从应用程序的前端实现开始我们的工作。接下来，我们将基于 Spring MVC 的、客户端应用程序的 Web 内容服务。

9.2.10　基于 Spring MVC 的 Web 内容服务

在 Spring MVC 中，HTTP 请求由控制器进行处理。这里，控制器定义为一个类，并通过@Controller 加以注解——这与@RestController 的注解方式十分类似。关于控制器的工作方式，一种较好的理解方式是考察一个具体的实例。下面将构建一个简单的 Spring MVC 控制器，并处理发送至/say/hello 路径的 HTTP GET 请求（通过返回一个视图），进而向用户显示一个 HTML 页面。

相应地，在 com.example.placereviewer 中创建一个 controller 包，并添加下列类定义：

```
package com.example.placereviewer.controller

import org.springframework.stereotype.Controller
import org.springframework.web.bind.annotation.GetMapping
import org.springframework.web.bind.annotation.RequestMapping

@Controller
@RequestMapping("/say")
class HelloController {

  @GetMapping("/hello")
  fun hello(): String {
    return "hello"
  }
}
```

不难发现，控制器的创建过程并不复杂。基于@Controller 的 HelloController 注解通知 Spring，该类定义为 Spring MVC 控制器，因而具备处理 HTTP 请求操作能力。此外，基于@RequestMapping("/say")的 HelloController 注解行为表明，该控制器处理包含/say 基路径的 HTTP 请求。当前控制器中定义了 hello()动作，由于该动作采用@GetMapping("/hello")加以注解，因而将处理/path/hello 路径的请求。hello()返回的字符串表示为视图资源名称——在将某个请求发送至该例程中时，将被显示于客户端上。

hello()要求名为 hello 的视图返回至客户端，那么，下一项任务则是将这一类视图添加至当前项目中。通常情况下，视图一般被添加至 Spring 项目 resources 目录下的 templates 文件夹中。在 templates 上单击鼠标右键，在弹出的快捷菜单中选择 New | HTML File 命令，进而将 hello.html 文件添加至当前项目中，如图 9.4 所示。

第 9 章 创建 Place Reviewer 后台应用程序

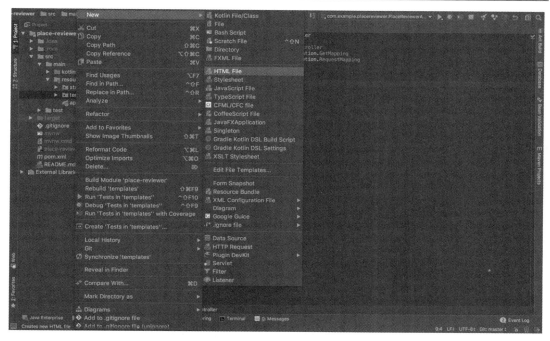

图 9.4 将 hello.html 文件添加至当前项目中

用户将被提示提供 HTML 页面的名称，这里输入 hello 作为页面名称，如图 9.5 所示。

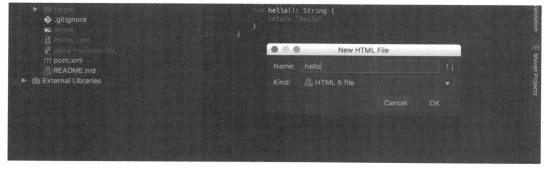

图 9.5 输入 hello 作为页面名称

IntelliJ IDEA 将在所选择的目录中生成 HTML 文件。待操作完成后，即可对其内容进行调整，进而包含基本的 HTML 内容，如下所示：

```
<!DOCTYPE html>
<html lang="en">
```

```html
<head>
  <meta charset="UTF-8">
  <title>Hello</title>
</head>
<body>
Hello world!
</body>
</html>
```

接下来将对所创建的控制器进行测试,以查看在向对应路径发送 GET 请求时,是否返回了一个包含"Hello World!"消息的 HTML 页面。相应地,需要将 GET 请求添加至 /say/hello 中,并作为 Spring Security 无须验证的请求。针对于此,可简单地调整 WebSecurityConfig 中的 configure(HttpSecurity),并支持/say/hello 路径的 GET 请求,如下所示:

```
@Throws(Exception::class)
override fun configure(http: HttpSecurity) {
    http.authorizeRequests()
        .antMatchers(HttpMethod.GET,"/say/hello").permitAll() // added line
        .antMatchers(HttpMethod.GET,"/register").permitAll()
        .antMatchers(HttpMethod.POST,"/users/registrations").permitAll()
        .antMatchers(HttpMethod.GET, "/css/**").permitAll()
        .antMatchers(HttpMethod.GET, "/webjars/**").permitAll()
        .anyRequest().authenticated()
        .and()
        .formLogin()
        .loginPage("/login")
        .successHandler { request, response, _ ->
            redirectStrategy.sendRedirect(request, response, "/home")
        }
        .permitAll()
        .and()
        .logout()
        .permitAll()
}
```

构建并运行 Spring 应用程序,随后开启 Web 浏览器并访问 URL:http://localhost:5000/say/hello,对应结果如图 9.6 所示。

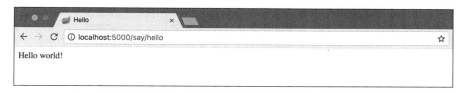

图 9.6　构建并运行 Spring 应用程序

9.3　利用 ELK 管理 Spring 应用程序日志

当构建发布系统时，服务器日志文件的管理方式是一项重点考察内容。服务区日志是一类日志文件，并通过服务器予以创建和管理。日志文件通常由服务器执行的活动列表构成。ELK（Elasticsearch、Logstash 和 Kibana）栈则强调了应用程序日志文件的管理方式，本节将学习如何利用 ELK 栈实现 Spring 应用程序日志文件的管理。

9.3.1　利用 Spring 生成日志

在开始设置 ELK 栈管理 Spring 日志之前，首先需要配置 Spring 以生成日志文件，这可通过 Spring 项目的 application.properties 文件予以实现。下面配置 Place Reviewer 后台应用程序并生成日志。

打开项目的 application.properties 文件，并添加下列代码：

```
logging.file=application.log
```

上述代码将配置 Spring，并在 application.log 文件中生成、保存服务器日志。在项目下一次启动时，该文件将在项目的根目录中创建并保存。这一步骤对于配置服务器日志来说不可或缺。下面将设置日志栈，并从安装 Elasticsearch 开始。

9.3.2　安装 Elasticsearch

安装 Elasticsearch 需要执行下列步骤：

（1）访问 https://www.elastic.co/downloads/elasticsearch，选择下载 ZIP 文件格式的 Elasticsearch 包。

（2）下载完毕后，解压 Elasticsearch ZIP 文件。

（3）在终端中运行 Elasticsearch，即运行 bin/elasticsearch（Windows 环境下为 bin/elasticsearch.bat），如图 9.7 所示。

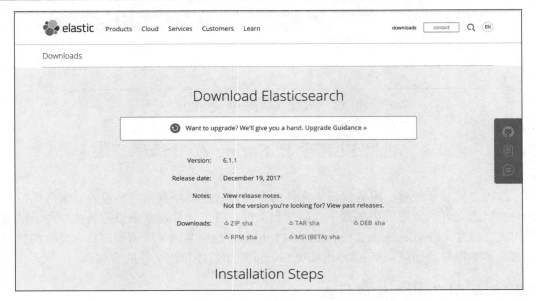

图 9.7 下载、安装 Elasticsearch

bin/elasticsearch 执行完毕后，Elasticsearch 将在系统中运行，如图 9.8 所示。

图 9.8 在系统中运行 Elasticsearch

在终端上运行下列命令，检测 Elasticsearch 是否可正常运行。

```
curl -XGET http://localhost:9200
```

如果一切就绪，即会得到下列响应结果。

```
{
  "name" : "Df8YuN2",
  "cluster_name" : "elasticsearch",
  "cluster_uuid" : "Z8SYAKLNSZaMiGkYz7ihfg",
  "version" : {
    "number" : "6.1.1",
    "build_hash" : "bd92e7f",
    "build_date" : "2017-12-17T20:23:25.338Z",
    "build_snapshot" : false,
    "lucene_version" : "7.1.0",
    "minimum_wire_compatibility_version" : "5.6.0",
    "minimum_index_compatibility_version" : "5.0.0"
  },
  "tagline" : "You Know, for Search"
}
```

9.3.3 安装 Kibana

Kibana 的安装过程与 Elasticsearch 类似，包括下列步骤：

（1）访问 https://www.elastic.co/downloads/kibana，下载 Kibana 压缩包。
（2）解压 Kibana。
（3）运行 bin/kibana。

在下载并运行 Kibana 后，可启动浏览器并访问 http://localhost:5601/。如果一切工作正常，将会看到如图 9.9 所示的 Kibana Web 页面。

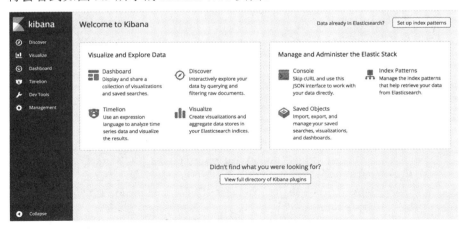

图 9.9　启动 Kibana

9.3.4 Logstash

安装 Logstash 需要执行下列步骤：
（1）访问 https://www.elastic.co/downloads/logstash 并下载 ZIP 压缩包。
（2）解压 Logstash 压缩包。

除了下载并运行 Logstash 之外，还需要对其进行适当配置，以理解 Spring 日志文件结构。因此，需要创建一个 Logstash 配置文件，其中包含了 3 个较为重要的部分，即输入、过滤和输出。每部分内容都设置了插件，并在日志文件处理中饰演了对应的角色。接下来，可在相关目录中创建 logstash.conf 文件，并添加下列代码。

```
input {
  file {
    type => "java"
    path => "/<path-to-project>/place-reviewer/application.log"
    codec => multiline {
      pattern => "^%{YEAR}-%{MONTHNUM}-%{MONTHDAY} %{TIME}.*"
      negate => "true"
      what => "previous"
    }
  }
}

filter {
  #Tag log lines containing tab character followed by 'at' as stacktrace.
  if [message] =~ "\tat" {
    grok {
      match => ["message", "^(\tat)"]
      add_tag => ["stacktrace"]
    }
  }
  #Grok Spring Boot's default log format
  grok {
    match => [ "message",
               "(?<timestamp>%{YEAR}-%{MONTHNUM}-%{MONTHDAY} %{TIME})
               %{LOGLEVEL:level} %{NUMBER:pid} --- \[(?<thread>
               [A-Za-z0-9-]+)\][A-Za-z0-9.]*\.(?<class>
               [A-Za-z0-9#_]+)\)\s*:\s+(?<logmessage>.*)",
               "message",
               "(?<timestamp>%{YEAR}-%{MONTHNUM}-%{MONTHDAY} %{TIME})
               %{LOGLEVEL:level} %{NUMBER:pid} --- .+?
```

```
                  :\s+(?<logmessage>.*)"
        ]
    }

    #Parsing timestamps in timestamp field
    date {
      match => [ "timestamp" , "yyyy-MM-dd HH:mm:ss.SSS" ]
    }
}

output {
    # Print each event to stdout and enable rubydebug.
    stdout {
      codec => rubydebug
    }
    # Send parsed log events to Elasticsearch
    elasticsearch {
      hosts => ["127.0.0.1"]
    }
}
```

解释全部插件超出了本书的讨论范围，读者可留意相关注释以理解代码的功能。另外，还应将输入部分文件插件中的 path 路径修改为 Place Reviewer 应用程序的 application.log 文件的绝对路径。

当 Logstash 配置文件处理完毕后，利用下列命令运行 Logstash：

`/bin/logstash -f logstash.conf`

如果一切配置顺利，Logstash 将开始存储日志事件。最后一项工作是配置 Kibana，并读取存储数据。

9.3.5 配置 Kibana

Kibana 的配置过程较为简单，进而可读取存储于 Elasticsearch 索引中的日志。对此，可访问 Kibana Web UI（http://localhost:5601/），单击左侧导航栏上的 Management，进而访问设置管理页面。在配置 Kibana 时，首先需要生成索引模式。对此，可单击管理页面中的 Index Patterns，并管理 Kibana 所识别的索引模式，如图 9.10 所示。

鉴于首次创建索引模式，因而将显示如图 9.11 所示的提示画面。

在 Index pattern 文本框中输入 Kibana 所识别的索引名（显示于当前页面中），随后可执行下一个步骤。此时需要选取 Time Filter field name，如图 9.12 所示。

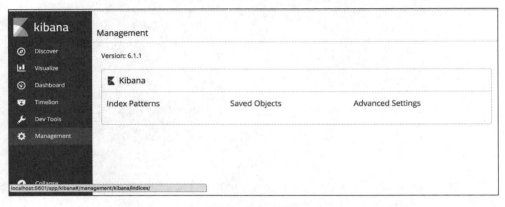

图 9.10 索引模式

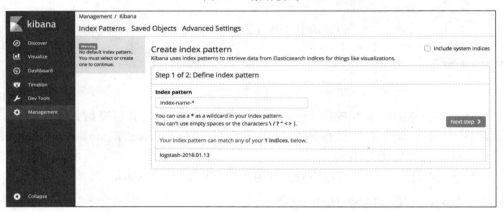

图 9.11 提示信息

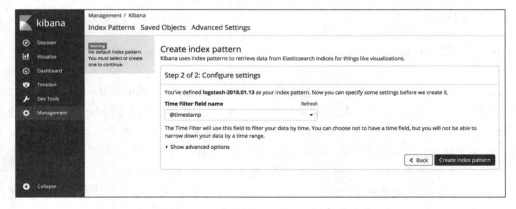

图 9.12 选取 Time Filter field name

此处可选取下拉菜单中的@timestamp。随后，单击 Create index pattern 按钮结束索引模式的创建过程。通过选择 Index Patterns，读者可从设置管理页面中随时管理索引模式。图 9.13 显示了保存后的模式。

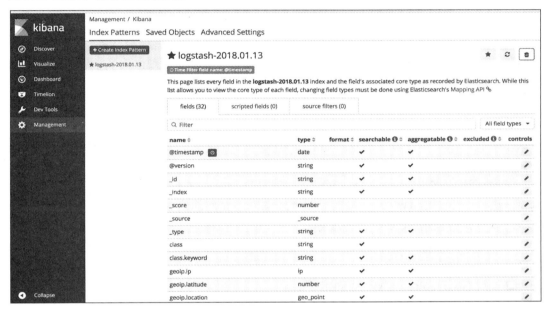

图 9.13　查看保存后的模式

至此，Kibana 的配置过程暂告一段落。

9.4　本章小结

本章深入讨论了 Web 开发中 Kotlin 语言及其应用，并实现了 Place Reviewer 应用程序。除此之外，我们还学习了如何配置 Spring 框架项目，并采用 Spring MVC 构建基于 MVC 设计模式的应用程序。进一步讲，本章介绍了 Spring Security 的配置方式，以防止 Spring Web 应用程序中未经验证的访问行为。最后，本章还考察了 ELK 栈，以及如何管理服务器栈。

第 10 章将完成 Place Reviewer 应用程序，并实现其前端内容。在前端实现过程中，读者将学习如何利用 Google Places API 构建富 Web 应用程序，并对 Spring 框架构建的 Web 应用程序进行测试。

第 10 章 实现 Place Reviewer 前端

第 9 章通过构建 Place Reviewer 网站，继续讨论了 Kotlin 语言在 Web 应用程序设计中的应用。其中涉及 MVC 模式，并从高层视角考察 MVC 应用程序中的主要组件，包括模型、视图以及控制器。在读者理解了 MVC 设计模式及其工作方式后，第 9 章还针对 Place Reviewer 应用程序介绍了其设计和实现方案。

在第 9 章中，首先介绍了应用程序用例，随后是构建应用程序所需的数据，在数据标识完毕后，接下来探讨了后端应用程序开发，同时设置了与应用程序通信的数据库，并实现了应用程序所需的实体和模型。

最后，我们还通过 Spring Security 进一步探讨了 Place Reviewer 应用程序的安全问题，即验证操作（未采用 JWT）。最后，读者还学习了如何创建 Spring MVC 应用程序的控制器，以及如何利用 ELK 栈管理服务器日志。

本章将介绍 Place Reviewer 应用程序的前端开发，进而完善该程序的所有内容。本章主要涉及以下内容：
- ❑ 与 Google Places API 协同工作。
- ❑ 应用程序测试。
- ❑ 向 AWS 部署 Web 应用程序。

下面开始着手讨论视图的实现过程。

10.1 利用 Thymeleaf 生成视图

如前所述，视图是应用程序创建的数据表达方式，也是用户与 MVC 模式应用程序之间的主要交互点。视图层利用多种不同的技术向用户显示相关信息。Spring 支持多种视图选择方案，即模板。Spring 应用程序中的模板功能通过模板引擎予以提供。简单地讲，模板引擎允许使用应用程序的视图层的静态模板文件。此外，模板引擎也称作模板库。Spring 支持下列模板库的应用：
- ❑ Thymeleaf。
- ❑ Freemaker。
- ❑ Tiles。

❑ Velocity。

实际内容远不止于此，还存在许多其他的模板库可供 Spring 使用。本章将利用 Thymeleaf 向应用程序提供模板处理支持。回忆一下，在项目的开始阶段，我们曾纳入了基于 Thymeleaf 的模板支持功能。具体来说，在项目 pom.xml 文件的依赖关系中，曾向项目中添加了 Thymeleaf，如下所示：

```
<dependencies>
...
<dependency>
 <groupId>org.springframework.boot</groupId>
 <artifactId>spring-boot-starter-thymeleaf</artifactId>
</dependency>
...
</dependencies>
```

关于 Thymeleaf 的正式定义，其官方网站中描述到：Thymeleaf 是一类针对 Web 和单机环境的服务器端 Java 模板引擎，旨在向开发工作流中添加模板——HTML 将能够正确地显示于浏览器中，并以静态原型工作，允许在开发团队中进行更强的协作。关于 Thymeleaf 及其设计目标，读者可访问 http://thymeleaf.org 以了解更多内容。

第 9 章曾讨论一个简单示例，并针对 HelloController 实现了一个 hello.html 视图，但所生成的视图仅向用户显示了"Hello world！"消息，本章将介绍一些更复杂的视图。我们首先创建一个视图，以方便用户在平台上进行注册。

10.1.1 实现用户注册视图

本节将完成两项任务。首先将创建一个视图层，并在 Place Reviewer 平台上实现新用户的注册过程。其次，将创建相应的控制器和动作，并利用注册视图显示用户，随后处理注册表单提交操作。对此，可在 Place Reviewer 项目中设置 register.html 模板。回忆一下，全部模板文件均位于 resources 的 templates 目录下。下面将下列模板 HTML 添加至当前文件中。

```
<!DOCTYPE html>
<html lang="en" xmlns:th="http://www.thymeleaf.org">
<head>
 <title>Register</title>
 <link rel="stylesheet" th:href="@{/css/app.css}"/>
 <link rel="stylesheet"
  href="/webjars/bootstrap/4.0.0-beta.3/css/bootstrap.min.css"/>
```

```html
    <script src="/webjars/jquery/3.2.1/jquery.min.js"></script>
    <script src="/webjars/bootstrap/4.0.0-beta.3/
     js/bootstrap.min.js"></script>
</head>
<body>
    <nav class="navbar navbar-default nav-enhanced">
      <div class="container-fluid container-nav">
        <div class="navbar-header">
          <div class="navbar-brand">
            Place Reviewer
          </div>
        </div>
        <ul class="navbar-nav" th:if="${principal != null}">
          <li>
            <form th:action="@{/logout}" method="post">
              <button class="btn btn-danger" type="submit">
                <i class="fa fa-power-off" aria-hidden="true"></i>
                Sign Out
              </button>
            </form>
          </li>
        </ul>
      </div>
    </nav>
    <div class="container-fluid" style="z-index: 2; position: absolute">
      <div class="row mt-5">
        <div class="col-sm-4 col-xs-2"> </div>
        <div class="col-sm-4 col-xs-8">
          <form class="form-group col-sm-12 form-vertical form-app"
            id="form-register" method="post"
            th:action="@{/users/registrations}">
            <div class="col-sm-12 mt-2 lead text-center text-primary">
              Create an account
            </div>
            <hr>
            <input class="form-control" type="text" name="username"
             placeholder="Username" required/>
            <input class="form-control mt-2" type="email" name="email"
             placeholder="Email" required/>
            <input class="form-control mt-2" type="password" name="password"
             placeholder="Password" required/>
```

```
            <span th:if="${error != null}" class="mt-2 text-danger"
             style="font-size: 10px" th:text="${error}"></span>
            <button class="btn btn-primary form-control mt-2 mb-3"
             type="submit">
              Sign Up!
            </button>
          </form>
        </div>
        <div class="col-sm-4 col-xs-2"></div>
      </div>
    </div>
  </body>
</html>
```

上述代码片段使用了 HTML 针对用户注册页面生成模板。Web 页面自身较为简单。仅包含了导航栏以及一个表单，用户可于其中输入提交所需的注册信息。作为 Thymeleaf 模板，使用一些基于 Thymeleaf 的特定属性也较为合理，下面考察其中的一些属性。

- ❑ th:href。该属性表示为修饰符属性。当模板引擎对此加以处理时，将计算所用的链接 URL，并将其设置于所用的标签中。例如，该属性所用的标签包括<a>和<link>。下列代码片段使用了 th:href 属性：

```
<link rel="stylesheet" th:href="@{/css/app.css}"/>
```

- ❑ th:action。该属性的工作方式类似于 HTML 动作属性，当提交表单时用于指定表单数据的发送位置。下列代码片段表明，表单数据应发送至包含 /users/registrations 路径的端点位置处。

```
<form class="form-group col-sm-12 form-vertical form-app"
 id="form-register" method="post"
 th:action="@{/users/registrations}">
  ...
</form>
```

- ❑ th:text。该属性用于确定 HTML 标签是否可根据条件测试结果予以显示，如下所示：

```
<span th:if="${error != null}" class="mt-2 text-danger"
 style="font-size: 10px" th:text="${error}"></span>
```

在上述代码中，如果存在模型属性错误，且对应值不等于 null，那么，span 标签将显示于 HTML 页面上；否则将不予显示。

除此之外，导航栏中还使用了 th:if，表明何时显示一个按钮，以使用户可注销其账

号,如下所示:

```html
<ul class="navbar-nav" th:if="${principal != null}">
  <li>
    <form th:action="@{/logout}" method="post">
      <button class="btn btn-danger" type="submit">
        <i class="fa fa-power-off" aria-hidden="true"></i> Sign Out
      </button>
    </form>
  </li>
</ul>
```

如果模板中设置了 principal 模型属性且不为 null,那么将会显示注销按钮。除非用户登录其账号,否则 principal 一直为 null。

初看之下,导航栏于模板间的直接添加方式尚且令人满意,但需要注意的是,在应用程序中多次使用导航条 DOM 元素是很常见的事情,因而不应在模板中多次重复编写相同的代码。为了避免不必要的重复性工作,需要将导航栏实现为片段,并可随时将其纳入模板中。

在 templates 中创建 fragments 目录,添加 navbar.html 文件并输入下列代码:

```html
<!DOCTYPE html>
<html lang="en" xmlns:th="http://www.thymeleaf.org">
  <head>
    <meta charset="UTF-8">
  </head>
  <body>
    <nav class="navbar navbar-default nav-enhanced" th:fragment="navbar">
      <div class="container-fluid container-nav">
        <div class="navbar-header">
          <div class="navbar-brand">
            Place Reviewer
          </div>
        </div>
        <ul class="navbar-nav" th:if="${principal != null}">
          <li>
            <form th:action="@{/logout}" method="post">
              <button class="btn btn-danger" type="submit">
                <i class="fa fa-power-off" aria-hidden="true"></i> Sign Out
              </button>
            </form>
          </li>
        </ul>
```

```
    </div>
  </nav>
 </body>
</html>
```

上述代码片段利用 th:fragment 属性定义了模板所支持的导航栏片段。当使用 th:insert 时，可随时将定义后的片段插入模板中。修改 register.html 文件中<body>标签的内部 HTML，并使用最新定义的判断，如下所示：

```
<!DOCTYPE html>
<html lang="en" xmlns:th="http://www.thymeleaf.org">
 <head>
   <title>Register</title>
   <link rel="stylesheet" th:href="@{/css/app.css}"/>
   <link rel="stylesheet"
    href="/webjars/bootstrap/4.0.0-beta.3/css/bootstrap.min.css"/>
   <script src="/webjars/jquery/3.2.1/jquery.min.js"></script>
   <script src="/webjars/bootstrap/4.0.0-beta.3/
    js/bootstrap.min.js"></script>
 </head>
 <body>
   <div th:insert="fragments/navbar :: navbar"></div>
   <!-- inserting navbar fragment -->
   <div class="container-fluid" style="z-index: 2; position: absolute">
     <div class="row mt-5">
       <div class="col-sm-4 col-xs-2">
       </div>
       <div class="col-sm-4 col-xs-8">
         <form class="form-group col-sm-12 form-vertical form-app"
          id="form-register" method="post"
          th:action="@{/users/registrations}">
           <div class="col-sm-12 mt-2 lead text-center text-primary">
             Create an account
           </div>
           <hr>
           <input class="form-control" type="text" name="username"
            placeholder="Username" required/>
           <input class="form-control mt-2" type="email" name="email"
            placeholder="Email" required/>
           <input class="form-control mt-2" type="password"
            name="password" placeholder="Password" required/>
           <span th:if="${error != null}" class="mt-2 text-danger"
            style="font-size: 10px" th:text="${error}"></span>
```

```html
            <button class="btn btn-primary form-control mt-2 mb-3"
                type="submit">
              Sign Up!
            </button>
         </form>
      </div>
      <div class="col-sm-4 col-xs-2"></div>
    </div>
  </div>
 </body>
</html>
```

不难发现,导航条 HTML 与片段的分离使代码更加简洁,并有助于提升模板的开发质量。

当对用户注册页面设置了相关模板后,即可确定控制器,并向站点的访问用户显示该模板。下面定义应用程序的控制器,并向请求用户显示 Place Reviewer 应用程序的 Web 页面。

向 controller 包中添加 ApplicationController 类,如下所示:

```kotlin
package com.example.placereviewer.controller

import org.springframework.stereotype.Controller
import org.springframework.web.bind.annotation.GetMapping

@Controller
class ApplicationController {

  @GetMapping("/register")
  fun register(): String {
    return "register"
  }
}
```

上述代码片段并无特别之处,其中创建了一个包含单一动作的 MVC 控制器,通过向用户显示 register.html 视图,处理/register 下的 HTTP GET 请求。

在查看新创建的注册页面之前,还需要添加 register.html 所需的 app.css 文件。诸如 CSS 文件这一类静态资源应添加至应用程序 resource 目录下的 static 目录中。相应地,在 static 目录下生成 css 目录,并加入包含下列代码的 app.css 文件。

```css
//app.css
.nav-enhanced {
```

```
  background-color: #00BFFF;
  border-color: blueviolet;
  box-shadow: 0 0 3px black;
}

.container-nav {
  height: 10%;
  width: 100%;
  margin-bottom: 0;
}

.form-app {
  background-color: white;
  box-shadow: 0 0 1px black;
  margin-top: 50px !important;
  padding: 10px 0;
}
```

下面可尝试运行 Place Reviewer 应用程序。当应用程序启动时，可打开浏览器并访问 http://localhost:5000/register 页面。

接下来，我们需要实现用户注册过程中的逻辑内容。针对于此，应声明一个动作，接收注册表单提供的表单数据，并对此类数据进行适当处理，旨在平台上成功地注册用户。如果读者还记得，表单数据应通过 POST 发送至/users/registrations。最终，我们需要定义一个动作，并处理此类 HTTP 请求。下面向 com.example.placereviewer.controller 包中添加 UserController 类，对应代码如下所示：

```
package com.example.placereviewer.controller

import com.example.placereviewer.component.UserValidator
import com.example.placereviewer.data.model.User
import com.example.placereviewer.service.SecurityService
import com.example.placereviewer.service.UserService
import org.springframework.stereotype.Controller
import org.springframework.ui.Model
import org.springframework.validation.BindingResult
import org.springframework.web.bind.annotation.GetMapping
import org.springframework.web.bind.annotation.ModelAttribute
import org.springframework.web.bind.annotation.PostMapping
import org.springframework.web.bind.annotation.RequestMapping

@Controller
@RequestMapping("/users")
```

```kotlin
class UserController(val userValidator: UserValidator,
    val userService: UserService, val securityService: SecurityService) {

    @PostMapping("/registrations")
    fun create(@ModelAttribute form: User, bindingResult:
            BindingResult, model: Model): String {
        userValidator.validate(form, bindingResult)

        if (bindingResult.hasErrors()) {
            model.addAttribute("error", bindingResult.allErrors.first()
                                                    .defaultMessage)
            model.addAttribute("username", form.username)
            model.addAttribute("email", form.email)
            model.addAttribute("password", form.password)

            return "register"
        }

        userService.register(form.username, form.email, form.password)
        securityService.autoLogin(form.username, form.password)

        return "redirect:/home"
    }
}
```

create()处理发送至/users/registrations 的 HTTP POST 请求，并接收 3 个参数。其中，第一个参数为表单，即 User 类对象。@ModelAttribute 用于注解 form。@ModelAttribute 表明，当前参数应通过模型获取。表单模型属性由表单提交给端点的数据填充。此外，username、email 和 password 均由注册表单所提交。所有的 User 类型对象均包含 username、email 和 password 属性。因此，表单提交的数据将赋予至对应的模型属性中。

函数的第二个参数表示为 BindingResult 实例。BindingResult 作为 DataBinder 的持有者，此处用于绑定 UserValidator 所执行的验证处理结果，稍后将对此加以定义。函数的第三个参数为 Model，据此可将属性添加至模型中，以供视图层的后续访问使用。

在进一步解释 create()动作的实现逻辑之前，需要实现 UserValidator 和 SecurityService。UserValidator 的唯一任务是验证提交至后台的用户数据。对此，可定义 com.example.placereviewer.component 包，并于其中定义 UserValidator 类，如下所示：

```kotlin
package com.example.placereviewer.component

import com.example.placereviewer.data.model.User
```

```
import com.example.placereviewer.data.repository.UserRepository
import org.springframework.stereotype.Component
import org.springframework.validation.Errors
import org.springframework.validation.ValidationUtils
import org.springframework.validation.Validator

@Component
class UserValidator(private val userRepository: UserRepository) : Validator {

  override fun supports(aClass: Class<*>?): Boolean {
    return User::class == aClass
  }

  override fun validate(obj: Any?, errors: Errors) {
    val user: User = obj as User
```

下列代码验证所提交的用户参数是否为空。对于空参数，将显示相应的错误代码和错误消息。

```
ValidationUtils.rejectIfEmptyOrWhitespace(errors, "username",
        "Empty.userForm.username", "Username cannot be empty")
ValidationUtils.rejectIfEmptyOrWhitespace(errors, "password",
        "Empty.userForm.password", "Password cannot be empty")
ValidationUtils.rejectIfEmptyOrWhitespace(errors, "email",
        "Empty.userForm.email", "Email cannot be empty")
```

下列代码用于验证所提交用户名的长度，对应长度不应小于 6。

```
if (user.username.length < 6) {
  errors.rejectValue("username", "Length.userForm.username",
                "Username must be at least 6 characters in length")
}
```

下列代码验证所提交的用户名是否已存在：

```
if (userRepository.findByUsername(user.username) != null) {
  errors.rejectValue("username", "Duplicate.userForm.username",
            "Username unavailable")
}
```

验证所提交密码长度。此处，密码长度不应小于 8 个字符。

```
if (user.password.length < 8) {
  errors.rejectValue("password", "Length.userForm.password",
```

```
            "Password must be at least 8 characters in length")
    }
  }
}
```

UserValidator 实现了 Validator 接口，并用于验证对象。因此，UserValidator 覆写了两个方法，即 supports(Class<*>?)和 validate(Any?, Errors)。supports()用于判断验证器是否可验证提供给它的对象。对于 UserValidator，supports()检测所提供的对象是否为 User 类实例。因此，全部 User 类型的对象均支持 UserValidator 验证。

validate()将对所提供的对象进行验证。如果验证失败，则利用 Error 对象注册错误信息。另外，读者还应留意 validate()方法中的注释内容，以进一步了解相关功能。

下面开始着手处理 SecurityService，并尝试实现 SecurityService，以简化登录用户的验证操作，以及注册用户的自动登录问题。

这里，可向 com.example.placereviewer.service 中添加 SecurityService 接口，如下所示：

```
package com.example.placereviewer.service

interface SecurityService {
  fun findLoggedInUser(): String?
  fun autoLogin(username: String, password: String)
}
```

接下来，向 com.example.placereviewer.service 中添加 SecurityServiceImpl 类。顾名思义，SecurityServiceImpl 实现了 SecurityService 接口，如下所示：

```
package com.example.placereviewer.service

import org.springframework.beans.factory.annotation.Autowired
import org.springframework.security.authentication.AuthenticationManager
import org.springframework.security.authentication.UsernamePasswordAuthenticationToken
import org.springframework.security.core.context.SecurityContextHolder
import org.springframework.security.core.userdetails.UserDetails
import org.springframework.stereotype.Service

@Service
class SecurityServiceImpl(private val userDetailsService: AppUserDetailsService)
        : SecurityService {

  @Autowired
```

```
lateinit var authManager: AuthenticationManager

override fun findLoggedInUser(): String? {
  val userDetails = SecurityContextHolder.getContext()
                                         .authentication.details

  if (userDetails is UserDetails) {
    return userDetails.username
  }

  return null
}

override fun autoLogin(username: String, password: String) {
  val userDetails: UserDetails = userDetailsService
                                 .loadUserByUsername(username)

  val usernamePasswordAuthenticationToken =
          UsernamePasswordAuthenticationToken(userDetails, password,
                          userDetails.authorities)

  authManager.authenticate(usernamePasswordAuthenticationToken)

  if (usernamePasswordAuthenticationToken.isAuthenticated) {
    SecurityContextHolder.getContext().authentication =
          usernamePasswordAuthenticationToken
  }
 }
}
```

findLoggedInUser()用于返回当前登录用户的用户名。Username 的检索借助于 Spring 框架的 SecurityContextHolder 类完成。UserDetails 实例的检索方式可描述为：调用 SecurityContextHolder.getContext().authentication.details，并访问登录用户的验证信息。需要注意的是，SecurityContextHolder.getContext().authentication.details 返回一个 Object，而非 UserDetails 实例。因此，此处需要执行类型检测，以确保当前所检索的对象也符合 UserDetails 类型。若是，则返回当前登录用户的用户名；否则返回 null。

在完成了平台注册后，autoLogin()方法用于执行简单的用户验证工作。其间，所提交的用户名和密码作为参数传递至 autoLogin()中；随后，将针对注册用户生成 UsernamePasswordAuthenticationToken。待 UsernamePasswordAuthenticationToken 实例创建完毕后，可利用 AuthenticationManager 验证用户令牌。若 UsernamePasswordAuthenticationToken 验

证成功,则将当前用户的验证属性设置为 UsernamePasswordAuthenticationToken。

在完成了必要的类定义后,下面返回至 UserController 以结束对 create()的讨论。在 create()中,首先利用 UserValidator 实例验证所提交的表单输入。此处,表单数据验证过程中出现的错误都与 Spring 置于控制器中的 BindingResult 实例绑定在一起。考察下列代码片段:

```
if (bindingResult.hasErrors()) {
 model.addAttribute("error", bindingResult.allErrors
                                        .first().defaultMessage)
 model.addAttribute("username", form.username)
 model.addAttribute("email", form.email)
 model.addAttribute("password", form.password)

 return "register"
}
```

代码首先检测 bindingResult,进而判断表单数据验证过程中是否产生错误。当发生错误时,将检索第一个错误的消息,并设置一个 model 属性错误,以保存错误消息,以便后续视图进行访问。除此之外,还可创建 model 属性,并加载用户提交的每项输入。最后,还需要重新向用户显示注册视图。

在上述代码片段中,应注意针对同一 Model 实例的多个方法调用。对此,一种较为清晰的方法是利用 Kotlin 中的 with 函数,如下所示:

```
if (bindingResult.hasErrors()) {
 with (model) {
   addAttribute("error", bindingResult.allErrors.first().defaultMessage)
   addAttribute("error", form.username)
   addAttribute("email", form.email)
   addAttribute("password", form.password)
 }
 return "register"
}
```

其中不难发现 with 函数的简单、方便之处。因此,可使用 with 函数,并对 UserController 进行适当调整。

此处读者可能会产生疑问,为何要将用户的提交数据存储于模型属性中呢?其原因是为了能够存在一种方法,可将注册表单中的数据重置为重新显示注册视图后所提交的内容;否则,即使仅存在一个无效表单输入,用户也必须反复输入所有表单数据,这无疑会令人感到沮丧。

若用户提交的输入内容均为有效,则执行下列代码:

```
userService.register(form.username, form.email, form.password)
securityService.autoLogin(form.username, form.password)
return "redirect:/home"
```

正如期望的那样,当用户提交的数据有效,将在平台上进行注册,并自动登录至其账号。最后,该用户将被重定向至其主页上。在体验注册表单之前,还需要完成以下两项工作:

(1) 使用定义于 register.html 中的模型属性。

(2) 创建 home.html 模板,以及显示该模板的控制器。

上述两项任务均较为简单。首先通过 model 属性修改 register.html 中的表单,如下所示:

```
<form class="form-group col-sm-12 form-vertical form-app"
 id="form-register" method="post" th:action="@{/users/registrations}">
  <div class="col-sm-12 mt-2 lead text-center text-primary">
    Create an account
  </div>
  <hr>
  <!-- utilized model attributes with th:value -->
  <input class="form-control" type="text" name="username"
   placeholder="Username" th:value="${username}" required/>
  <input class="form-control mt-2" type="email" name="email"
   placeholder="Email" th:value="${email}" required/>
  <input class="form-control mt-2" type="password" name="password"
   placeholder="Password" th:value="${password}" required/>
  <span th:if="${error != null}" class="mt-2 text-danger"
   style="font-size: 10px" th:text="${error}"></span>
  <button class="btn btn-primary form-control mt-2 mb-3" type="submit">
    Sign Up!
  </button>
</form>
```

上述代码使用了 Thymeleaf 的 th:value 模板属性,并将表单输入中的值预置于相应的 model 属性值中。下面定义一个简单的 home.html 模板,并将 home.html 模板添加至 templates 目录中,如下所示:

```
<html>
  <head>
    <title> Home</title>
  </head>
  <body>
```

```
    You have been successfully registered and are now in your home page.
  </body>
</html>
```

更新 ApplicationController，并包含一个动作，以处理针对/home 的 GET 请求，如下所示：

```
package com.example.placereviewer.controller

import com.example.placereviewer.service.ReviewService
import org.springframework.stereotype.Controller
import org.springframework.ui.Model
import org.springframework.web.bind.annotation.GetMapping
import java.security.Principal
import javax.servlet.http.HttpServletRequest

@Controller
class ApplicationController(val reviewService: ReviewService) {

  @GetMapping("/register")
  fun register(): String {
    return "register"
  }

  @GetMapping("/home")
  fun home(request: HttpServletRequest, model: Model,
           principal: Principal): String {
    val reviews = reviewService.listReviews()

    model.addAttribute("reviews", reviews)
    model.addAttribute("principal", principal)

    return "home"
  }
}
```

home 动作检索存储于数据库中的评论列表。除此之外，home 动作设置了一个模型属性，用于加载包含当前登录用户的信息主体结构。最后，home 动作向用户显示主页内容。

在完成了必要的准备工作后，下面在 Place Reviewer 平台上注册用户，构建并运行应用程序，并从浏览器（http://localhost:5000/register）中访问注册页面。首先，可输入并提交无效表单数据，进而执行表单的验证工作，如图 10.1 所示。

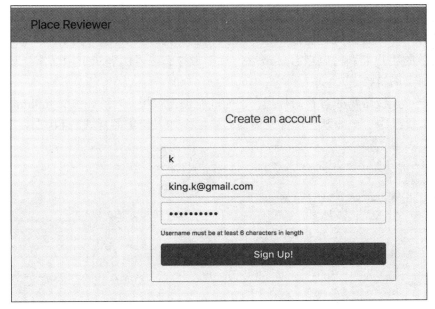

图 10.1　执行表单的验证工作

不难发现，错误被 UserValidator 检测到，并成功地绑定至 BindingResult，随后作为错误信息在视图中予以显示。读者可针对表单尝试输入其他无效数据，以确保验证行为的准确性。下面验证注册逻辑方面的工作。分别在用户名、电子邮件、密码文本框中输入 king.kevin、king.k@gmail.com 以及 Kingsman406，并于随后单击 "Sign Up!" 按钮。此时将生成一个新的账号，并显示主页画面，如图 10.2 所示。

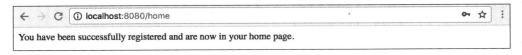

图 10.2　生成一个新的账号，并显示主页画面

当然，后续内容还将对主页进行适当调整，现在让我们将注意力转向创建一个合适的用户登录页面。

10.1.2　实现登录视图

类似于实现用户注册视图，当前，首先需要围绕视图模板展开工作。登录视图所需的模板应包含一个表单，并作为输入内容接收登录用户的用户名和密码。除此之外，还需设置一个按钮，以实现登录表单的提交操作——毕竟，若无法提交，表单将变得毫无意义。此外，如果登录过程出现错误，还应存在某种方式对用户予以提示。例如，用户输入了无效的用户名和密码组合。最后，对于未持有账号的登录页面访客，还应提供一个账号注册页面链接。

在制定了模板所需的各种条件后，下面开始创建模板。对此，将 login.html 文件添加至 template 目录中。一如既往，首先需要向模板中加入所需的样式表和脚本，如下所示：

```html
<!DOCTYPE html>
<html lang="en" xmlns:th="http://www.thymeleaf.org">
  <head>
    <title>Login</title>
    <link rel="stylesheet" th:href="@{/css/app.css}"/>
    <link rel="stylesheet"
     href="/webjars/bootstrap/4.0.0-beta.3/css/bootstrap.min.css"/>

    <script src="/webjars/jquery/3.2.1/jquery.min.js"></script>
    <script src="/webjars/bootstrap/4.0.0-beta.3/
     js/bootstrap.min.js"></script>
  </head>
  <body>
  </body>
</html>
```

在加入了模板所需的样式和 JavaScript 脚本后，下面考察模板的\<body\>。如前所述，当载入页面时，HTML 模板的\<body\>包含了 DOM 元素，并向用户予以显示。在 login.html 的\<body\>标签中，添加下列代码：

```html
<div th:insert="fragments/navbar :: navbar"></div>

<div class="container-fluid" style="z-index: 2; position: absolute">
  <div class="row mt-5">
    <div class="col-sm-4 col-xs-2"></div>
    <div class="col-sm-4 col-xs-8">
     <form class="form-group col-sm-12 form-vertical form-app"
       id="form-login" method="post" th:action="@{/login}">
       <div class="col-sm-12 mt-2 lead text-center text-primary">
         Login to your account
       </div>
       <hr>
       <input class="form-control" type="text" name="username"
```

```
         placeholder="Username" required/>
       <input class="form-control mt-2" type="password"
         name="password" placeholder="Password" required/>
       <span th:if="${param.error}" class="mt-2 text-danger"
         style="font-size: 10px">
         Invalid username and password combination
       </span>
       <button class="btn btn-primary form-control mt-2 mb-3"
         type="submit">
         Go!
       </button>
     </form>
     <div class="col-sm-12 text-center" style="font-size: 12px">
       Don't an account? Register <a href="/register">here</a>
     </div>
   </div>
   <div class="col-sm-4 col-xs-2">
     <div th:if="${param.logout}"
       class="col-sm-12 text-success text-right">
       You have been logged out.
     </div>
   </div>
 </div>
</div>
```

在<body>添加完毕后，所创建的 HTML 页面依然可以描述登录页面所需的结构。除了加入所需的表单之外，我们还向页面中添加了之前创建的导航栏片段——此处无须编写样板代码。此外，若在登录过程中出现错误，用户还可提供与此相关的反馈方式。对应代码如下所示：

```
<span th:if="${param.error}" class="mt-2 text-danger"
 style="font-size: 10px">
 Invalid username and password combination
</span>
```

若 param.error 被设置，则表明用户登录过程中出现错误，同时会向用户显示 "Invalid username and password combination" 消息。需要注意的是，登录页面一般是用户与 Web 应用程序的首个接触点，同时也是交互会话过程中的最后一个交互点，例如用户的注销操作。在用户与应用程序交互完毕并注销时，应被重定向至登录页面。对此，可添加一些文本信息，以提示用户已退出当前账号，如下所示：

```html
<div th:if="${param.logout}" class="col-sm-12 text-success text-right">
  You have been logged out.
</div>
```

在成功从当前账号注销后,将向用户显示<div>标签。此时,理想状态下应实现一个控制器并显示 login.html。但不要忘记,我们已通过 MvcConfig 且利用自定义 Spring MVC 配置实现了这一项内容,如下所示:

```
...
override fun addViewControllers(registry: ViewControllerRegistry?) {
  registry?.addViewController("/login")?.setViewName("login")
}
...
```

此处利用 ViewControllerRegistry 实例添加了一个视图控制器,处理/login 的请求,并将所用视图设置为刚刚实现的登录模板。构建并运行应用程序,并查看最新生成的视图。Web 页面可通过 http:localhost:5000/login 进行访问,如图 10.3 所示。

图 10.3　最新生成的视图

当输入无效的用户信息后,将会显示错误消息,如图 10.4 所示。

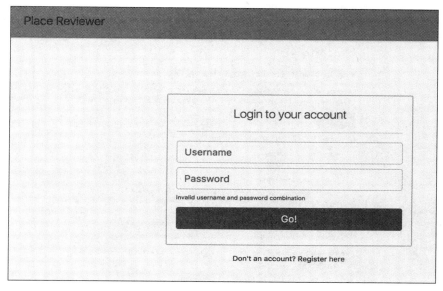

图 10.4　无效输入信息验证

另外一方面，在输入了有效信息后，用户将被转至应用程序的主页。对于主页，我们需要完善其视图层，并自此开始与 Google Places API 协同工作。因此，在执行后续处理之前，需要对应用程序进行配置。

10.1.3　Google Places API Web 服务

Google Places API 的配置过程十分简单，仅涉及以下两个步骤。

（1）获取 API 密钥。

（2）将 Google Places API 添加至 Web 应用程序中。

1．获取 API 密钥

读者可访问 https://developers.google.com/places/web-service/get-api-key，定位至 Get an API key 部分，并单击 GET A KEY 按钮，如图 10.5 所示。

随后将显示相关模态框，读者可以此选取或创建与 Google Places API Web 服务集成的项目。单击下拉菜单并选择 Create a new project。此时，读者将被提示输入项目名称，此处可输入 Place Reviewer 作为项目名称，如图 10.6 所示。

在输入了项目名称后，单击 NEXT 按钮执行后续操作。随后，当前项目将通过 API 予以设置，同时还将显示所用的 API 密钥，如图 10.7 所示。

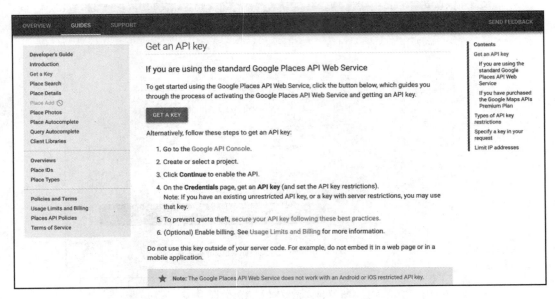

图 10.5　获取 API 密钥

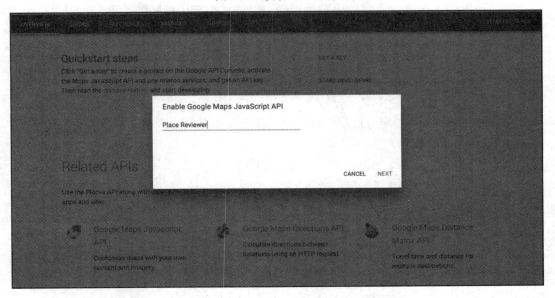

图 10.6　输入 Place Reviewer 作为项目名称

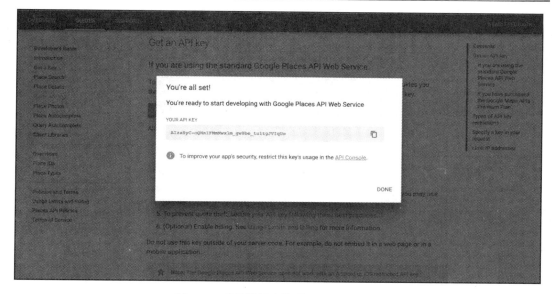

图 10.7 项目设置完毕并显示 API 密钥

在持有 API 密钥后,下面考察如何使用 API 密钥。

2. 在 Web 应用程序中设置 Google Places

对于 Google Places API Web 服务,API 的应用十分简单,且与生成 API 密钥十分相似。当在 Web 应用程序中使用生成的 API 密钥时,仅需在使用 Web 服务的页面标记中包含以下 HTML 代码:

```
<script type="text/javascript"
src="https://maps.googleapis.com/maps/api/js?key={{API_KEY}}&libraries=places"></script>
```

其中,确保利用所生成的 API 密钥替换{{API_KEY}}。

10.1.4 实现主视图

在开始编写代码之前,可针对所创建的视图设置一个粗略的图形化模型。从长远来看,这将为构建过程提供一个清晰的方向,从而节省大量的时间。

主页的创建过程涉及以下几个步骤:

(1)显示最新的地点评论。

(2)可直接访问发表评论的 Web 页面。

（3）提供一种方式，用户可退出当前账号。

（4）借助于地图，用户可查看评论地点的准确位置。

综上所述，图10.8显示了最终模板的粗略示意图。

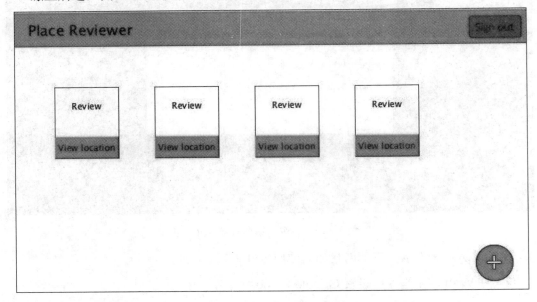

图10.8 最终模板的示意图

图10.8中的各项内容实现起来并不复杂。单击View location将显示应用程序用户，其间，地图将显示评论地点的准确位置。

在明晰了即将创建的模板后，下面开始对其进行编码。与以往一样，首先需要向模板中加入外部样式表和脚本。打开home.html文件并添加下列代码：

```html
<!DOCTYPE html>
<html lang="en" xmlns:th="http://www.thymeleaf.org">
 <head>
  <title>Home</title>
  <!-- Addition of external stylesheets -->
  <link rel="stylesheet" th:href="@{/css/app.css}"/>
  <link rel="stylesheet"
    href="https://cdnjs.cloudflare.com/ajax/libs/toastr.js
        /latest/toastr.min.css">
  <link rel="stylesheet"
    href="/webjars/bootstrap/4.0.0-beta.3/css/bootstrap.min.css"/>
  <link rel="stylesheet"
```

```html
    href="https://maxcdn.bootstrapcdn.com/font-awesome
        /4.7.0/css/font-awesome.min.css"/>
<link href="https://fastcdn.org/Buttons/2.0.0/css/buttons.css"
 rel="stylesheet">

<!-- Inclusion of external Javascript -->
<script src="/webjars/jquery/3.2.1/jquery.min.js"></script>
<script src="https://cdnjs.cloudflare.com/ajax/libs
            /toastr.js/latest/toastr.min.js">
</script>
<script src="https://cdnjs.cloudflare.com/ajax/libs
            /popper.js/1.12.6/umd/popper.min.js">
</script>
<script src="/webjars/bootstrap/4.0.0-beta.3/
            js/bootstrap.min.js"></script>
<script src="https://fastcdn.org/Buttons/2.0.0/js/buttons.js"></script>
<script type="text/javascript"
 src="https://maps.googleapis.com/maps/api/js?key={{API_KEY}}
        &libraries=places">
</head>
</html>
```

除了外部样式表之外，还将使用到模板中的内部样式。当在 HTML 文件中定义内部样式表时，可在 HTML 头中简单地添加<sstyle>，并输入所期望的 CSS 规则即可。接下来，可向 home.html 中添加下列样式：

```html
</script>

    <!-- Definition of internal styles -->
    <style>
      #map {
        height: 400px;
      }

      .container-review {
        background-color: white;
        border-radius: 2px;
        font-family:sans-serif;
        box-shadow: 0 0 1px black;
        border-color: black;
        padding: 0;
        min-width: 250px;
        height: 230px;
```

```css
       }

       .review-author {
         font-size: 15px
       }

       .review-location {
         font-size: 12px
       }

       .review-title {
         font-size: 13px;
         text-decoration-style: dotted;
         height: calc(20 / 100 * 230px);
       }

       .review-content {
         font-size: 12px;
         height: calc(40 / 100 * 230px);
       }

       .review-header {
         height: calc(20 / 100 * 230px)
       }

       hr {
         margin: 0;
       }

       .review-footer {
         height: calc(20 / 100 * 230px);
       }
    </style>
```

下面考察页面的主体内容。读者已经了解到，构成 HTML 模板主体的所有元素须位于<body>标签内。相应地，可向模板文件中添加下列 HTML 代码：

```html
<!-- Invokes the showNoReviewNotification() function defined in -->
<!-- internal Javascript of this file upon document load. -->
<body
  th:onload="'javascript:showNoReviewNotification(
    ' + ${reviews.size() == 0} + '
  )'">
```

```html
<div th:insert="fragments/navbar :: navbar"></div>
<div class="container">
  <div class="row mt-5">
    <!-- Creates view containers for each review retrieved -->
    <!-- Distinct <div> containers are created for the -->
    <!-- review author, location, title and body. -->
    <div th:each="review: ${reviews}"
      class="col-sm-2 container-review mt-4 mr-2">
      <div class="review-header pt-1">
        <div class="col-sm-12 review-author text-success">
          <b th:text="${review.reviewer.username}"></b>
        </div>
        <div th:text="${review.placeName}"
         class="col-sm-12 review-location">
        </div>
      </div>
      <hr>
      <b>
        <div th:text="${review.title}"
          class="col-sm-12 review-title pt-1">
        </div>
      </b>
      <hr>
      <div th:text="${review.body}"
       class="col-sm-12 review-content pt-2">
      </div>
      <div class="review-footer">
        <!-- Creation of distinct DOM
        <button> elements for the display of reviewed locations. -->
          <!-- Upon button click, the application renders a modal
            showing the reviewed location on a map -->
          <button class="col-sm-12 button button-small button-primary"
            type="button" data-toggle="modal" data-target="#mapModal"
            style="height: inherit; border-radius: 2px;"
            th:onclick="'javascript:showLocation(
              ' + ${review.latitude} + ','
                + ${review.longitude} + ',\''
                + ${review.placeId} + '\'
            )'">
            <i class="fa fa-map-o" aria-hidden="true"></i>
            View location
          </button>
```

```html
        </div>
      </div>
    </div>
  </div>
```

读者不必为代码块的具体功能而担心,稍后将会逐一加以解释。下面继续向 home.html 中添加下列代码:

```html
<!-- Modal creation -->
<div class="modal fade" id="mapModal">
  <div class="modal-dialog modal-lg" role="document">
    <div class="modal-content">
      <div class="modal-header">
        <h5 class="modal-title">Reviewed location</h5>
        <button type="button" class="close"
         data-dismiss="modal" aria-label="Close">
          <span aria-hidden="true">&times;</span>
        </button>
      </div>
      <div class="modal-body">
        <div class="container-fluid">
          <div id="map"> </div>
        </div>
      </div>
      <div class="modal-footer">
        <button type="button" class="btn btn-primary"
         data-dismiss="modal">
          Done
        </button>
      </div>
    </div>
  </div>
</div>
```

上述代码声明了一种模态框,用于加载显示评论地址的地图,稍后将对此予以实现。下面继续讨论\<body\>,并添加下列代码。

```html
<span style="bottom: 20px; right: 20px; position: fixed">
  <form method="get" th:action="@{/create-review}">
    <button class="button button-primary button-circle
     button-giant navbar-bottom" type="submit">
      <i class="fa fa-plus"></i>
    </button>
```

```
</form>
</span>
```

最后，还需针对 HTML 页加入内部 JavaScript，如下所示：

```
<script>
 //Shows a toast notification to the user when no review is present
 function showNoReviewNotification(show) {
   if (show) {
     toastr.info('No reviews to see');
   }
 }
```

下列函数初始化并显示地图，进而显示评论的地理位置。

```
function showLocation(latitude, longitude, placeId) {
 var center = new google.maps.LatLng(latitude, longitude);

 var map = new google.maps.Map(document.getElementById('map'), {
   center: center,
   zoom: 15,
   scrollwheel: false
 });
 var service = new google.maps.places.PlacesService(map);

 loadPlaceMarker(service, map, placeId);
}
```

加载位置标记将在评论地址上创建地图标记，如下所示：

```
      function loadPlaceMarker(service, map, placeId) {
        var request = {
          placeId: placeId
        };

        service.getDetails(request, function (place, status) {
          if (status === google.maps.places.PlacesServiceStatus.OK) {
            new google.maps.Marker({
              map: map,
              title: place.name,
              place: {
                placeId: place.place_id,
                location: place.geometry.location
              }
```

```
            })
        }
    });
  }
</script>
</body>
```

下面从<head>标签开始,并着手实现视图方面的内容。之前曾添加了主页所需的样式表和脚本,下列代码添加了 CSS 方面的内容。

```
<link rel="stylesheet" th:href="@{/css/app.css}"/>
<link rel="stylesheet" href="https://cdnjs.cloudflare.com/ajax/libs/toastr.js/latest/toastr.min.css">
<link rel="stylesheet" href="/webjars/bootstrap/4.0.0-beta.3/css/bootstrap.min.css"/>
<link rel="stylesheet" href="https://maxcdn.bootstrapcdn.com/font-awesome/4.7.0/css/font-awesome.min.css"/>
<link href="https://fastcdn.org/Buttons/2.0.0/css/buttons.css" rel="stylesheet">
```

上述代码针对应用程序的 CSS 加入了外部样式表。其中,Toastr 库用于生成 JavaScript 通知消息;Bootstrap 库在设置站点和 Web 应用程序方面表现得十分强大;Font Awesome 则是针对站点和 Web 应用程序的图标工具;Buttons 则是一个高度自定义 Web 和 CSS 按钮库。

在包含了 CSS 后,还需要设置一些外部 JavaScript 方面的内容,如下所示:

```
<script src="/webjars/jquery/3.2.1/jquery.min.js"></script>
<script src="https://cdnjs.cloudflare.com/ajax/libs/toastr.js/latest/toastr.min.js"></script>
<script src="https://cdnjs.cloudflare.com/ajax/libs/popper.js/1.12.6/umd/popper.min.js"></script>
<script src="/webjars/bootstrap/4.0.0-beta.3/js/bootstrap.min.js"></script>
<script src="https://fastcdn.org/Buttons/2.0.0/js/buttons.js"></script>
<script type="text/javascript" src="https://maps.googleapis.com/maps/api/js?key={{API_KEY}}&libraries=places"></script>
```

上述脚本所包含的顺序依次为:JQuery,针对简化客户端脚本处理而设计的 JavaScript

库；Toastr 和 Popper，用于管理 Web 应用程序弹出消息的库；随后是 Bootstrap、Buttons 和 Google Places API Web 服务。再次强调，对于 Google Places API Web 服务，需要利用 API 密钥替换{{API_KEY}}，这一点十分重要。

JavaScript 之后，我们将针对定义内部样式表。然而，样式表及其创建过程的讨论超出了本书的范围，读者可回顾 CSS 方面的内容。对于 home.html，下面添加<body>：

```
<body th:onload="'javascript:showNoReviewNotification(' +
${reviews.size()== 0} + ')'">
```

其中，th:onload 用于指定页面完全加载后需要运行的 JavaScript，即产生 onload 事件后所需执行的代码。此处，所运行的脚本为模板中定义的 JavaScript 函数 showNoReviewNotification(boolean)。若模型提供的评论列表为空，该函数则显示相关消息，表明当前不存在可查看的评论内容。showNoReviewNotification(boolean)在模板中的声明如下所示：

```
function showNoReviewNotification(show) {
  if (show) {
    toastr.info('No reviews to see');
  }
}
```

showNoReviewNotification(boolean)函数接收单一参数 show。若 show 为 true，Toastr 库将向用户显示通知消息。

对于可显示的评论内容，则针对每条评论创建一个容器，如下所示：

```
<!-- Creates view containers for each review retrieved -->
<!-- Distinct <div> containers are created for the -->
<!-- review author, location, title and body. -->
<div th:each="review: ${reviews}" class="col-sm-2 container-review mt-4 mr-2">
  <div class="review-header pt-1">
    <div class="col-sm-12 review-author text-success">
      <b th:text="${review.reviewer.username}"></b>
    </div>
    <div th:text="${review.placeName}" class="col-sm-12 review-location">
    </div>
  </div>
  <hr>
  <b>
    <div th:text="${review.title}" class="col-sm-12 review-title pt-1">
    </div>
```

```
    </b>
    <hr>
    <div th:text="${review.body}" class="col-sm-12 review-content pt-2">
    </div>
    <div class="review-footer">
      <!-- Creation of distinct DOM <button> elements for the
      display of reviewed locations. -->
      <!-- Upon button click, the application renders a modal
      showing the reviewed location on a map -->
      <button class="col-sm-12 button button-small button-primary"
        type="button" data-toggle="modal" data-target="#mapModal"
        style="height: inherit; border-radius: 2px;"
        th:onclick="'javascript:showLocation('
          + ${review.latitude} + ','
          + ${review.longitude} + ',\''
          + ${review.placeId} + '\
        ')'">
        <i class="fa fa-map-o" aria-hidden="true"></i>
        View location
      </button>
    </div>
</div>
```

每个评论容器将显示评论者的用户名、评论地址名称、评论标题、评论内容，以及可供用户查看评论位置的按钮。这里，Thymeleaf 的 th:each 用于遍历 reviews 列表中的每条评论。如下所示：

```
<div th:each="review: ${reviews}" class="col-sm-2 container-review mt-4 mr-2">
```

当理解遍历处理过程时，一种较好的方法是将 th:each="review:${reviews}" 读作 "For each review in reviews"。此处，当前所遍历的评论由 review 遍历加载。因此，被遍历的评论所持有的数据可像其他对象那样予以访问，如下所示：

```
<div th:text="${review.placeName}" class="col-sm-12 review-location">
</div>
```

其中，th:text 负责将<div>持有的文本设置为 review.placeName 的赋值结果。另外，这里有必要解释一下位置地图相对于用户的显示方式。考察下列代码：

```
<button class="col-sm-12 button button-small button-primary" type="button"
  data-toggle="modal" data-target="#mapModal" style="height: inherit;
  border-radius: 2px;"
```

```
th:onclick="'javascript:showLocation('
 + ${review.latitude} + ','
 + ${review.longitude} + ',\''
 + ${review.placeId} + '\
 ')'">
 <i class="fa fa-map-o" aria-hidden="true"></i>
 View location
</button>
```

上述代码块定义了一个按钮，并在产生单击事件时执行两项任务。首先向用户显示 ID mapModal 标识的模态框；其次将初始化并显示地图，进而显示评论的准确地理位置。其中，地图的显示过程可通过定义于模板文件中的 showLocation() JavaScript 函数完成。

showLocation()函数接收 3 个参数。第一个参数表示经度坐标，第二个参数表示维度坐标，第三个参数表示评论地理位置的唯一标识符，即位置 ID。地理位置的位置 ID 可通过 Google Places API 获得。首先，showLocation()检索地理位置坐标的中心点——可使用 Google Places API 中的 google.maps.LatLng 类。简单地讲，LatLng 表示地理坐标点（精度和维度）。当检索到中心点后，新地图将通过 Map 类（由 Google Places API 提供）创建，如下所示：

```
var map = new google.maps.Map(document.getElementById('map'), {
  center: center,
  zoom: 15,
  scrollwheel: false
});
```

生成后的地图将置于包含 ID map 的 DOM 容器元素中。在地图构建完毕后，可借助于 loadPlaceMarker() 函数在准确位置处生成位置标记。loadPlaceMarker() 函数接收 google.maps.places.PlacesService、Map 以及一个位置 ID 作为其参数。其中，PlacesService 类包含了检索位置信息和搜索位置的相关方法。

google.maps.places.PlacesService 实例首先用于检索包含特定位置 ID（评论的地理位置）的位置信息。如果位置信息被成功检索，status ===google.maps.places.PlacesServiceStatus.OK 的计算结果为 true，位置标记将置于当前地图上。相应地，该标记通过 google.maps.Marker 类予以创建。Marker()接收一个可选的 Options 对象作为其唯一参数。若存在该对象，那么，位置标记将通过特定的可选项予以创建。此时，地图定义于 Options 对象中。因此，标记在生成后即添加至地图中。

最后，我们还向模板中添加了一个表单，并在提交后发送/reviews/new 路径上的 GET 请求。此外，还设置了一个按钮——单击该按钮将提交表单。对应代码如下所示：

```html
<form method="get" th:action="@{/reviews/new}">
  <button class="button button-primary button-circle button-giant
    navbar-bottom" type="submit"><i class="fa fa-plus"></i></button>
</form>
```

至此,主页设计暂告一段落。读者可尝试重新构建并运行应用程序、注册账号并查看如图 10.9 所示的主页画面。

图 10.9　主页画面显示效果

当前,平台中尚未显示任何评论,我们需要围绕主页画面执行相关操作,进而生成评论内容。

10.1.5　生成评论

截至目前,前述内容构建了用户注册视图、登录视图,以及登录用户主页,以查看所发表的评论内容。下面考察与发表评论相关的视图。像以往一样,在创建视图之前,首先定义相关动作,以向用户显示相关视图。在 Application Controller 类中添加下列代码:

```kotlin
@GetMapping("/create-review")
fun createReview(model: Model, principal: Principal): String {
  model.addAttribute("principal", principal)
  return "create-review"
}
```

通过向客户端返回 create-review.html 模板,createReview()动作处理/create-review 路径的 HTTP GET 请求。下面向项目 template 目录中添加 create-review.html 文件。

同样,需要向 createreview.html 中添加外部样式和脚本,如下所示:

```html
<!DOCTYPE html>
<html lang="en" xmlns:th="http://www.thymeleaf.org">
```

```html
<head>
  <title>New review</title>
  <!-- Addition of external stylesheets -->
  <link rel="stylesheet" th:href="@{/css/app.css}"/>
  <link rel="stylesheet" href="/webjars/bootstrap/4.0.0-beta.3
                /css/bootstrap.min.css"/>
  <link rel="stylesheet" href="https://maxcdn.bootstrapcdn.com
                /font-awesome/4.7.0/css/font-awesome.min.css"/>
  <link href="https://fastcdn.org/Buttons/2.0.0/css/buttons.css"
   rel="stylesheet">

  <!-- Inclusion of external Javascript -->
  <script src="/webjars/jquery/3.2.1/jquery.min.js"></script>
  <script src="https://cdnjs.cloudflare.com/ajax/libs/popper.js
                /1.12.6/umd/popper.min.js"></script>
  <script src="/webjars/bootstrap/4.0.0-beta.3/
                js/bootstrap.min.js"></script>
  <script src="https://fastcdn.org/Buttons/2.0.0/js/buttons.js"> </script>
  <script type="text/javascript" src="https://maps.googleapis.com/
                maps/api/js?key={{API_KEY}}&libraries=places">
  </script>
```

下列代码针对 Web 页面添加了所需的内部样式表。

```html
  <!-- Definition of internal styles -->
  <style>
    #map {
      height: 400px;
    }

    #container-place-data {
      height: 0;
      visibility: hidden;
    }

    #container-place-info {
      font-size: 14px;
    }

    #container-selection-status {
      visibility: hidden;
    }
  </style>
</head>
```

接下来需要构建表单,用于评论数据的输入。在 create-review.html 模板中添加下列代码:

```html
<body>
  <div th:insert="fragments/navbar :: navbar"> </div>
  <div class="container-fluid">
    <div class="row">
      <div class="col-sm-12 col-xs-12">
        <!-- Review form creation -->
        <form class="form-group col-sm-12 form-vertical form-app"
          id="form-login" method="post" th:action="@{/reviews}">
          <div class="col-sm-12 mt-2 lead">Write your review</div>
          <div th:if="${error != null}" class="text-danger"
            th:text="${error}"> </div>
          <hr>
          <input class="form-control" type="text" name="title"
            placeholder="Title" th:value="${title}" required/>
          <textarea class="form-control mt-4" rows="13" name="body"
            placeholder="Review" th:value="${body}" required></textarea>
          <div class="form-group" id="container-place-data">
            <!-- Input fields for location specific form data -->
            <!-- Form input data for the fields below are
              provided by the Google Places API -->
            <input class="form-control" id="place_address"
              th:value="${placeAddress}" type="text" name="placeAddress"
              required/>
            <input class="form-control" id="place_name" type="text"
              name="placeName" th:value="${placeName}" required/>
            <input class="form-control" id="place_id" type="text"
              name="placeId" th:value="${placeId}" required/>
            <input id="location-lat" type="number" name="latitude"
              step="any" th:value="${latitude}" required/>
            <input id="location-lng" type="number" name="longitude"
              step="any" th:value="${longitude}" required/>
          </div>
          <div class="form-group mb-3">
            <button class="button button-pill" type="button"
              data-toggle="modal" data-target="#mapModal">
              <i class="fa fa-map-marker" aria-hidden="true"></i>
              Select Location
            </button>
            <button class="button button-pill button-primary">
```

```html
      Submit Review</button>
    </div>
    <div class="text-success ml-2" id="container-selection-status">
      Location selected</div>
  </form>
 </div>
</div>
```

下面添加模态框，以使用户可从地图中选择评论地理位置，代码如下所示：

```html
<!-- Map Modal -->
<div class="modal fade" id="mapModal">
  <div class="modal-dialog modal-lg" role="document">
    <div class="modal-content">
      <div class="modal-header">
        <h5 class="modal-title">Select place to review</h5>
        <button type="button" class="close" data-dismiss="modal"
         aria-label="Close">
          <span aria-hidden="true">&times;</span>
        </button>
      </div>
      <div class="modal-body">
        <div class="container-fluid">
          <div id="map"> </div>
            <div class="row mt-2" id="container-place-info">
              <div class="col-sm-12" id="container-place-name">
                <b>Place Name:</b>
              </div>
              <div class="col-sm-12" id="container-place-address">
                <b>Place Address:</b>
              </div>
            </div>
        </div>
      </div>
      <div class="modal-footer">
        <button type="button" class="btn btn-primary"
         data-dismiss="modal">Done</button>
      </div>
    </div>
  </div>
</div>
```

最后，还需要添加内部 JavaScript 以完成当前模板，代码如下所示：

```
<script>
 // form field reference creation
 var formattedAddressField = document
                     .getElementById('place_address');
 var placeNameField = document.getElementById('place_name');
 var placeIdField = document.getElementById('place_id');
 var latitudeField = document.getElementById('location-lat');
 var longitudeField = document.getElementById('location-lng');

 // container reference creation
 var containerPlaceName = document.getElementById
                     ('container-place-name');
 var containerPlaceAddress = document.getElementById
                     ('container-place-address');
 var containerSelectionStatus = document.getElementById
                     ('container-selection-status');
```

在上述代码片段中，创建了指向当前页面中 DOM 元素的引用，包括位置输入框（针对地址、名称、ID 以及地址的经纬度坐标）。除此之外，还添加了指向显示所选地址信息的容器的引用，例如地址名称和地址。对此，需要声明相应的函数，其中包括 initialize()、getPlaceDetailsById()、updateViewData()、setFormValues()、showSelectionsStatusContainer() 以及 setContainerText()。

接下来首先向模板中加入 initialize() 和 getPlaceDetailsById() 函数，如下所示：

```
//invoked to initialize Google map
function initialize() {

 navigator.geolocation.getCurrentPosition(function(location) {
   var latitude = location.coords.latitude;
   var longitude = location.coords.longitude;

   var center = new google.maps.LatLng(latitude, longitude);

   var map = new google.maps.Map(document.getElementById('map'), {
     center: center,
     zoom: 15,
     scrollwheel: false
   });

   var service = new google.maps.places.PlacesService(map);
```

```
    map.addListener('click', function(data) {
      getPlaceDetailsById(service, data.placeId);
    });
  });
}
```

下列函数可通过 Google Places API 获取特定位置，进而检索具体的地理位置。

```
function getPlaceDetailsById(service, placeId) {
  var request = {
    placeId: placeId
  };

  service.getDetails(request, function (place, status) {
    if (status === google.maps.places.PlacesServiceStatus.OK) {
      updateViewData(place)
    }
  });
}
```

下面是 updateView() 和 setFormValues() 函数。

```
//Invoked to update view information
function updateViewData(place) {
  setFormValues(
    place.formatted_address,
    place.name,
    place.place_id,
    place.geometry.location.lat(),
    place.geometry.location.lng()
  );

  setContainerText('<b>Place Name: </b>' + place.name,
    '<b>Place Address: </b>' + place.formatted_address);

  showSelectionStatusContainer();
}
```

当调用下列函数时，将更新视图表单数据。

```
function setFormValues(formattedAddress, placeName, placeId,
                      latitude, longitude) {
```

```
            formattedAddressField.value = formattedAddress;
            placeNameField.value = placeName;
            placeIdField.value = placeId;
            latitudeField.value = latitude;
            longitudeField.value = longitude;
        }
```

最后,添加下列代码以完成当前模板。

```
        function showSelectionStatusContainer() {
            containerSelectionStatus.style.visibility = 'visible'
        }

        function setContainerText(placeNameText, placeAddressText) {
            containerPlaceName.innerHTML = placeNameText;
            containerPlaceAddress.innerHTML = placeAddressText;
        }
        // Initializes map upon window load completion
        google.maps.event.addDomListener(window, 'load', initialize);
    </script>
  </body>
</html>
```

与上述模板类似,create-review.html 中包含了外部和内部 CSS,以及 HTML<head>标签中模板所需的 JavaScript。进一步讲,需要创建一个表单,并接收下列表单数据作为其输入内容:

- title:针对所发表的评论定义的用户标题。
- body:评论内容,即主要的评论文本。
- placeAddress:评论的地理位置。
- placeName:评论的地名。
- placeId:评论位置的唯一 ID。
- latitude:评论位置的经度坐标。
- longitude:评论位置的纬度坐标。

此处,无须针对用户提供 placeAddress、placeName、placeId、latitud 以及 longitude 的表单输入。因此,可隐藏前述输入元素的父<div>。此外,还需使用 Google Places API 检索位置信息。需要注意的是,在模板中,我们利用模态框显示所选地址的地图。该模态框可通过添加至模板中的 button 进行切换,如下所示:

```
<button class="button button-pill" type="button" data-toggle="modal"
datatarget="#mapModal">
```

```html
<i class="fa fa-map-marker" aria-hidden="true"></i> Select Location
</button>
```

单击按钮将向用户显示地图模态框。当显示地图时,用户可从地图中单击评论地点。相应地,执行这一类单击动作将触发地图的单击事件,并通过定义于模板中的监听器进行处理,如下所示:

```
map.addListener('click', function(data) {
  getPlaceDetailsById(service, data.placeId);
});
```

getPlacesDetailsById()函数接收两个参数,即 google.maps.places.PlacesService 实例,以及评论信息的位置 ID。随后,PlacesService 实例用于检索位置信息。在获取了相关信息后,视图也据此进行更新:设置特定位置的表单数据、更新地图模态框中的位置名称和地址容器,并向用户显示位置被成功选定这一消息。当位置选取以及全部所需表单数据输入操作完毕后,用户即可提交其评论内容。

在查看评论发表页面之前,还需创建一个评论验证器以及一个控制器,以处理发送至/reviews 路径中的 POST 请求。对此,可向 com.example.placereviewer.component 中添加 ReviewValidator 类,如下所示:

```kotlin
package com.example.placereviewer.component

import com.example.placereviewer.data.model.Review
import org.springframework.stereotype.Component
import org.springframework.validation.Errors
import org.springframework.validation.ValidationUtils
import org.springframework.validation.Validator

@Component
class ReviewValidator: Validator {

  override fun supports(aClass: Class<*>?): Boolean {
    return Review::class == aClass
  }

  override fun validate(obj: Any?, errors: Errors) {
    val review = obj as Review

    ValidationUtils.rejectIfEmptyOrWhitespace(errors, "title",
                    "Empty.reviewForm.title", "Title cannot be empty")
```

```
        ValidationUtils.rejectIfEmptyOrWhitespace(errors, "body",
                        "Empty.reviewForm.body", "Body cannot be empty")
        ValidationUtils.rejectIfEmptyOrWhitespace(errors, "placeName",
                        "Empty.reviewForm.placeName")
        ValidationUtils.rejectIfEmptyOrWhitespace(errors,
                        "placeAddress","Empty.reviewForm.placeAddress")
        ValidationUtils.rejectIfEmptyOrWhitespace(errors, "placeId",
                        "Empty.reviewForm.placeId")
        ValidationUtils.rejectIfEmptyOrWhitespace(errors, "latitude",
                        "Empty.reviewForm.latitude")
        ValidationUtils.rejectIfEmptyOrWhitespace(errors, "longitude",
                        "Empty.reviewForm.longitude")

        if (review.title.length < 5) {
          errors.rejectValue("title", "Length.reviewForm.title",
                        "Title must be at least 5 characters long")
        }

        if (review.body.length < 5) {
          errors.rejectValue("body", "Length.reviewForm.body",
                        "Body must be at least 5 characters long")
        }
    }
}
```

前述内容曾解释了自定义验证器的工作机制,下面通过代码方式对此稍作回顾,即针对与评论相关的 HTTP 请求实现控制器类。相应地,在 com.example.placereviewer.controller 中定义 ReviewController 类,并添加下列代码:

```
package com.example.placereviewer.controller

import com.example.placereviewer.component.ReviewValidator
import com.example.placereviewer.data.model.Review
import com.example.placereviewer.service.ReviewService
import org.springframework.stereotype.Controller
import org.springframework.ui.Model
import org.springframework.validation.BindingResult
import org.springframework.web.bind.annotation.ModelAttribute
import org.springframework.web.bind.annotation.PostMapping
import org.springframework.web.bind.annotation.RequestMapping
import javax.servlet.http.HttpServletRequest
```

```kotlin
@Controller
@RequestMapping("/reviews")
class ReviewController(val reviewValidator: ReviewValidator,
                      val reviewService: ReviewService) {

 @PostMapping
 fun create(@ModelAttribute reviewForm: Review, bindingResult: BindingResult,
            model: Model, request: HttpServletRequest): String {
   reviewValidator.validate(reviewForm, bindingResult)

   if (!bindingResult.hasErrors()) {
     val res = reviewService.createReview(request.userPrincipal.name,
                                          reviewForm)
     if (res) {
       return "redirect:/home"
     }
   }

   with (model) {
     addAttribute("error", bindingResult.allErrors.first().defaultMessage)
     addAttribute("title", reviewForm.title)
     addAttribute("body", reviewForm.body)
     addAttribute("placeName", reviewForm.placeName)
     addAttribute("placeAddress", reviewForm.placeAddress)
     addAttribute("placeId", reviewForm.placeId)
     addAttribute("longitude", reviewForm.longitude)
     addAttribute("latitude", reviewForm.latitude)
   }

   return "create-review"
 }
}
```

在定义了 ReviewValidator 和 ReviewController 类之后，构建并运行当前项目，执行用户登录操作，并从浏览器中访问 http://localhost:5000/create-review。

页面加载完毕后，将显示一个表单，用户可于其中添加新的评论内容，如图 10.10 所示。

用户需要在提交评论之前选取评论地理位置。对此，可单击 Select Location 按钮，如图 10.11 所示。

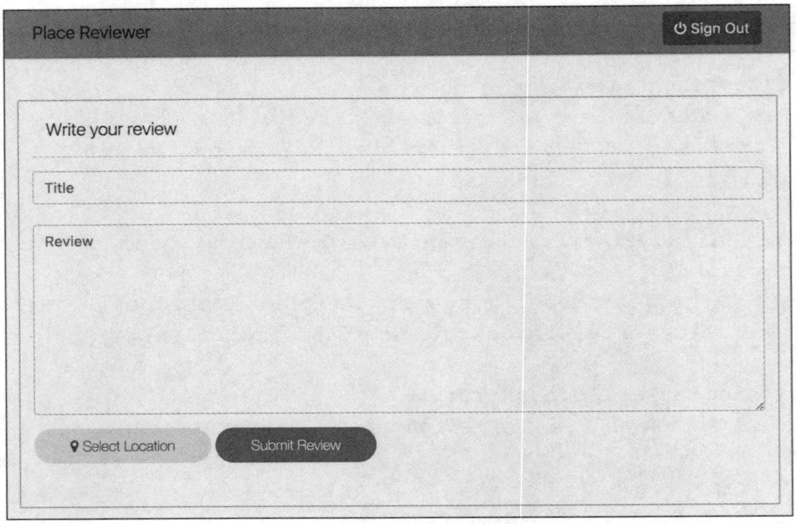

图 10.10 发表评论内容

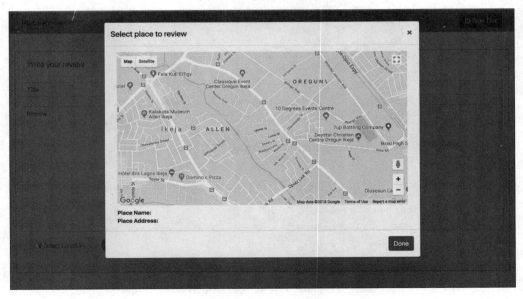

图 10.11 选取地理位置

单击 Select Location 按钮将向用户显示一个模态框,其中包含了选取评论地理位置的地图。从地图上单击某个位置将显示地图上的一个信息窗口,其中包含了与单击位置相关的信息。除此之外,加载所选位置名称和地址的模态框容器将被更新,如图 10.12 所示。

第 10 章 实现 Place Reviewer 前端

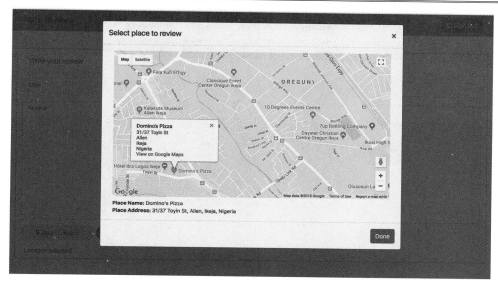

图 10.12 选取评论位置

在用户选择了评论位置后,即可单击 Done 按钮关闭模态框,并继续填写评论的标题和内容,如图 10.13 所示。

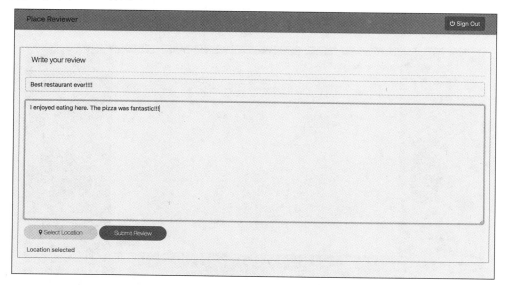

图 10.13 填写评论的标题和内容

需要注意的是,评论表单现在表明已成功地选择了评论位置。当用户填写所有的评论信息后,即可单击 Submit Review 按钮提交评论内容,如图 10.14 所示。

图 10.14　提交评论内容

在提交了评论后,用户将被重定位至其主页,并可查看所提交的评论。单击主页中评论上的 View location 按钮将显示一个模态框,其中包含了评论地理位置的地图,如图 10.15 所示。

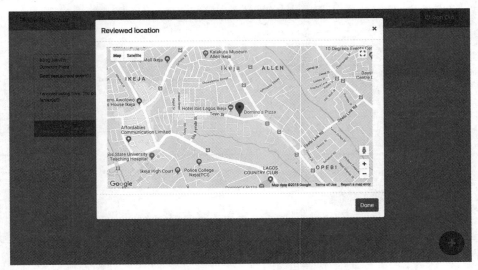

图 10.15　显示评论地理位置的地图

显示于用户的地图包含了一个标记,表示评论的准确地理位置。

至此,我们已实现 Place Reviewer 应用程序的全部核心功能。下面讨论 Spring 应用程序的测试行为。

10.2 Spring 应用程序测试

前述内容曾简要地介绍了应用程序的测试操作,及其在软件设计过程中的必要性。Spring 应用程序可通过 4 个步骤进行测试。

(1) 向项目中添加必要的测试依赖关系。
(2) 定义配置类。
(3) 配置测试类并使用自定义配置。
(4) 编写所需的测试程序。

下面将对上述步骤逐一加以讨论。

10.2.1 添加测试依赖关系

首先需要向项目中添加测试依赖关系。打开 Place Reviewer 项目的 pom.xml 文件,并添加下列依赖关系:

```xml
<dependency>
 <groupId>junit</groupId>
 <artifactId>junit</artifactId>
 <version>4.12</version>
 <scope>test</scope>
 <exclusions>
   <exclusion>
     <groupId>org.hamcrest</groupId>
     <artifactId>hamcrest-core</artifactId>
   </exclusion>
 </exclusions>
</dependency>
<dependency>
 <groupId>org.hamcrest</groupId>
 <artifactId>hamcrest-library</artifactId>
 <version>1.3</version>
```

```
<scope>test</scope>
</dependency>
```

后续各节将学习如何利用 jUnit 和 Hamcrest 编写测试程序。JUnit 是针对 Java 编程语言的测试框架；而 Hamcrest 库则提供了匹配器，二者结合使用后可创建有效的 Intent 表达式。

10.2.2 定义配置类

定义测试配置类有助于测试的正常运行。在 Place Reviewer 项目的 src/test/kotlin 目录中，将 config 包添加至 com.example.placereviewer 中，并向该包中加入 TestConfig 类，如下所示：

```
package com.example.placereviewer.config

import org.springframework.context.annotation.ComponentScan
import org.springframework.context.annotation.Configuration

@Configuration
@ComponentScan(basePackages = ["com.example.placereviewer"])
class TestConfig
```

10.2.3 利用自定义配置设置配置类

对此，可打开 Spring 应用程序的测试类，并使用@ContextConfiguration 注解指定测试类所用的配置类。相应地，打开 PlaceReviewerApplicationTests.kt 文件（位于 src/test/kotlin 目录下的 com.example.placereviewer 包中），并按照下列方式设置配置类：

```
package com.example.placereviewer

import com.example.placereviewer.config.TestConfig
import org.junit.runner.RunWith
import org.springframework.boot.test.context.SpringBootTest
import org.springframework.test.context.ContextConfiguration
import org.springframework.test.context.junit4.SpringRunner

@RunWith(SpringRunner::class)
@SpringBootTest
```

```
@ContextConfiguration(classes = [TestConfig::class])
class PlaceReviewerApplicationTests
```

下面尝试编写应用程序测试内容。

10.2.4 编写第一个测试程序

编写应用程序测试与编写 Spring 应用程序其他部分的代码基本类似，并可利用应用程序其他部分中的组件和服务，下面将对此予以介绍。

向 com.example.placereviewer.service 中添加 TestUserService 接口，如下所示：

```
package com.example.placereviewer.service

import com.example.placereviewer.data.model.User

interface TestUserService {
  fun getUser(): User
}
```

随后向包中添加 TestUserServiceImpl 类，如下所示：

```
package com.example.placereviewer.service

import com.example.placereviewer.data.model.User
import org.springframework.stereotype.Service

@Service
internal class TestUserServiceImpl : TestUserService {

  //Test stub mimicking user retrieval
  override fun getUser(): User {
    return User(
      "user@gmaiil.com",
      "test.user",
      "password"
    )
  }
}
```

返回至 PlaceReviewerApplicationTests.kt 文件中，对其进行修改以反映此类变化，如下所示：

```kotlin
package com.example.placereviewer

import com.example.placereviewer.config.TestConfig
import com.example.placereviewer.data.model.User
import com.example.placereviewer.service.TestUserService
import org.hamcrest.Matchers.instanceOf
import org.hamcrest.MatcherAssert.assertThat
import org.junit.Test
import org.junit.runner.RunWith
import org.springframework.beans.factory.annotation.Autowired
import org.springframework.boot.test.context.SpringBootTest
import org.springframework.test.context.ContextConfiguration
import org.springframework.test.context.junit4.SpringRunner

@RunWith(SpringRunner::class)
@SpringBootTest
@ContextConfiguration(classes = [TestConfig::class])
class PlaceReviewerApplicationTests {

  @Autowired
  lateinit var userService: TestUserService

  @Test
  fun testUserRetrieval() {

    val user = userService.getUser()

    assertThat(user, instanceOf(User::class.java))
  }
}
```

testUserRetrieval()测试方法在执行时使用定义于 TestUserServiceImpl 中的存根（stub）方法检索用户，并检测该方法返回的对象是否为 User 类实例。

当运行测试程序时，可单击 IDE 中的 Run Test 按钮，如图 10.16 所示。

此时，testUserRetrieval 将被运行，运行测试结果将显示于 IDE 窗口的下方，如图 10.17 所示。

运行结果表明，当前程序已通过测试。当开发更大、更复杂的应用程序，并为应用程序模块编写测试时，读者将会发现，测试过程常常会失败。此时，读者应保持冷静并对程序进行调试。随着时间的推移，读者将能够创建更加可靠的软件。

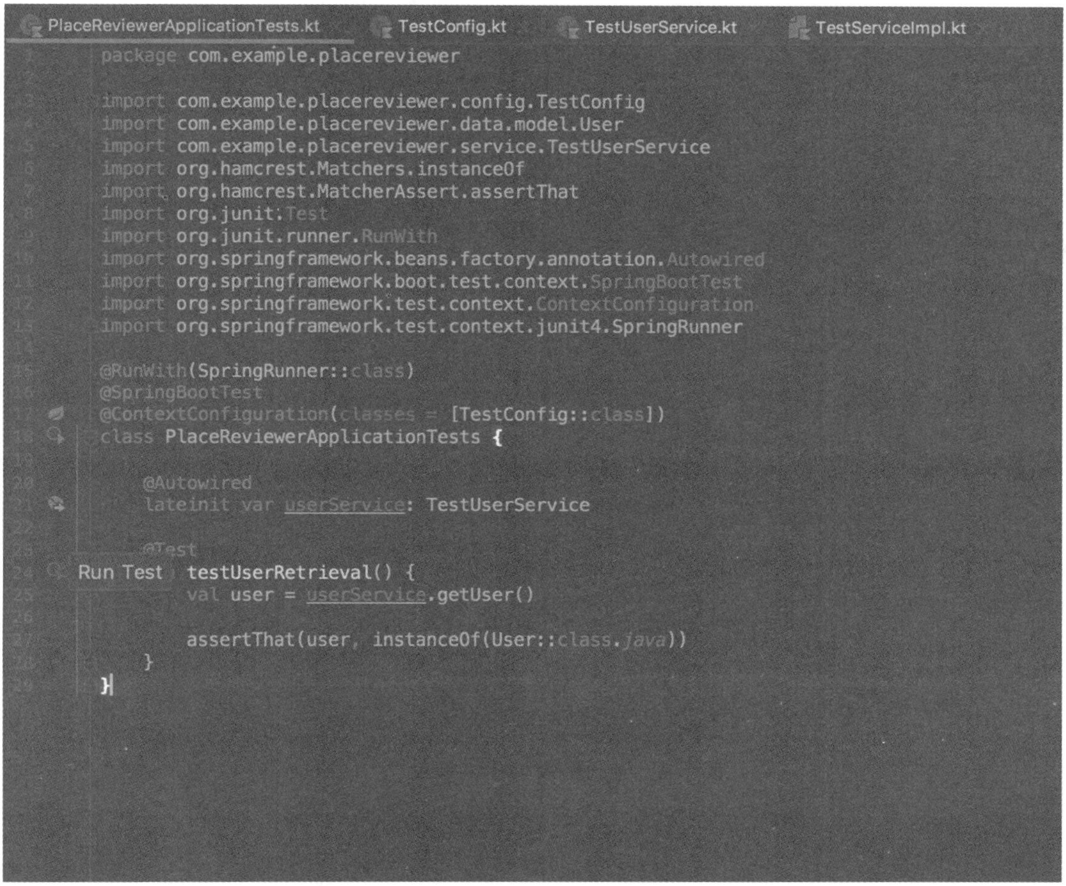

图 10.16 运行测试程序

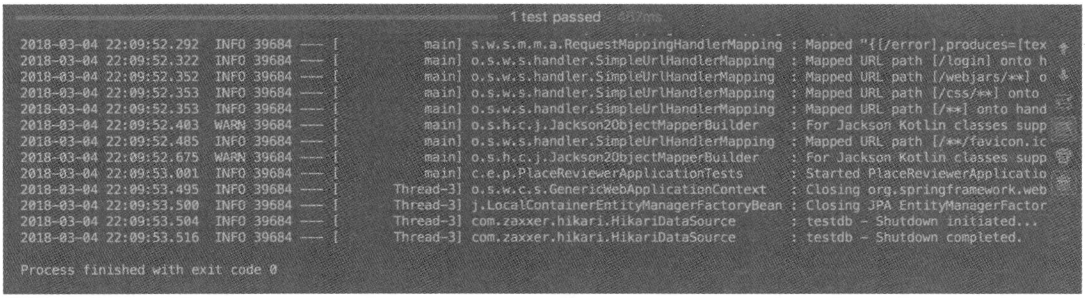

图 10.17 显示测试结果

10.3 本章小结

本章完成了 Place Reviewer 应用程序,其间详细讨论了 Spring MVC 应用程序视图层的构建过程。进一步讲,我们学习了如何将应用程序与 Google Places API Web 服务进行整合,进而实现了包含定位感知特性的应用程序。

除此之外,本章还借助于 Validator 类和 BindingResult,介绍了表单输入验证方案。最后,本章还探讨了测试的配置操作,并针对 Spring 应用程序编写相应的测试方法。

后　　记

　　如果读者正在阅读这一部分内容，相信您已经完整地阅读了本书。首先，请允许我向读者表示衷心的祝贺！您对 Kotlin 这门语言的投入和执着的确令人称赞。那么，接下来的问题是什么？这一问题很可能会在读者的脑海中闪过。本节尝试对此予以解答。

　　首先，读者必须努力掌握 Kotlin 这门语言，这需要大量的实践——相信我，这完全是可行的。以下几点建议仅供读者参考：

- ❑ 每天使用 Kotlin 语言进行程序设计，这一点非常重要，并可确保读者不断巩固所学到的知识。此外，这将有助于读者发现新的结构、模式、数据结构和范例，从而提升读者的编程技能。
- ❑ 广泛地阅读。这一点无论如何强调都不过分。为了掌握相关技能，并尽可能地获取更多的知识，读者必须养成阅读 Kotlin 相关话题的习惯。在开始阶段，读者可尝试阅读 Kotlin 参考文档，对应网址为 https://kotlinlang.org/docs/。此外，读者还可阅读其他与 Kotlin 相关的优秀专著。
- ❑ 咨询与 Kotlin 有关的问题。在学习 Kotlin 这门语言的过程中，很多时候都会经历各种问题。这里的建议是，不要对这些问题置之不理，否则读者将会错过一次美好的学习经历。针对于此，Stack Overflow 和 Quora 等平台是专门为知识共享而打造的，读者应确保养成使用这些平台的习惯，并及时寻求问题的答案。同样，若读者具有较高的水平，也可以自由地回答问题。
- ❑ 尝试使用工具编写程序。熟练地掌握一种工具对于编写 Kotlin 应用程序来说是不可或缺的。
- ❑ 如果读者保持对 Kotlin 的热情并遵循这些建议，那么，掌握这门语言绝非难事。这里，也祝您一路顺风！